Edexcel AS Chemistry

Ann Fullick Bob McDuell

STUDENTS' BOOK

A PEARSON COMPANY

How to use this book

This book contains a number of great features that will help you find your way around your AS Chemistry course and support your learning.

Introductory pages

Each unit has two introductory pages, giving you an idea of the chemistry to come and linking the content to the chemical ideas, how chemists work and chemistry in action that you will encounter in the unit.

HSW boxes

How Science Works is a key feature of your course. The many HSW boxes within the text will help you cover all the new aspects of How Science Works that you need. These include how scientists investigate ideas and develop theories, how to evaluate data and the design of studies to test their validity and reliability, and how science affects the real world including informing decisions that need to be taken by individuals and society.

Main text

The main part of the book covers all you need to learn for your course. The text is supported by many diagrams and photographs that will help you understand the concepts you need to learn.
Key terms in the text are shown in bold type. These terms are defined in the interactive glossary that can be found on the software using the 'search glossary' feature.

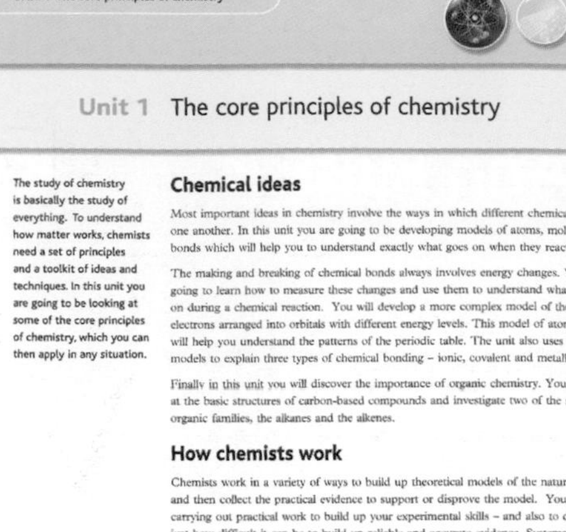

Introductory pages

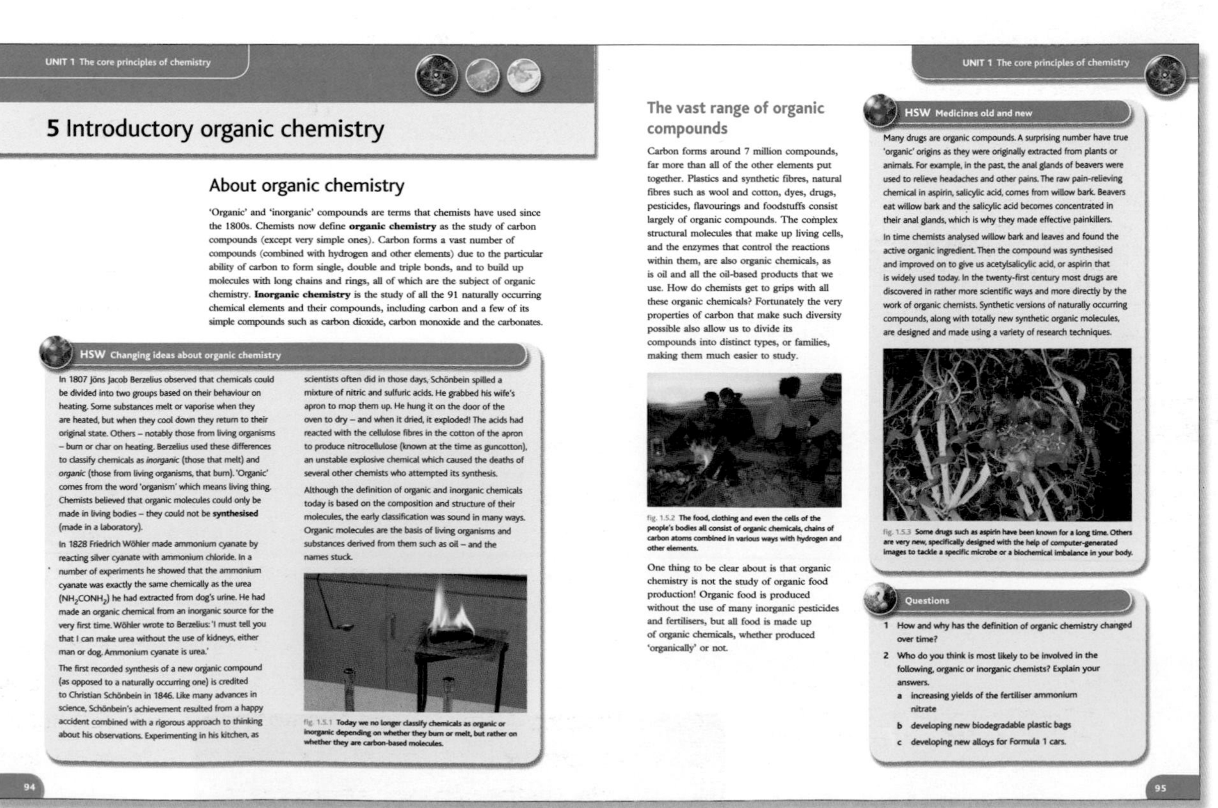

Main text

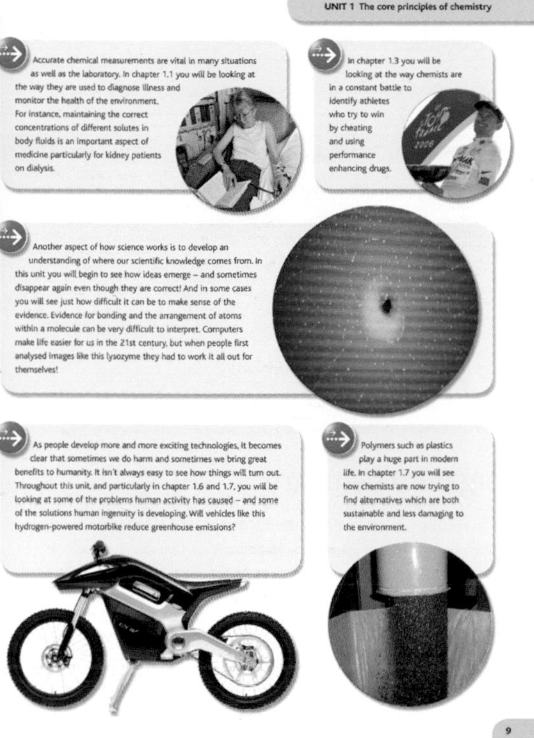

UNIT 1 The core principles of chemistry

Accurate chemical measurements are vital in many situations as well as the laboratory. In chapter 1.1 you will be looking at the way they are used to diagnose illness and monitor the health of the environment. For instance, maintaining the correct concentrations of different solutes in body fluids is an important aspect of medicine particularly for kidney patients on dialysis.

In chapter 1.3 you will be looking at the way chemists are in a constant battle to identify athletes who try to win by cheating and using performance enhancing drugs.

Another aspect of how science works is to develop an understanding of where our scientific knowledge comes from. In this unit you will begin to see how ideas emerge – and sometimes disappear again even though they are correct! And in some cases you will see just how difficult it can be to make sense of the evidence. Evidence for bonding and the arrangement of atoms within a molecule can be very difficult to interpret. Computers make life easier for us in the 21st century, but when people first analysed images like this lysozyme they had to work it all out for themselves!

As people develop more and more exciting technologies, it becomes clear that sometimes we do harm and sometimes we bring great benefits to humanity. It isn't always easy to see how things will turn out. Throughout this unit, and particularly in chapter 1.6 and 1.7, you will be looking at some of the problems human activity has caused – and some of the solutions human ingenuity is developing. Will vehicles like this hydrogen-powered motorbike reduce greenhouse emissions?

Polymers such as plastics play a huge part in modern life. In chapter 1.7 you will see how chemists are now trying to find alternatives which are both sustainable and less damaging to the environment.

9

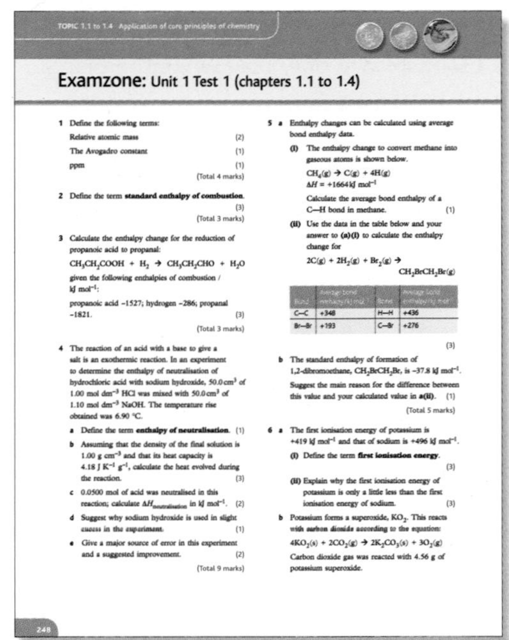

TOPIC 1.1 to 1.4 Application of core principles of chemistry

Examzone: Unit 1 Test 1 (chapters 1.1 to 1.4)

1 Define the following terms:
Relative atomic mass (2)
The Avogadro constant (1)
ppm (1)
(Total 4 marks)

2 Define the term **standard enthalpy of combustion**. (3)
(Total 3 marks)

3 Calculate the enthalpy change for the reduction of propanoic acid to propanal:
$CH_3CH_2COOH + H_2 \rightarrow CH_3CH_2CHO + H_2O$
given the following enthalpies of combustion / kJ mol^{-1}:
propanoic acid −1527; hydrogen −286; propanal −1821. (3)
(Total 3 marks)

4 The reaction of an acid with a base to give a salt is an exothermic reaction. In an experiment to determine the enthalpy of neutralisation of hydrochloric acid with sodium hydroxide, 50.0 cm^3 of 1.00 mol dm^{-3} HCl was mixed with 50.0 cm^3 of 1.10 mol dm^{-3} NaOH. The temperature rise obtained was 6.90 °C.
a Define the term **enthalpy of neutralisation**. (1)
b Assuming that the density of the final solution is 1.00 g cm^{-3} and that its heat capacity is 4.18 J K^{-1} g^{-1}, calculate the heat evolved during the reaction. (3)
c 0.0500 mol of acid was neutralised in this reaction; calculate $\Delta H_{neutralisation}$ in kJ mol^{-1}. (2)
d Suggest why sodium hydroxide is used in slight excess in the experiment. (1)
e Give a major source of error in this experiment and a suggested improvement. (2)
(Total 9 marks)

5 a Enthalpy changes can be calculated using average bond enthalpy data.
(i) The enthalpy change to convert methane into gaseous atoms is shown below.
$CH_4(g) \rightarrow C(g) + 4H(g)$
$\Delta H = +1664 \text{ kJ mol}^{-1}$
Calculate the average bond enthalpy of a C—H bond in methane. (1)
(ii) Use the data in the table below and your answer to **(a)(i)** to calculate the enthalpy change for
$2C(g) + 2H_2(g) + Br_2(g) \rightarrow CH_2BrCH_2Br(g)$

Bond	Average bond enthalpy / kJ mol^{-1}	Bond	Average bond enthalpy / kJ mol^{-1}
C—C	+348	H—H	+436
Br—Br	+193	C—Br	+276

(3)
b The standard enthalpy of formation of 1,2-dibromoethane, CH_2BrCH_2Br, is −37.8 kJ mol^{-1}. Suggest the main reason for the difference between this value and your calculated value in **a(ii)**. (1)
(Total 5 marks)

6 a The first ionisation energy of potassium is +419 kJ mol^{-1} and that of sodium is +496 kJ mol^{-1}.
(i) Define the term **first ionisation energy**. (3)
(ii) Explain why the first ionisation energy of potassium is only a little less than the first ionisation energy of sodium. (3)
b Potassium forms a superoxide, KO_2. This reacts with carbon dioxide according to the equation:
$4KO_2(s) + 2CO_2(g) \rightarrow 2K_2CO_3(s) + 3O_2(g)$
Carbon dioxide gas was reacted with 4.56 g of potassium superoxide.

248

Examzone page

Practical boxes

Your course contains a number of core practicals that you may be tested on. These boxes indicate links to practical work. Your teacher will give you opportunities to cover these investigations.

Worked example boxes

These boxes give example questions and answers, taking you through the calculations step-by-step and showing how you could set out your answers.

Question boxes

At the end of each section of text you will find a box containing questions that cover what you have just learnt. You can use these questions to help you check whether you have understood what you have just read, and whether there is anything that you need to look at again.

Examzone pages

At the end of the book you will find 9 pages of exam questions from past papers. You can use these questions to test how fully you have understood the units, as well as to help you practise for your exams. There are 2 tests for each unit

The contents list shows you that there are two units in the book, matching the Edexcel AS specification for Chemistry. There are 7 chapters in unit 1 (chapters 1.1 to 1.7) and 11 chapters in unit 2 (chapters 2.1 to 2.11). Page numbering in the contents list, and in the index at the back of the book, will help you find what you are looking for.

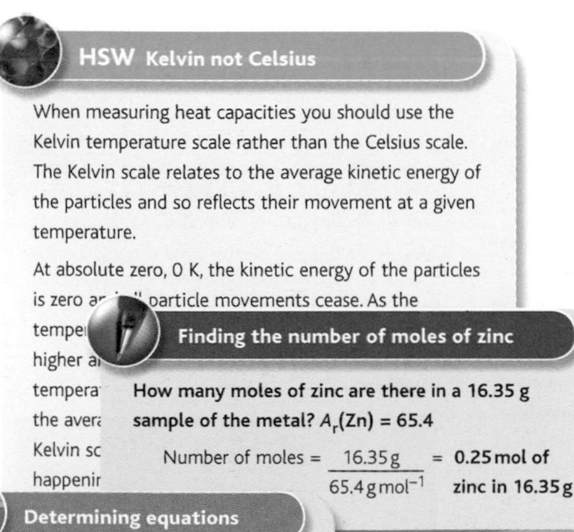

HSW Kelvin not Celsius

When measuring heat capacities you should use the Kelvin temperature scale rather than the Celsius scale. The Kelvin scale relates to the average kinetic energy of the particles and so reflects their movement at a given temperature.

At absolute zero, 0 K, the kinetic energy of the particles is zero and all particle movements cease. As the tempe… higher a… tempera… the avera… Kelvin sc… happenin…

Finding the number of moles of zinc

How many moles of zinc are there in a 16.35 g sample of the metal? $A_r(Zn)$ = 65.4

$$\text{Number of moles} = \frac{16.35\text{ g}}{65.4\text{ g mol}^{-1}} = \textbf{0.25 mol of zinc in 16.35 g}$$

Determining equations

You can carry out a similar investigation yourself to determine the equation for the reaction between potassium chroma…

Questions

1 What is ionisation energy and why is it so important in the development of a useful model of the electronic structure of an atom?

2 Rank the following orbitals in order of energy level, starting with the lowest (the one from which it would take most energy to remove an electron): 3p, 1s, 2s, 3d, 4s.

How to use your ActiveBook

The ActiveBook is an electronic copy of the book, which you can use on a compatible computer. The CD-ROM will only play while the disc is in the computer. The ActiveBook has these features:

Find Resources

Click on this tab to see menus which list all the electronic files on the ActiveBook.

Student Book tab

Click this tab at the top of the screen to access the electronic version of the book.

Key words

Click on any of the words in **bold** to see a box with the word and what it means. Click 'play' to listen to someone read it out for you to help you pronounce it.

Interactive view

Click this button to see all the icons on the page that link to electronic files, such as documents and spreadsheets. You have access to all of the features that are useful for you to use at home on your own. If you don't want to see these links you can return to **Book view**.

Student Book | Find Resources | Glossary | Help

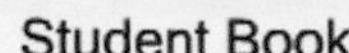

UNIT 1 The core principles of chemistry

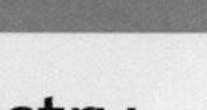

5 Introductory organic chemistry

About organic chemistry

'Organic' and 'inorganic' compounds are terms that chemists have used since the 1800s. Chemists now define **organic chemistry** as the study of carbon compounds (except very simple ones). Carbon forms a vast number of compounds (combined with hydrogen and other elements) due to the particular ability of carbon to form single, double and triple bonds, and to build up molecules with long chains and rings, all of which are the subject of organic chemistry. **Inorganic chemistry** is the study of all the 91 naturally occurring chemical elements and their compounds, including carbon and a few of its simple compounds such as carbon dioxide, carbon monoxide and the carbonates.

HSW Changing ideas about organic chemistry

In 1807 Jöns Jacob Berzelius observed that chemicals could be divided into two groups based on their behaviour on heating. Some substances melt or vaporise when they are heated, but when they cool down they return to their original state. Others – notably those from living organisms – burn or char on heating. Berzelius used these differences to classify chemicals as *inorganic* (those that melt) and *organic* (those from living organisms, that burn). 'Organic' comes from the word 'organism' which means living thing. Chemists believed that organic molecules could only be made in living bodies – they could not be **synthesised** (made in a laboratory).

In 1828 Friedrich Wöhler made ammonium cyanate by reacting silver cyanate with ammonium chloride. In a number of experiments he showed that the ammonium cyanate was exactly the same chemically as the urea (NH_2CONH_2) he had extracted from dog's urine. He had made an organic chemical from an inorganic source for the very first time. Wöhler wrote to Berzelius: 'I must tell you that I can make urea without the use of kidneys, either man or dog. Ammonium cyanate is urea.'

The first recorded synthesis of a new organic compound (as opposed to a naturally occurring one) is credited to Christian Schönbein in 1846. Like many advances in science, Schönbein's achievement resulted from a happy accident combined with a rigorous approach to thinking about his observations. Experimenting in his kitchen, as scientists often did in those days, Schönbein spilled a mixture of nitric and sulfuric acids. He grabbed his wife's apron to mop them up. He hung it on the door of the oven to dry – and when it dried, it exploded! The acids had reacted with the cellulose fibres in the cotton of the apron to produce nitrocellulose (known at the time as guncotton), an unstable explosive chemical which caused the deaths of several other chemists who attempted its synthesis.

Although the definition of organic and inorganic chemicals today is based on the composition and structure of their molecules, the early classification was sound in many ways. Organic molecules are the basis of living organisms and substances derived from them such as oil – and the names stuck.

fig. 1.5.1 Today we no longer classify chemicals as organic or inorganic depending on whether they burn or melt, but rather on whether they are carbon-based molecules.

94

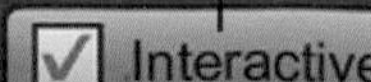

Page turn | Interactive

Glossary

Click this tab to see all of the key words and what they mean. Click 'play' to listen to someone read them out to help you pronounce them.

Help

Click on this tab at any time to search for help on how to use the ActiveBook.

The vast range of organic compounds

Carbon forms around 7 million compounds, far more than all of the other elements put together. Plastics and synthetic fibres, natural fibres such as wool and cotton, dyes, drugs, pesticides, flavourings and foodstuffs consist largely of organic compounds. The complex structural molecules that make up living cells, and the enzymes that control the reactions within them, are also organic chemicals, as is oil and all the oil-based products that we use. How do chemists get to grips with all these organic chemicals? Fortunately the very properties of carbon that make such diversity possible also allow us to divide its compounds into distinct types, or families, making them much easier to study.

fig. 1.5.2 **The food, clothing and even the cells of the people's bodies all consist of organic chemicals, chains of carbon atoms combined in various ways with hydrogen and other elements.**

One thing to be clear about is that organic chemistry is not the study of organic food production! Organic food is produced without the use of many inorganic pesticides and fertilisers, but all food is made up of organic chemicals, whether produced 'organically' or not.

HSW Medicines old and new

Many drugs are organic compounds. A surprising number have true 'organic' origins as they were originally extracted from plants or animals. For example, in the past, the anal glands of beavers were used to relieve headaches and other pains. The raw pain-relieving chemical in aspirin, salicylic acid, comes from willow bark. Beavers eat willow bark and the salicylic acid becomes concentrated in their anal glands, which is why they made effective painkillers.

In time chemists analysed willow bark and leaves and found the active organic ingredient. Then the compound was synthesised and improved on to give us acetylsalicylic acid, or aspirin that is widely used today. In the twenty-first century most drugs are discovered in rather more scientific ways and more directly by the work of organic chemists. Synthetic versions of naturally occurring compounds, along with totally new synthetic organic molecules, are designed and made using a variety of research techniques.

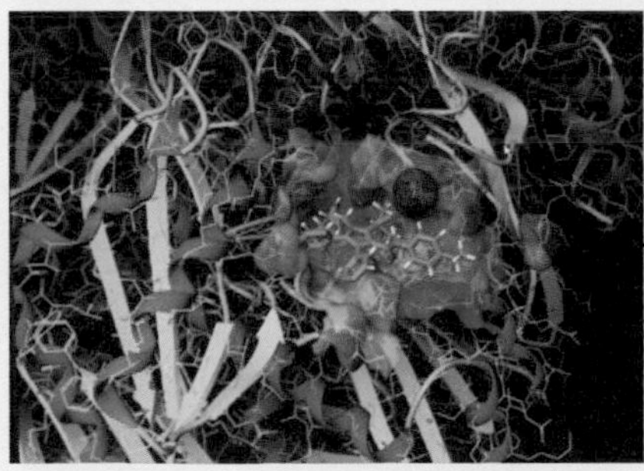

fig. 1.5.3 **Some drugs such as aspirin have been known for a long time. Others are very new, specifically designed with the help of computer-generated images to tackle a specific microbe or a biochemical imbalance in your body.**

Questions

1 How and why has the definition of organic chemistry changed over time?

2 Who do you think is most likely to be involved in the following, organic or inorganic chemists? Explain your answers.

a increasing yields of the fertiliser ammonium nitrate

b developing new biodegradable plastic bags

c developing new alloys for Formula 1 cars.

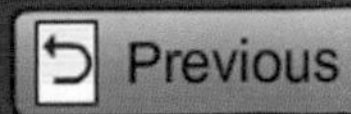 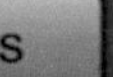  

Zoom feature

Just click on a section of the page and it will magnify so that you can read it easily on screen. This also means that you can look closely at photos and diagrams.

CONTENTS

Unit 1 The core principles of chemistry – p8

Unit 2 Application of core principles of chemistry – p142

Unit 1 The core principles of chemistry

The study of chemistry is basically the study of everything. To understand how matter works, chemists need a set of principles and a toolkit of ideas and techniques. In this unit you are going to be looking at some of the core principles of chemistry, which you can then apply in any situation.

Chemical ideas

Most important ideas in chemistry involve the ways in which different chemicals react with one another. In this unit you are going to be developing models of atoms, molecules and bonds which will help you to understand exactly what goes on when they react.

The making and breaking of chemical bonds always involves energy changes. You are going to learn how to measure these changes and use them to understand what is going on during a chemical reaction. You will develop a more complex model of the atom with electrons arranged into orbitals with different energy levels. This model of atomic structure will help you understand the patterns of the periodic table. The unit also uses these models to explain three types of chemical bonding – ionic, covalent and metallic.

Finally in this unit you will discover the importance of organic chemistry. You will look at the basic structures of carbon-based compounds and investigate two of the simplest organic families, the alkanes and the alkenes.

How chemists work

Chemists work in a variety of ways to build up theoretical models of the nature of matter and then collect the practical evidence to support or disprove the model. You will be carrying out practical work to build up your experimental skills – and also to demonstrate just how difficult it can be to build up reliable and accurate evidence. Systematic and random errors all too often interfere with the results! Increasingly chemists rely on instrumentation to get the evidence they need to support their theories and you will be looking at some examples, particularly the mass spectrometer.

The difference between ideal (theoretical) measurements and real, practically based results can be seen clearly when you consider the lattice energies of different compounds using Born Haber cycles calculated from theory and by experimentation!

Chemistry in action

Although a lot of the chemistry you learn in school is about models and theory, chemistry plays an enormous role in the real world. In this unit you will be looking at a variety of examples of chemistry in action which range from detecting athletes who are cheating by using drugs through to recycling polymers and trying to build a more sustainable world.

Chemistry and chemists do not exist in a vacuum. The work of chemists has a big effect on the life you lead and on the well-being and the economy of the whole planet. Some of the developments and discoveries of chemists raise ethical problems which scientists cannot answer. Society as a whole has to decide what is acceptable and what is not. So you will be looking at chemical ideas, practical chemistry and the role of chemistry in society.

Accurate chemical measurements are vital in many situations as well as the laboratory. In chapter 1.1 you will be looking at the way they are used to diagnose illness and monitor the health of the environment. For instance, maintaining the correct concentrations of different solutes in body fluids is an important aspect of medicine particularly for kidney patients on dialysis.

In chapter 1.3 you will be looking at the way chemists are in a constant battle to identify athletes who try to win by cheating and using performance enhancing drugs.

Another aspect of how science works is to develop an understanding of where our scientific knowledge comes from. In this unit you will begin to see how ideas emerge – and sometimes disappear again even though they are correct! And in some cases you will see just how difficult it can be to make sense of the evidence. Evidence for bonding and the arrangement of atoms within a molecule can be very difficult to interpret. Computers make life easier for us in the 21st century, but when people first analysed images like this lysozyme they had to work it all out for themselves!

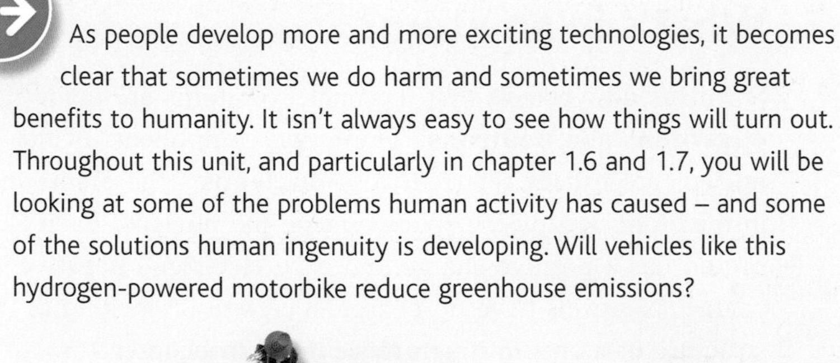

As people develop more and more exciting technologies, it becomes clear that sometimes we do harm and sometimes we bring great benefits to humanity. It isn't always easy to see how things will turn out. Throughout this unit, and particularly in chapter 1.6 and 1.7, you will be looking at some of the problems human activity has caused – and some of the solutions human ingenuity is developing. Will vehicles like this hydrogen-powered motorbike reduce greenhouse emissions?

Polymers such as plastics play a huge part in modern life. In chapter 1.7 you will see how chemists are now trying to find alternatives which are both sustainable and less damaging to the environment.

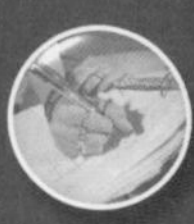

1 Formulae, equations and amounts of substance

The foundations of chemistry

fig 1.1.1 Everything in and on the Earth is made up of chemicals. Chemists investigate the way different elements and compounds react together in both living and non-living material.

At the heart of chemical theory is the atom as the building block of all matter. The first recorded reference we have to the idea of atoms as a basis of matter comes from around 2600 years ago in ancient India! The idea of atoms in the West can be traced back about 2500 years, to the Greek philosophers Leucippus and his pupil Democritus. The word **atom** comes from the Greek word meaning 'indivisible'. It is likely that a similar theory was held by some of the great Arab scientists such as Jabir ibn Hayyan during the ninth and tenth centuries, but the idea was not accepted in Europe until the start of the nineteenth century when a Manchester schoolteacher called John Dalton made the case for atoms based on evidence from experiments with gases.

Throughout the nineteenth and early twentieth centuries a body of theory about the nature of atoms and elements and the way they react together gradually built up. This was based on careful observation and experimentation by scientists such as Berzelius, Kekulé, Rutherford and Thomson. From the 1860s onwards Dmitri Mendeleev and others developed the periodic table, which mapped the known elements according to their properties.

HSW Evidence for the existence of atoms

The conclusive proof that matter is made up of atoms was provided by Albert Einstein in 1905. His calculations showed how Brownian motion is best explained in terms of a particle theory of matter. The random movement of microscopic smoke or pollen particles suspended in a gas or liquid is due to collisions between these particles and the particles of the gas or liquid.

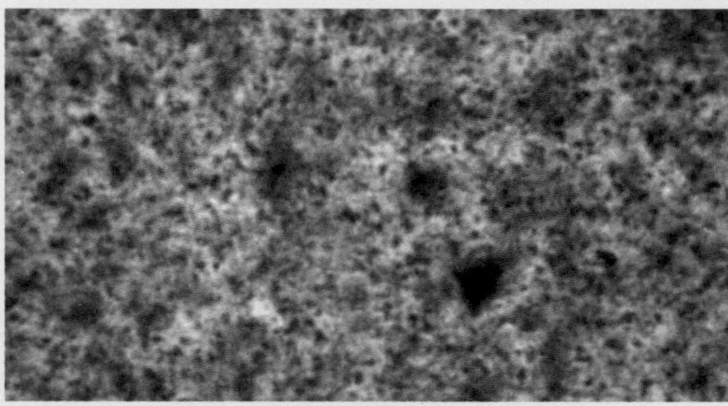

fig 1.1.2 The dancing, zig-zag motion of smoke particles in air or of pollen grains in water is known as Brownian motion, after the botanist Robert Brown who first observed it in 1827.

What's in an atom?

Scientists now believe that the nuclei of atoms are made up of **protons** and **neutrons**. These two components of the nucleus are usually referred to as **nucleons**. The **electrons** in the atom occupy the space outside the nucleus. Each proton has a positive charge and each electron a negative charge. Neutrons have no charge. You will be looking at the structure of atoms in much more detail in chapter 1.3.

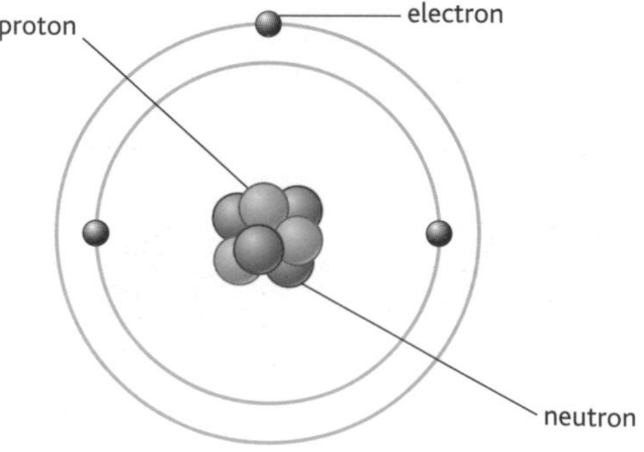

fig. 1.1.3 Atoms consist of protons and neutrons in a nucleus, surrounded by electrons.

An **element** is a substance that cannot be broken down chemically into simpler substances. The atoms of an element all contain the same number of protons. The number of protons in the nucleus of an atom is the **atomic number** of the element, given the symbol *Z*. The number of protons is equal to the number of electrons – an atom is electrically neutral. The protons and neutrons in an atom have mass, but the electron's mass is negligible. The number of protons (*Z*) plus the number of neutrons (*N*) in an atom is known as the **mass number** (*A*). **Figure 1.1.4** shows how the data for a nucleus are presented.

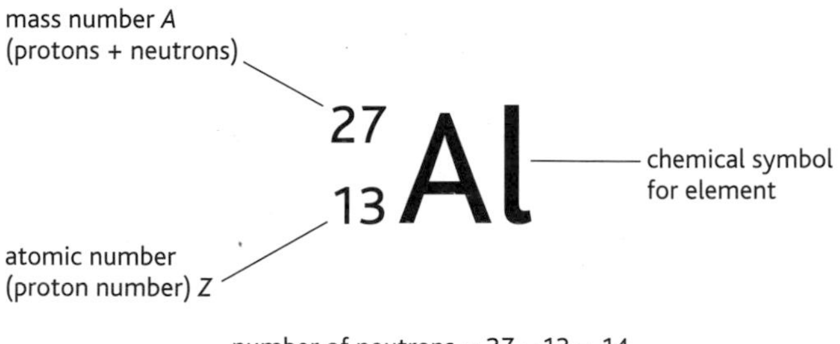

fig. 1.1.4 How the details of the nucleus are written with the chemical symbol, for aluminium. The number of neutrons is the mass number minus the atomic number

Isotopes

All the atoms of a particular element have the same atomic number – they all have the same number of protons and electrons. It is the number of electrons in the atom that determines its chemical properties.

However, atoms within the same element may have different numbers of neutrons. Neutrons have mass, so this means that the atoms making up an element may have different masses. Atoms with the same atomic number but different numbers of neutrons are called **isotopes**. Most elements are a mixture of isotopes. Sometimes the presence of extra neutrons makes the nucleus unstable, so some 'heavy' isotopes are radioactive. Although different isotopes of the same element may have different **physical properties** – they have a different mass and some may be radioactive – they always have the same **chemical properties** because the number of electrons stays the same.

For example, hydrogen has three isotopes, hydrogen-1, hydrogen-2 and hydrogen-3, shown in **fig. 1.1.5**. Around 99.985% of the atoms in a container of hydrogen will be hydrogen-1. About 0.015% are hydrogen-2, with twice the mass of hydrogen-1. The number of hydrogen-3 atoms in a sample of hydrogen is variable – these nuclei are **unstable**, and undergo radioactive decay. A shorthand way of representing the isotopes is ^1H, ^2H and ^3H (see **fig. 1.1.5**) or ^1_1H, ^2_1H and ^3_1H.

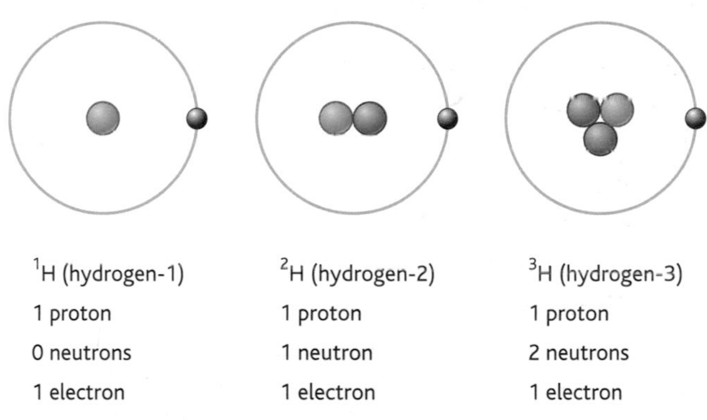

^1H (hydrogen-1)	^2H (hydrogen-2)	^3H (hydrogen-3)
1 proton	1 proton	1 proton
0 neutrons	1 neutron	2 neutrons
1 electron	1 electron	1 electron

fig. 1.1.5 The three isotopes of hydrogen.

	Lithium-7	Silicon-28	Copper-65	Silver-109	Uranium-238
Atomic number *Z*	3	14	29	47	92
Mass number *A*	7	28	65	109	238
Number of neutrons *N* (= *A* – *Z*)	4	14	36	62	146
Symbol	^7_3Li	$^{28}_{14}Si$	$^{65}_{29}Cu$	$^{109}_{47}Ag$	$^{238}_{92}U$

table 1.1.1 This table shows how many protons and neutrons combine to make up the nuclei of different atoms.

Questions

1 An atom has atomic number 26 and mass number 56. How many protons, neutrons and electrons does it have?

2 Work out the number of protons, electrons and neutrons present in the following atoms:

a $^{23}_{11}Na$ b $^{127}_{53}I$ c $^{84}_{36}Kr$

Formulae and equations

Holding atoms together

In compounds formed between metals and non-metals, the atoms are usually joined by **ionic bonding**. The atoms lose or gain electrons so that they have a complete outer shell, which is the most stable state for an ion. An atom that loses or gains one or more electrons forms an **ion**. An atom that loses an electron becomes a positive ion, and an atom that gains an electron becomes a negative ion. Strong forces of attraction (**ionic bonds**) hold the oppositely charged ions in a giant lattice structure.

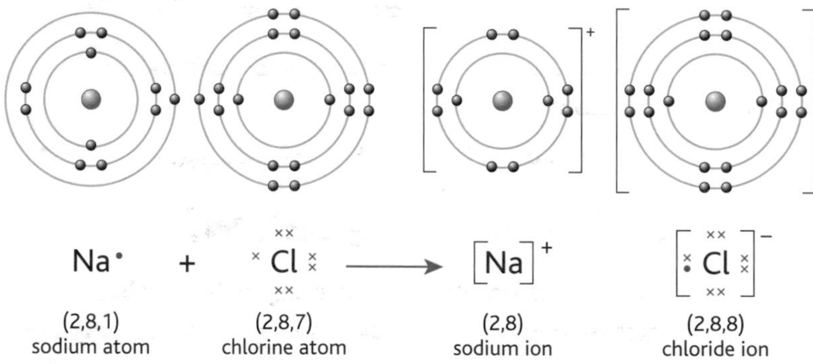

fig. 1.1.6 The formation of sodium chloride, NaCl – an example of ionic bonding. The dot and cross diagrams show the full outer shells in the ions produced.

Many chemical reactions do not involve ions. Non-metals generally gain electrons to achieve a stable outer shell. When non-metals react together, the atoms achieve full outer shells by sharing electrons – this is known as **covalent bonding**. The atoms in the molecules are then held together because they are sharing pairs of electrons. These strong bonds between the atoms are known as **covalent bonds**.

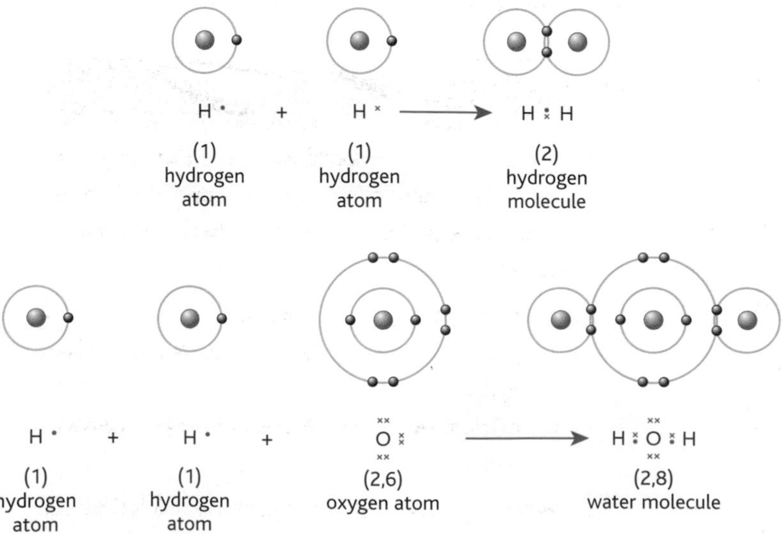

fig. 1.1.7 Examples of covalent bonding. Dot and cross diagrams show how electrons are shared to achieve a full outer shell.

Representing chemical reactions

Chemists need to explain what is happening during a chemical reaction as fully, yet as simply as possible. Equations help them calculate the amounts of reactants they need, predict the yields of products and tell other chemists exactly what experiment they did.

When magnesium burns in air, one atom of magnesium reacts with one atom of oxygen to form magnesium oxide. In theory this reaction can be represented by a simple chemical equation:

$$Mg + O \rightarrow MgO$$

However, oxygen normally exists as **diatomic** molecules, each containing two oxygen atoms, so this equation should read:

$$Mg + O_2 \rightarrow MgO + O$$

But now you have an oxygen atom left over! You need to produce a **balanced equation** showing all the reactants in the form in which they take part in the reaction:

$$2Mg + O_2 \rightarrow 2MgO$$

This kind of balancing act is performed for every chemical equation. The total mass of the product or products of a chemical reaction is always equal to the total mass of the reactants. The products have the same atoms as the reactants – but arranged in a different way.

To check if an equation is balanced, count the number of each type of atom on both sides. If the numbers are equal, and all the atoms are part of a complete molecule, then the equation is balanced.

For example, when nitrogen reacts with hydrogen, ammonia is produced:

$$N + H \rightarrow NH_3$$

But nitrogen and hydrogen are diatomic gases, N_2 and H_2, so, assuming there is one nitrogen molecule:

$$N_2 + 1\frac{1}{2}H_2 \rightarrow NH_3 + N$$

But you can't have half molecules or a single nitrogen atom. To balance the equation you need three molecules of hydrogen, which make two ammonia molecules:

$$N_2 + 3H_2 \rightarrow 2NH_3$$

Finally add up the different atoms – there are two nitrogen atoms and six hydrogen atoms on both sides of the equation, all in complete molecules, so it is balanced.

The state symbols

It is also important to show the **state symbols** for the reacting chemicals – whether each is a solid (**s**), a liquid (**l**), a gas (**g**) or in solution in water (aqueous solution, **aq**).

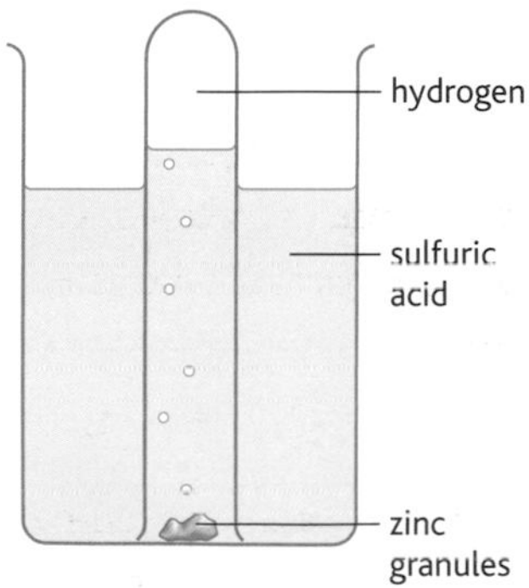

$$Zn(s) + H_2SO_4(aq) \rightarrow ZnSO_4(aq) + H_2(g)$$

fig. 1.1.8 Zinc reacts with sulfuric acid to produce hydrogen. It helps to know the state of all the reactants and products when planning the apparatus you need.

Ions in solution

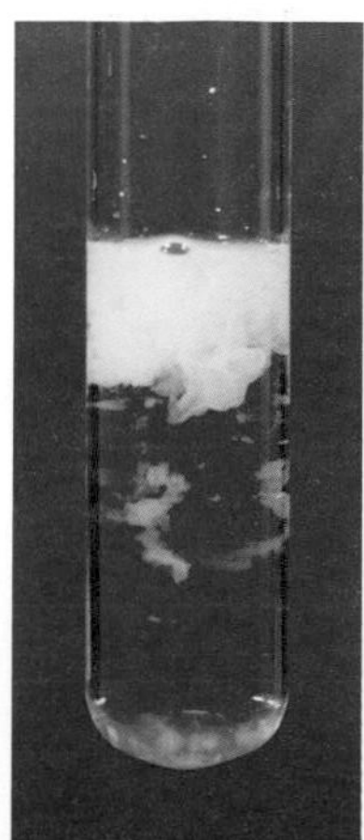

Many reactions involve ions in solution. When substances react in this way an ionic equation is useful. For example, when a solution of sodium hydroxide is added to a solution of magnesium chloride, sodium chloride, NaCl and a solid precipitate of magnesium hydroxide, $Mg(OH)_2$ are formed:

$$2NaOH(aq) + MgCl_2(aq) \rightarrow 2NaCl(aq) + Mg(OH)_2(s)$$

fig. 1.1.9 Magnesium hydroxide is present in 'Milk of Magnesia'. It can be made by adding a solution of sodium hydroxide to a solution of magnesium chloride.

This is a **molecular equation**, because it shows the complete formula of every substance. However, none of the substances consists of molecules – both reactants dissolve in water to form ions, so an **ionic equation** gives a more accurate picture:

$$2Na^+(aq) + 2OH^-(aq) + Mg^{2+}(aq) + 2Cl^-(aq) \rightarrow 2Na^+(aq) + 2Cl^-(aq) + Mg(OH)_2(s)$$

The sodium ions and chloride ions appear in exactly the same way on each side of the equation. Ions like this are often called **spectator ions**, and can be left out of the equation:

$$\cancel{2Na^+(aq)} + 2OH^-(aq) + Mg^{2+}(aq) + \cancel{2Cl^-(aq)} \rightarrow \cancel{2Na^+(aq)} + \cancel{2Cl^-(aq)} + Mg(OH)_2(s)$$

This produces the much simpler **overall ionic equation**:

$$Mg^{2+}(aq) + 2OH^-(aq) \rightarrow Mg(OH)_2(s)$$

Which type of equation?

The type of equation you use depends on what you want to show. A molecular equation gives a complete description of the reactants and the amounts you need in a practical experiment. An ionic equation shows what is actually happening in solution, eg when sodium hydroxide and magnesium chloride react together. The power of overall ionic equations is that they let you make generalisations, eg any solution containing magnesium ions will react with any other solution containing hydroxide ions to form a precipitate of magnesium hydroxide. As you continue through this book you will discover more ways of representing reactions.

Questions

1 Write balanced chemical equations including state symbols for the following word equations. Use the data in the periodic table at the back of the book to help you:

a zinc + oxygen → zinc oxide

b potassium + water → hydrogen and potassium hydroxide

c calcium carbonate → calcium oxide + carbon dioxide

2 When aqueous solutions of potassium chloride, KCl and silver nitrate, $AgNO_3$ are mixed, a precipitate of silver chloride forms. Write a molecular and an ionic equation for this reaction.

How big and heavy are atoms?

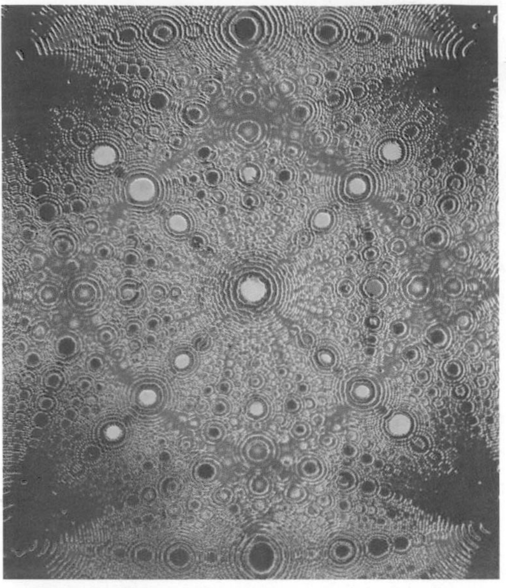

Atoms are far too small to be seen with an ordinary optical microscope. However, it is possible to see atoms using more sophisticated techniques. **Figure 1.1.10** shows the tip of a platinum needle seen through an **ion microscope**. This instrument uses helium ions instead of light, giving much higher magnifications. The radius of an individual atom is around 1×10^{-10} m.

fig. 1.1.10 The ion microscope reveals that platinum consists of layers of platinum atoms, arranged in a highly symmetrical crystal lattice. The image produced is striking.

HSW Working with numbers

When dealing with atoms, you need to use both very small and very large numbers. Powers of 10 are often used to express numbers like this. For example, the approximate distance travelled by light in one year is 1×10^{16} m, or 10 000 000 000 000 000 m. The 16 is called the **index** – the power to which 10 is raised – and 10 is the **base**.

Similarly, the approximate radius of a platinum atom is 1×10^{-10} m, or 0.000 000 000 1 m. The sign of the index tells you which way the decimal point moves.

Numbers are often written in **standard form** as a number between 1 and 10 multiplied by 10 raised to the appropriate power. So for example, the boiling point of nickel is 3.0×10^3 K, and the wavelength of red light is 6.3×10^{-7} m. (Both of these examples are quoted to 2 significant figures.)

Relative atomic mass

The mass of an atom is also tiny, but calculating the masses and numbers of atoms is essential in practical chemistry. Using actual mass would be cumbersome, so chemists use a *relative* scale which is easier to handle.

Whenever you measure something, you do it by comparing what you are measuring with a known quantity. For example, if you buy a kilogram of apples, the mass of the apples has been determined by weighing it against a standard 1 kg mass. The cylinder in **fig. 1.1.11** is defined as having a mass of exactly 1 kg. You are effectively comparing the mass of the apples with the mass of this cylinder!

fig. 1.1.11 This cylinder of metal in Paris is the standard kilogram. It will be replaced in the future because it is very slowly losing mass as metal atoms evaporate from its surface.

In the same way, chemists compare the masses of atoms with the mass of a standard atom – the carbon-12 isotope. This is called the **relative atomic mass scale**. The **relative atomic mass** (**RAM**, given the symbol A_r) of an element is often shown on the periodic table. Relative atomic masses have no units – they compare the mass of one atom with that of another.

- The relative atomic mass of an element is defined as the average mass of its isotopes compared with the mass of an atom of the carbon-12 isotope.

The mass number of an element tells you the number of protons and neutrons in a typical atom of that element. However, as you have seen, the atoms in an element do not necessarily all have the same mass number if there are different isotopes. This is why the relative atomic mass scale is based on one isotope. On this scale, atoms of the carbon-12 isotope are assigned a relative atomic mass of 12. An atom of the hydrogen-1 isotope has a relative atomic mass of 1, one-twelfth of the RAM for the carbon-12 isotope.

Mixtures of isotopes

If you look up the relative atomic mass of carbon in a data book, you will find it given as 12.011 not 12.000. This is because naturally occurring carbon contains mainly atoms of carbon-12, but also some atoms of carbon-13 and carbon-14, making the average relative atomic mass of a carbon atom slightly greater than 12. The relative atomic mass of every element is an average which reflects the mix of isotopes and their differing masses within the element. Most elements have one major isotope and very small proportions of other isotopes, and for these it is usually accurate enough to assume that the value of the relative atomic mass is the same as its mass number.

Finding the relative atomic mass of iron

A sample of iron was analysed in a mass spectrometer (see chapter 1.3). Table 1.1.2 shows the isotopes that it contains. Find the relative atomic mass of this sample of iron.

	Iron-54	Iron-56	Iron-57	Iron-58
Relative abundance (%)	5.84	91.68	2.17	0.31

table 1.1.2 The **relative abundance** of isotopes in a sample of iron.

To find the relative atomic mass, you need to calculate the average of these four isotopes. However, they are present in different proportions, so just adding them together and dividing by 4 would give a misleading average. So you multiply the relative atomic mass of each isotope by its relative abundance to calculate a **weighted mean** which will give you the overall relative atomic mass:

$$A_r(\text{Fe}) = (54 \times 5.84\%) + (56 \times 91.68\%) + (57 \times 2.17\%) + (58 \times 0.31\%)$$

$A_r(\text{Fe})$ = **55.91**

For many calculations 56 is a close enough approximation to the relative atomic mass of iron. (This is the relative atomic mass of the most common isotope of iron.)

Questions

1 The element bromine has atomic number 35 and has two isotopes of relative atomic mass 79 and 81. Copy and complete table 1.1.3.

Atom	Number of protons	Number of neutrons
^{79}Br		
^{81}Br		

table 1.1.3 Two isotopes of bromine.

2 Calculate the relative atomic mass of each of the following elements:

a bromine (50.5% bromine-79, 49.5% bromine-81)

b silver (51.3% silver-107, 48.7% silver-109)

c chromium (4.3% chromium-50, 83.8% chromium-52, 9.6% chromium-53, 2.3% chromium-54)

The mole

Counting and weighing atoms

As a chemist you often need to 'count' and 'weigh' atoms. You need to mix chemicals together in the correct ratio to carry out a reaction. Or you might want to find out the composition and formula of the compound you have made, by calculating the ratio of the numbers of different atoms in your product from the amounts of the reactants. The **mole** provides a way of doing this.

Just like relative atomic mass, the mole is based on the carbon-12 isotope as a standard measure. The relative atomic mass of carbon-12 in grams (12.00 g) contains a set number of atoms. From this it follows that for any element, the relative atomic mass in grams contains this same number of atoms. This is the basis of the definition of the mole, the SI unit of the **amount of substance**. Like other units, the mole has an abbreviation, **mol**.

- A mole of any substance is defined as the amount of substance that contains as many particles (atoms, ions or molecules) as there are atoms in exactly 12 g of carbon-12.

The number of atoms in exactly 12 g of carbon-12 is found from experiments to be 6.02×10^{23} (to 3 significant figures). This number is known as the **Avogadro constant**, after the nineteenth-century Italian chemist Amedeo Avogadro. The Avogadro constant has the unit particles per mole (mol^{-1}) and is sometimes given the symbol L or N_A. One mole of any substance contains 6.02×10^{23} particles.

You can work out the number of moles in any amount of an element:

$$\text{Number of moles} = \frac{\text{mass in g}}{\text{molar mass in g mol}^{-1}}$$

The **molar mass** (symbol $\boldsymbol{M}$) is just the numerical value of the relative atomic mass, but with the unit grams per mole.

You can also use the Avogadro constant to work out how many atoms there are in a certain mass of a substance. For example, imagine you have 6 g of magnesium and 6 g of carbon-12 and you want to know how many atoms of each you have.

Finding the number of moles of zinc

How many moles of zinc are there in a 16.35 g sample of the metal? $A_r(Zn) = 65.4$

$$\text{Number of moles} = \frac{16.35\,\text{g}}{65.4\,\text{g mol}^{-1}} = \textbf{0.25 mol of zinc in 16.35 g}$$

You would divide the mass of each element by its molar mass and then multiply by the Avogadro constant:

$A_r(^{12}C) = 12$

$$\text{Number of moles} = \frac{6\text{ g}}{12\text{ g mol}^{-1}} = 0.5\text{ mol}$$

Therefore the number of carbon-12 atoms $= 0.5 \times 6.02 \times 10^{23} = 3.01 \times 10^{23}$

$A_r(Mg) = 24.31$

$$\text{Number of moles} = \frac{6\text{ g}}{24.3\text{ g mol}^{-1}} = 0.25\text{ mol}$$

Therefore the number of magnesium atoms $= 0.25 \times 6.02 \times 10^{23} = 1.505 \times 10^{23}$

You can see from this that 6 g of magnesium has half as many atoms as 6 g of carbon-12. This is logical given that the magnesium atom has twice the mass of a carbon-12 atom. We can also say that 6 g of magnesium is 0.25 of a mole of magnesium and 6 g of carbon-12 is 0.5 of a mole of carbon-12.

fig. 1.1.12 Counting atoms is like counting these sweets. The quickest way is to weigh them – but we need to know how much each individual sweet, or atom, weighs first

HSW Units

A great deal of science is based on measuring physical quantities, such as length and mass, using a number combined with a unit. For example, a length may be quoted as 2.5 km or 2500 m. To make communication between scientists and engineers easier, a common system of units is now in use in the world of science. This system is called the **Système International (SI)**, and it consists of a set of seven **base units**. The base SI units are the metre (m), the kilogram (kg), the second (s), the ampere (A), the kelvin (K), the candela (cd) and the mole (mol). All other measurements relate to these in some way, and so they are known as **derived units**.

fig. 1.1.13 **All these clocks measure time – but do they all measure using the exact SI standard second?**

The units of mass and density illustrate how base units and derived units are related.

- Mass has the unit kilograms (kg) in the SI system.
- Density measures the mass per unit volume of a substance.
- Volume is expressed as metres cubed (m^3).

So the units of density are kilograms per metre cubed. This may be written as kilograms/metre^3, kg/m^3 or kg m^{-3}. (In chemistry you are more likely to use the smaller derived unit of g cm^{-3} for density.)

It is tempting to assume that base units such as length and mass are absolute and unchanging, but it isn't that simple. For example, as you saw in **fig. 1.1.11**, the standard kilogram is constantly losing mass. There is an identical cylinder stored in London and the two are compared regularly. The Paris standard has lost 50 parts per billion compared with the London cylinder. Scientists don't know if the loss is due to pollutants when it was forged, solvents used to clean it or because on rare occasions it has been handled. The quantity is very small (less than the mass of a grain of sugar) so it doesn't matter when you buy a kilogram of apples, but in the scientific world it is very important. Now scientists are looking at new ways to define the kilogram, including relating it to the mass of an atom. Progress will be reviewed in 2011.

Questions

1 Calculate the number of moles in:
   - a 8.0 g of oxygen-16 atoms
   - b 48 g of carbon-12 atoms
   - c 8.88 g of chlorine-35 atoms.

2 Calculate the number of atoms in:
   - a 36 g of carbon-12
   - b 4 g of magnesium-24
   - c 532 g of caesium-133
   - d 57 g of lead-208

Using moles

So far you have used relative atomic mass, the Avogadro constant and the mole for calculations involving atoms and elements. Now you can apply these principles to compounds as well.

Relative formula mass and relative molecular mass

For a compound, the sum of the relative atomic masses of all the atoms in the chemical formula is called the **relative formula mass**. In covalent compounds that are made up of molecules, this is referred to as the **relative molecular mass** ($\boldsymbol{M_r}$).

The relative molecular mass can apply to elements as well as compounds. Most gaseous elements exist as diatomic molecules, eg oxygen, O_2. Relative molecular mass is used rather than relative atomic mass for practical chemistry: $M_r(O_2) = 32$.

Finding the relative molecular mass

Calculate the relative molecular mass of carbon dioxide. $A_r(C) = 12.0$ and $A_r(O) = 16.0$ (to 1 decimal place).

$M_r(CO_2) = 12.0 + (2 \times 16.0) = \mathbf{44.0}$

For ionic compounds the relative formula mass is calculated in the same way, using the formula.

Relative formula mass for NaCl $= A_r(Na) + A_r(Cl) = 23 + 35.5 = 58.5$

The relative atomic mass can be used for ions because the atoms have simply lost or gained electrons, which have virtually no mass.

More about molar mass

Like relative atomic mass, the relative molecular mass or relative formula mass does not have units. By definition, the relative molecular mass of a substance in grams contains one mole of molecules of the substance – so 44 g of carbon dioxide contains 6.02×10^{23} molecules of CO_2.

Once again the **molar mass** is the relative molecular or formula mass in grams per mole. It tells you the number of grams of that substance that makes up one mole, eg the molar mass of carbon dioxide is 44 g mol^{-1} and the molar mass of sodium chloride is 58.5 g mol^{-1}.

fig. 1.1.14 One mole looks very different depending on which compound you are considering.

Calculating reacting quantities

A student plans an experiment that uses 0.2 mol of calcium phosphate, $Ca_3(PO_4)_2$. How many grams of calcium phosphate need to be weighed out?

Relative formula mass $[Ca_3(PO_4)_2] = A_r(Ca) \times 3 + 2 \times [A_r(P) + 4 \times A_r(O)]$

$= (40.1 \times 3) + 2 \times [31.0 + (4 \times 16.0)]$

$= 120.3 + 2 \times (31 + 64) = 310.3$

$M[Ca_3(PO_4)_2] = 310.3$ g mol^{-1}

1 mol of calcium phosphate has a mass of 310.3 g, so 0.2 mol would weigh:

310.3 g mol^{-1} × 0.2 mol = **62.1 g (to 1 decimal place)**

Always check that your answer makes sense!
A tenth (0.1) of a mole would weigh about 31 g. 0.2 is twice as much as this, so an answer of 62.1 g seems reasonable.

More about moles and equations

In an equation, the symbol for an element stands for two things – an atom of the element and a mole of atoms of the element. For example, when carbon burns in a limited supply of oxygen to produce carbon monoxide, the equation tells you the reacting quantities:

$2C(s) + O_2(g) \rightarrow 2CO(g)$

2 atoms + 1 molecule → 2 molecules

2 mol of atoms + 1 mol of molecules → 2 mol of molecules

It also tells you the state (solid, liquid, gas or aqueous ions) of the reactants and products.

It is important to specify precisely which particle you mean whenever you use moles, eg does 'a mole of oxygen' mean oxygen atoms or oxygen molecules? Giving the formula or clearly stating the particles involved avoids this confusion.

From equation to reaction

Using moles you can work out from a chemical equation the quantities of reactants you need to carry out the reaction. For example, how much magnesium and oxygen do you need to make one mole of magnesium oxide? The equation is:

$2Mg(s) + O_2(g) \rightarrow 2MgO(s)$

2 mol + 1 mol → 2 mol

$(2 \times 24.3\ g\ mol^{-1}) + (32\ g\ mol^{-1}) \rightarrow (2 \times 40.3\ g\ mol^{-1})$

So 48.6 g of magnesium needs to be reacted with 32 g of oxygen (in practice, oxygen would be in excess) to make 80.6 g of magnesium oxide.

fig. 1.1.15 The bright white light produced when magnesium reacts with oxygen is used in fireworks and flares.

Calculations like these are carried out at an industrial level as well as in the school laboratory to determine the quantities of reacting chemicals needed. For example, in a recent year one UK-based fertiliser company produced 650 000 tonnes of ammonium nitrate (1 tonne = 1000 kg). This is one of the world's leading fertilisers. The final stage involves reacting ammonia with nitric acid:

$NH_3(g) + HNO_3(aq) \rightarrow NH_4NO_3(aq)$

fig. 1.1.16 Chemistry on this scale needs careful calculations so no reactants are wasted.

The industrial chemists need to know exactly how much ammonia and nitric acid are needed to make 650 000 tonnes of ammonium nitrate. They start off with lab-sized quantities and work up! 1 mol of ammonia reacts with 1 mol of nitric acid to produce 1 mol of ammonium nitrate. Calculating molar masses gives:

$M_r(NH_3) = 17\ g\ mol^{-1}$, $M_r(HNO_3) = 63\ g\ mol^{-1}$, $M_r(NH_4NO_3) = 80\ g\ mol^{-1}$

17 g of NH_3 + 63 g of HNO_3 → 80 g of NH_4NO_3

17 kg of NH_3 + 63 kg of HNO_3 → 80 kg of NH_4NO_3

138 125 tonnes of NH_3 + 511 875 tonnes of HNO_3 → 650 000 tonnes of NH_4NO_3

The mass of nitric acid needed will actually be even larger because the acid is in aqueous solution.

Reactions with gases

Many chemical reactions take place between gases, or involve gases as one of the reactants or products. Rather than weighing gases, it is more convenient to measure their volume. Provided that you are careful to state the temperature and pressure of the gas when you give its volume, this is quite straightforward since:

- One mole of any gas occupies 24 dm^3 at 25°C (298 K) and 1 atmosphere pressure.

HSW Avogadro and the gas laws

At the end of the eighteenth century, the English scientist Henry Cavendish established that when the gases hydrogen and oxygen react to form water, they always react in the ratio of 2:1 by volume. This work was repeated and extended by the French chemist Joseph-Louis Gay-Lussac. In 1805, Gay-Lussac published his findings. He discovered that when gases react, their volumes are simple ratios of whole numbers, provided that they are measured under the same conditions of temperature and pressure.

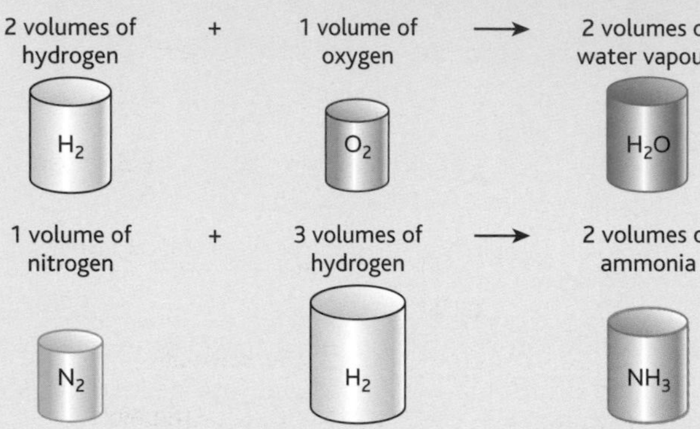

fig. 1.1.17 The volumes of reacting gases are always simple ratios of whole numbers.

The Italian chemist Amedeo Avogadro seized on Gay-Lussac's work and in 1811 published his law (now named as **Avogadro's law**). Avogadro's law states that:

- Equal volumes of all gases contain equal numbers of molecules, provided that they are at the same temperature and pressure.

By interpreting Gay-Lussac's work in this way, Avogadro provided chemists with a very powerful practical tool for measuring gases.

Avogadro's theory was not generally accepted for many years, for a number of reasons. Dalton's atomic theories, on which Avogadro built his ideas, were still very new. Avogadro's ideas were largely theoretical, with relatively little laboratory work to back them up – and he was not known as a careful researcher! He published his ideas in relatively minor scientific journals which many people did not read, and he didn't write particularly clearly so not everyone understood what he was trying to say. Finally, his ideas were in conflict with some of the best-known scientists of the day so many people ignored him to avoid rocking the boat. It was not until the middle of the nineteenth century that Avogadro's ideas were accepted by the majority of chemists.

fig. 1.1.18 Amedeo Avogadro, count of Quaregna (1776–1856). Avogadro started out as a lawyer but quickly turned to chemistry, applying a mathematical approach to combining volumes of gases, among other things.

How does Avogadro's law explain how gases react? Take the synthesis of ammonia from nitrogen and hydrogen as an example. Experiment showed that:

1 volume of nitrogen + 3 volumes of hydrogen → 2 volumes of ammonia

Using Avogadro's ideas, it follows from this that if one volume of any gas contains n molecules, then:

$1n$ molecules of nitrogen + $3n$ molecules of hydrogen → $2n$ molecules of ammonia

From which:

1 molecule of nitrogen + 3 molecules of hydrogen → 2 molecules of ammonia

So the chemical equation for the reaction can be written as:

$N_2(g) + 3H_2(g) \rightarrow 2NH_3(g)$

The power of Avogadro's law lies in the fact that it allows the masses of the molecules in two gases to be compared simply by taking equal volumes of the gases and comparing their masses. This enables chemists to measure the relative molecular masses of gases, and so to determine their formulae.

Molar volume

One of the major implications of Avogadro's law is that one mole of any gas must occupy the same volume under the same conditions – this is the **molar volume** of the gas, V_m. Scientists use the **standard temperature and pressure** (**STP**) to compare molar volumes. These are 1 atm pressure and 298 K (25 °C). Under these conditions, the volume of 1 mol of any gas is 24 dm^3, or very close to it. How can we use this information about gases and molar volumes?

Using molar volumes

Burning fuels 1

Coal is a fossil fuel which is made up of carbon. If 6 g of carbon is burned, what volume of carbon dioxide would you expect to be produced at 25 °C and 1 atm? What mass of carbon dioxide would this be?

$C(s) + O_2(g) \rightarrow CO_2(g)$

1 mol of carbon atoms + 1 mol of oxygen molecules
→ 1 mol of carbon dioxide molecules

A_r(C) = 12.0 (to 1 decimal place) so:

$$\frac{6\text{ g}}{12\text{ g mol}^{-1}} = 0.5\text{ mol}$$

So 0.5 mol of carbon reacts; this means that 0.5 mol of carbon dioxide is produced. $V_m(CO_2)$ = 24 dm^3 at 25 °C and 1 atm. Therefore:

Volume of carbon dioxide produced = 24 dm^3 × 0.5
= 12 dm^3

$M(CO_2)$ = 12 + 2(16) g mol^{-1} = 12 + 32 = 44 g mol^{-1}

Mass of carbon dioxide produced
= 44 g mol^{-1} × 0.5 mol = 22 g

Burning fuels 2

When propane, C_3H_8 is burned in oxygen, carbon dioxide and water are produced:

$C_3H_8(g) + 5O_2(g) \rightarrow 3CO_2(g) + 4H_2O(g)$
1 mol + 5 mol → 3 mol + 4 mol

If 30 dm^3 of propane (measured at 25 °C and 1 atm) is burned in air, what mass of carbon dioxide would you expect to be produced?

24 dm^3 of propane = 1 molar volume of propane

30 dm^3 of propane is 30/24 = 1.25 molar volumes

1 molar volume of propane burns to produce 3 molar volumes of carbon dioxide so:

1.25 molar volumes of propane → 3 × 1.25 molar volumes of carbon dioxide

= 3.75 molar volumes of carbon dioxide

This is equivalent to 3.75 molar masses, therefore

Mass of carbon dioxide produced = 3.75 (12.0 + 32.0) = 3.75 × 44.0 = **165 g**

Questions

1 Define carefully relative formula mass, relative molecular mass, molar mass and molar volume.

2 a Summarise Avogadro's law and explain why it is useful.
  b Explain why it took 50 years for Avogadro's ideas to become generally accepted by scientists.

3 Methane (CH_4) burns in oxygen (O_2) to form carbon dioxide (CO_2) and water (H_2O).
  a Write a balanced equation for this reaction.
  b How many moles of each substance are involved in the reaction?
  c What volume of methane, oxygen and carbon dioxide would you expect at 25 °C and 1 atm?
  d What mass of oxygen would you need to react with 8 dm^3 of methane, and what volume of carbon dioxide would be produced (All volumes measured at 25 °C and 1 atm)?

Calculating formulae using moles

Moles allow you to calculate quantities of reactants and products if you know their formulae and the equation. You can also use moles to calculate the formulae of substances involved in a reaction, and therefore write the equation for the reaction.

The empirical formula

The formula of a compound gives the ratio of the different atoms present. Chemical analysis can tell you the mass of each element present in a sample of a compound. From this you can find the number of moles of each element, and the ratio of one to the other. In this way you can work out the **empirical formula** of a compound. This is the simplest formula for a compound, showing the whole-number ratio of the numbers of atoms of each element present.

Finding an empirical formula

Analysis of a sample of aluminium chloride shows that it contains 5.8 g of aluminium and 22.9 g of chlorine. **Work out the empirical formula of aluminium chloride.**

	Al	Cl
Mass found by analysis	5.8 g	22.9 g
Molar mass	27.0 g mol^{-1}	35.5 g mol^{-1}
Moles of atoms in sample	$\frac{5.8\ g}{27.0\ g\ mol^{-1}}$ = 0.215 mol	$\frac{22.9\ g}{35.5\ g\ mol^{-1}}$ = 0.645 mol
Ratio of moles in sample (= ratio of atoms in sample)	1	3

table 1.1.4 **How to calculate the ratio of the moles of atoms in this sample.**

The ratio of the numbers of aluminium atoms to chlorine atoms in this sample of aluminium chloride is 1:3, so the empirical formula is $\mathbf{AlCl_3}$. However, other formulae such as Al_2Cl_6 or Al_3Cl_9 are also possible, and the analysis does not give any guide as to which of these is correct.

Finding the molecular formula

The empirical formula $AlCl_3$ does not tell us how many atoms of aluminium and chlorine are bound together in one molecule of aluminium chloride. The formula that describes this is the **molecular formula**. In the case of aluminium chloride, both $AlCl_3$ and Al_2Cl_6 are correct as the molecular formula! This is because aluminium chloride can exist in different forms depending on the conditions (see chapter 1.4). For many substances the empirical formula and the molecular formula are the same – water, H_2O and methane, CH_4 are good examples. To check this for your compound, you need to find the relative molecular mass of the compound. This may be found experimentally by measuring reacting quantities of chemicals and using this in calculations, or using a mass spectrometer (see chapter 1.3).

Giant structures of atoms and ions are described by their empirical formula. A molecular formula has no meaning for these substances because they are not made up of molecules.

Finding the molecular formula from the empirical formula

Styrene has the empirical formula CH. Its relative molecular mass is 104. What is its molecular formula?

The relative formula mass of the empirical formula CH (to 1 decimal place) is 12.0 + 1.0 = 13. Divide the relative molecular mass by this. If they are the same (result is 1) then the empirical formula and the molecular formula are the same. If the result is more than 1, multiply the empirical formula by the factor you have calculated to find the molecular formula. For styrene:

$$\frac{104}{13} = 8$$

The molecular formula of styrene is 8 times the empirical formula: $\mathbf{C_8H_8}$ These syrene molecules can join together to form polymers known as polystrene. These molecules contain hundreds of carbon and hydrogen atoms – so the empirical formula is useful again as polystrene has the formula $(C_8H_8)_n$ where $n > 50$

Finding the equation

Calculations with moles allow you to work out formulae and equations from measured masses or volumes of reactants and products. The worked example below shows the steps in this process.

Confirming an equation

Tomi thinks that the reaction between lithium and water at 25 °C is:

$$Li(s) + 2H_2O(l) \rightarrow Li(OH)_2(aq) + H_2(g)$$

Will disagrees. He thinks it should be:

$$2Li(s) + 2H_2O(l) \rightarrow 2LiOH(aq) + H_2(g)$$

They carried out the reaction in the lab and measured the quantities of some of the reacting materials.

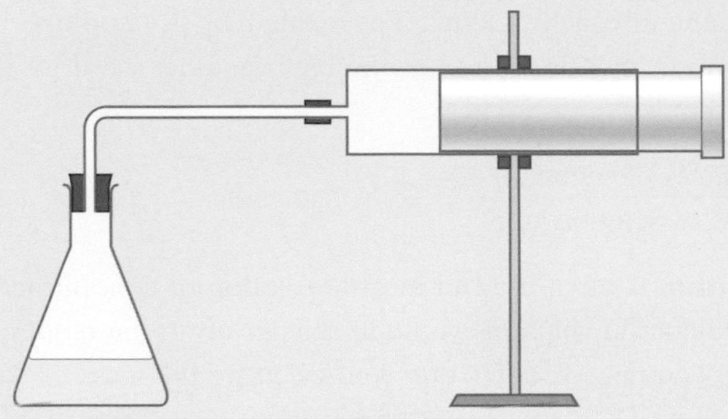

fig. 1.1.19 This apparatus can be used to collect the gas produced when lithium reacts with water, or any other reaction that gives off a gas.

They reacted 0.035 g of lithium with water and 60 cm^3 of hydrogen was collected in a gas syringe.

Use this information to confirm the correct equation for the reaction.

$A_r(Li) = 6.9$ g

$$\frac{0.035}{6.9} = 0.005 \text{ mol of lithium atoms}$$

1 mol of hydrogen molecules has a volume of 24 dm^3 (= 24 000 cm^3) at 25 °C

60 cm^3 of hydrogen molecules

$$= \frac{60}{24\,000} = 0.0025 \text{ mol of hydrogen molecules}$$

From this you can see that when 0.005 mol of lithium reacts with water, 0.0025 mol of hydrogen gas is formed. So when 2 mol of lithium reacts with water, 1 mol of hydrogen gas is formed. Therefore **Will's equation is correct.**

Questions

1 Calculate the mass of one mole of the following:

a As_2O_3 (the 'arsenic' of detective stories)

b PbN_6 (used in explosives to prime the charge)

c $Ca(NO_3)_2$ (used in matches)

d $Ca(C_6H_{12}NSO_3)_2$ (calcium cyclamate, an artificial sweetener).

2 Aluminium oxide, Al_2O_3, and hydrogen iodide, HI, react together as follows:

$$Al_2O_3(s) + 6HI(aq) \rightarrow 2AlI_3(aq) + 3H_2O(l)$$

a How many moles of hydrogen iodide would react completely with 0.5 mol of aluminium oxide?

b How many moles of aluminium iodide would this reaction form?

c What mass of aluminium iodide is this?

d What masses of aluminium oxide and hydrogen iodide would be required to produce 102 g of aluminium iodide?

3 The substance ATP is important for transporting energy in living cells. A sample of 1.6270 g of ATP was analysed and found to contain 0.3853 g of carbon, 0.05178 g of hydrogen, 0.2247 g of nitrogen and 0.2981 g of phosphorus. The rest was oxygen.

a Calculate the number of moles of each element present in the sample of ATP.

b Calculate the ratio of the numbers of moles of each element using the nearest whole numbers.

c Write down the empirical formula of ATP.

d The relative molecular mass of ATP is 507. What is its molecular formula?

Measuring concentration

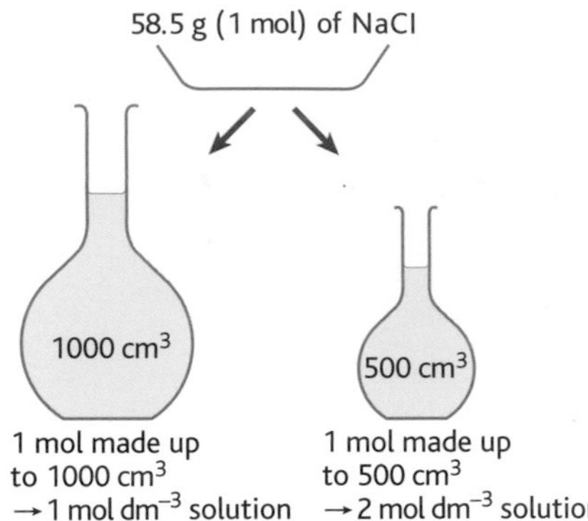

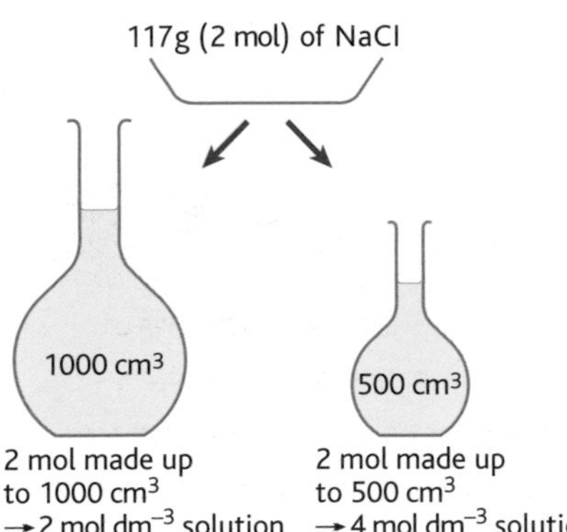

fig. 1.1.20 Making up solutions of different molarities.

Moles in solution

Many chemical reactions occur in **solution** – that is, with a **solute** (solid, liquid or gas) dissolved in a **solvent**. Scientists measure the **concentration** of a solution using **moles per cubic decimetre** (**mol dm^{-3}**).

The dm^3 is the unit of volume used in chemistry, rather than the litre used in everyday life, though the two units are effectively equivalent.

$$1 \text{ litre} = 1000 \text{ cm}^3 = (10 \text{ cm})^3 = 1 \text{ dm}^3$$

To make a solution of accurate concentration, you do not simply calculate the mass of solute you need then add it to 1 dm^3 of solvent – this would make a solution with a volume greater than 1 dm^3. Instead you dissolve the weighed solute in a small quantity of solvent, then add more solvent until the solution has a volume of exactly 1 dm^3. The symbol M (for **molar**) is sometimes used to mean mol dm^{-3}, and **molarity** is another word for concentration in mol dm^{-3}.

$$\text{Molarity} = \frac{\text{moles of solute}}{\text{volume of solution (dm}^3\text{)}}$$

The term **molar solution** is often used to mean a solution of concentration 1 mol dm^{-3}. When making up solutions you add solvent up to the required volume. If the volume you need is not 1 dm^3 you calculate the mass of solute accordingly.

Concentration calculations

Finding the mass of solute

In an experiment you need a 0.15 mol dm^{-3} solution of sodium chloride. How many grams of sodium chloride would you need to make 500 cm^3 of 0.15 mol dm^{-3} sodium chloride solution?

M(NaCl) = 23.0 + 35.5 g mol^{-1} = 58.5 g mol^{-1}

0.150 mol of NaCl is 58.5 g mol^{-1} × 0.150 mol = 8.8 g (to 1 decimal place)

Therefore to make 1 dm^3 of 0.150 M NaCl solution, you need to make 8.8 g of NaCl up to 1 dm^3 with water.

To make 500 cm^3 of 0.15 mol dm^{-3} NaCl solution you need **4.4 g of NaCl.**

Finding the molarity

If you weigh out 10.60 g of sodium carbonate, Na_2CO_3 and make it up to 250 cm^3 in a volumetric flask, what will be the molarity of the resulting solution?

First work out how many moles of Na_2CO_3 you have:

$M(Na_2CO_3)$ = (23.0 × 2) + 12.0 + (16 × 3) g mol^{-1}

= 46.0 + 12.0 + 48.0 g mol^{-1} = 106 g mol^{-1}

$$\text{Number of moles} = \frac{10.6 \text{ g}}{106 \text{ g mol}^{-1}} = 0.1 \text{ mol}$$

Now substitute into the equation:

$$\text{Molarity} = \frac{\text{moles of solute}}{\text{volume of solution (dm}^3\text{)}} = \frac{0.1 \text{ mol}}{0.25 \text{ dm}^3} = \mathbf{0.4 \text{ mol dm}^{-3}}$$

HSW Molarity and methodology

In any scientific investigation you need to be aware of any limitations in your technique and the types of errors and inaccuracies that can creep in. Then when you evaluate your method you can make realistic suggestions to improve it in future. There are a number of sources of inaccuracies and errors when making up solutions:

- measuring out the mass of solute – inaccuracies in the balance, mistakes in reading the scale, spilling solute after weighing, not transferring all the solute to the solution
- making up the solution – reading the level on the volumetric flask inaccurately so not adding exactly the right volume of solvent, not washing all of the solute into the solution
- changes in temperature – the volume of a solution changes as temperature goes up and down. You should note down the temperature when you make up a solution and its concentration will only be accurate at this temperature.

fig. 1.1.21 When working with solutions you need to remember all the possible sources of error.

HSW Molarity in medicine

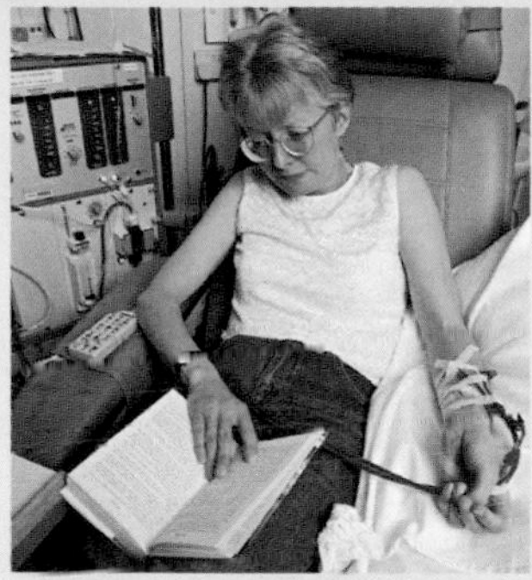

fig. 1.1.22 For kidney patients, solutes are kept at the correct concentrations in their blood using an artificial blood-filtering system known as dialysis.

Exact concentrations are very important in medicine. For example, the concentrations of sodium ions and potassium ions in your blood are vital for the regulation of your blood pressure and the working of your nervous system. Your kidneys control these concentrations – if levels in your blood are not normal this shows that your kidneys are not working properly.

The team at the renal unit at the Royal Infirmary of Edinburgh have developed a website which allows patients to see what doctors are looking for in a blood test. **Table 1.1.5** shows the normal ranges for a number of chemicals found in your blood. The quantities measured are very small. They include **micromoles** (1 μmol = 10^{-6} mol) and **millimoles** (1 mmol = 10^{-3} mol).

Solute	Normal blood level
Creatinine (a waste product of muscle metabolism)	60–120 μmol dm^{-3}
Urea (a waste product of protein metabolism)	3.5–6.5 mmol dm^{-3}
Na^+ (sodium ions)	135–145 mmol dm^{-3}
K^+ (potassium ions)	3.5–5.0 mmol dm^{-3}

table 1.1.5 Normal ranges for some blood solutes.

You can use table 1.1.5 to calculate the amount of the different substances in your body. For example, an average person has 5 dm^3 of blood. What is the minimum mass of sodium ions in the blood of an average person?

I mol of Na^+ has a mass of 23 g

I mmol of Na^+ has a mass of 23/1000 g

Minimum Na^+ concentration is 135 mmol dm^{-3}

So in 1 dm^3 of blood there will be $\frac{135 \times 23 \text{ g}}{1000}$ of sodium ions = 3.105 g

So the minimum mass of sodium ions in 5 dm^3 of blood is 3.105 × 5 = **15.525 g**

Other units of concentration

Percentage by mass

Sometimes molarity is not the best measure of concentration to use, eg you may need a quick comparison rather than calculating the number of atoms or moles involved.

A simple way of expressing concentration is as a percentage by mass or volume. To find the percentage by mass, you divide the mass of the solute by the mass of the whole solution and multiply by 100:

$$\text{Percentage by mass} = \frac{\text{mass of solute}}{\text{mass of solution}} \times 100\%$$

For example, if 100 g of solution contains 15 g of sodium chloride, the percentage of sodium chloride by mass is:

$$\frac{15\text{ g}}{100\text{ g}} \times 100\% = 15\% \text{ by mass}$$

Percentage by volume

For a mixture of two liquids or two gases, the percentage by volume can be useful:

$$\text{Percentage by volume} = \frac{\text{volume of one component}}{\text{total volume}} \times 100\%$$

For example, if you mix 5 cm^3 of ethanol with 45 cm^3 of water, the percentage by volume of ethanol would be:

$$\frac{5\text{ cm}^3}{5 + 45\text{ cm}^3} \times 100\% = \frac{5}{50} \times 100\% = 10\% \text{ by volume}$$

NUTRITIONAL INFORMATION

Typical Values	Per Biscuit	Per 100g
Energy	173kJ 41kcal	1734kJ 412kcal
Protein	0.9g	8.6g
Available Carbohydrate	6.5g	65.3g
of which sugars	1.9g	19.1g
Fat	1.3g	12.9g
of which saturates	0.5g	5.2g
Fibre	0.8g	8.0g

fig. 1.1.23 Food labels can give a lot of information – but how useful is it?

HSW Informing decisions about food

In recent years people have become more aware of the role of a healthy diet in reducing the risk of problems such as heart disease, diabetes or tooth decay. Food manufacturers now provide lots of information – a range of food labels may tell you the percentage of certain nutrients in the food, or the percentage of your daily requirements of different nutrients the food provides, or simply the mass in grams of different nutrients. This information may be presented as a table, a pie chart or a simple list – but how scientific is it, and how much help does it give?

Gill Cowburn and Lynn Stockley from the British Heart Foundation Health Promotion Research Group at the University of Oxford looked at 103 research papers investigating how useful people found food labelling. They discovered that many people claim to look at the labels, but relatively few actually understand all the information. Pie charts are of limited use unless the label explains what they mean. A comparison with the recommended dietary intake was helpful. Many of the studies were poorly carried out, and provided only self-reported data – in other words what people said about themselves and how they read and use labels, rather than the scientists observing what they actually do. They found very little existing evidence that food labelling has any impact on the quality of our diets.

Parts per million

fig. 1.1.24 An idyllic scene – but how clean is the air? Some local authorities in the UK, such as Tameside, provide air pollution levels on their websites.

For very small concentrations, percentages become less useful, eg a 0.01% solution is difficult to visualise. **Parts per million (ppm)** can be useful for low concentrations.

To find the concentration in parts per million you divide the mass of the solute by the total mass of the solution (or the gas mixture) and multiply by a million (10^6).

$$\text{Concentration} = \frac{\text{mass of component}}{\text{mass of solution}} \times 1\,000\,000 \text{ ppm}$$

For example, if there is 2 g of organophosphate pollution in 1.5 million grams of solution (eg river water), there will be:

$$\frac{2 \text{ g}}{1\,500\,000 \text{ g}} \times 1\,000\,000 \text{ ppm} = 1.3 \text{ ppm of organophosphate pollution}$$

Parts per million are often used for levels of pollutants in the air or in water. Parts per billion and parts per trillion are also widely used when measuring polluting chemicals present at very low levels (see **table 1.1.6**).

Description of air pollution	Low	Moderate	High	Very high
Sulfur dioxide (ppb, 15-minute average)	less than 100	100–199	200–399	400 or more
Ozone (ppb)	less than 50 (8-hour running average)	50–89 (hourly average)	90–179 (hourly average)	180 or more (hourly average)
Carbon monoxide (ppm, 8-hour running average)	less than 10	10–14	15–19	20 or more
Nitrogen dioxide (ppb, hourly average)	less than 150	150–299	300–399	400 or more
Fine particles ($\mu g\ m^{-3}$, 24-hour running average)	less than 50	50–74	75–99	100 or more

table 1.1.6 Parts per million (ppm) and parts per billion (ppb) are often used for the levels of pollutant gases, as you can see in these figures from Tameside.

Questions

1 A stock solution contains 10 mol dm^{-3} of hydrochloric acid. What volume of this solution should be dissolved in water to make 6 dm^3 of acid of concentration 0.1 mol dm^{-3}?

2 When solutions of potassium chloride, KCl and silver nitrate, $AgNO_3$ are mixed together, a precipitate of silver chloride forms.

a Write down a balanced molecular equation for this reaction.

b In an experiment, 24.0 cm^3 of 0.05 mol dm^{-3} silver nitrate solution exactly reacts with 15.0 cm^3 of potassium chloride solution. Calculate the molar concentration of the potassium chloride solution.

c How much KCl would be needed to produce 100 cm^3 of a solution of this concentration?

3 Using the data for the Royal Infirmary of Edinburgh (**table 1.1.5**), calculate the maximum and minimum mass of potassium ions you would expect to find in 1 dm^3 of the blood of a healthy individual.

4 a Why are self-reported data a poor basis for research evidence?

b What studies would be needed to determine if food labelling has an impact on dietary choices?

5 Using **table 1.1.6**:

a What is the minimum mass of CO you would expect to collect over 24 hours of very high air pollution?

b It is difficult to assess which pollutant is the biggest problem in Tameside. What other factors make the data more complex?

More ways to calculate equations

Equations from precipitation reactions

Another way of using calculations to confirm chemical equations is using precipitation reactions. For example, when you mix colourless solutions of potassium iodide, KI and lead nitrate, $Pb(NO_3)_2$ a bright yellow precipitate forms. This is lead iodide, PbI_2. The other product of the reaction is soluble potassium nitrate, KNO_3. You can filter out the yellow precipitate, then evaporate the solution to dryness to obtain the white solid potassium nitrate. The reaction can be represented by this unbalanced equation:

$$KI(aq) + Pb(NO_3)_2(aq) \rightarrow PbI_2(s) + KNO_3(aq)$$

How can you work out experimentally the proportions in which these compounds react together?

- Put 5 cm^3 of 1.0 mol dm^{-3} KI(aq) into each of seven different test tubes.
- Add 0.5 cm^3 of 1.0 mol dm^{-3} $Pb(NO_3)_2$(aq) to tube 1.
- Add 1.0 cm^3 of 1.0 mol dm^{-3} $Pb(NO_3)_2$(aq) to tube 2.
- Add 1.5 cm^3 of 1.0 mol dm^{-3} $Pb(NO_3)_2$(aq) to tube 3, etc.
- A yellow precipitate of lead iodide forms in each tube. Place each tube in a centrifuge and spin for the same length of time.
- Measure the depth of the yellow precipitate in each tube. The depth is a measure of the mass of the solid lead iodide precipitated in each tube.

As the volume of lead nitrate added increases, the amount of lead iodide produced increases too, shown by the increasing depth of the precipitate. This continues until 2.5 cm^3 of lead nitrate has been added, after which the amount of precipitate produced stays the same.

This tells you that all the reacting particles in the potassium iodide have been used up. So 5×10^{-3} mol of KI just reacts with 2.5×10^{-3} mol of $Pb(NO_3)_2$.

The reactants are in the proportion 2:1 – so 2 mol of potassium iodide reacts with 1 mol of lead nitrate. You know the overall equation for the reaction, so from this information you can work out the balanced equation.

$$2KI(aq) + Pb(NO_3)_2(aq) \rightarrow PbI_2(s) + 2KNO_3(aq)$$

This is a very simple example for which you could have written a balanced equation from first principles. However, it shows how the method is used for more complicated reactions when you do not know the products or the proportions in which the reactants react.

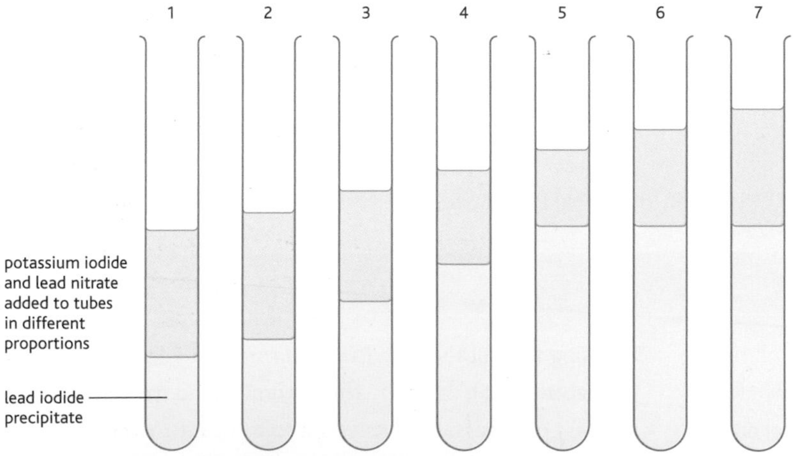

fig. 1.1.25 You can use the depth of precipitate produced as a measure of the mass of insoluble product.

Test tube	1	2	3	4	5	6	7
Volume of 1.0 mol dm^{-3} KI (cm^3)	5	5	5	5	5	5	5
Number of moles of KI	5×10^{-3}	5×10^{-3}	5×10^{-3}	5×10^{-3}	5×10^{-3}	5×10^{-3}	5×10^{-3}
Volume of 1.0 mol dm^{-3} $Pb(NO_3)_2$ (cm^3)	0.5	1.0	1.5	2.0	2.5	3.0	3.5
Number of moles of $Pb(NO_3)_2$	0.5×10^{-3}	1.0×10^{-3}	1.5×10^{-3}	2.0×10^{-3}	2.5×10^{-3}	3.0×10^{-3}	3.5×10^{-3}
Depth of precipitate (mm)	5	6	8	10	12	12	12

table 1.1.7 The depths of precipitate produced when different volumes of 1.0 M $Pb(NO_3)_2$(aq) are added to 5 cm^3 of 1.0 M KI(aq).

Determining equations

You can carry out a similar investigation yourself to determine the equation for the reaction between potassium chromate and barium chloride.

HSW More problems of methodology

Many errors might cast doubt on the validity of the results of this experiment, eg the height of a precipitate is affected by the shape of the tube (see **fig. 1.1.26**). Once the precipitate reaches the straight part of the tube, the results become much more reliable – but even then factors such as packing of the precipitate and whether the test tube is upright could cause unreliable results to creep in.

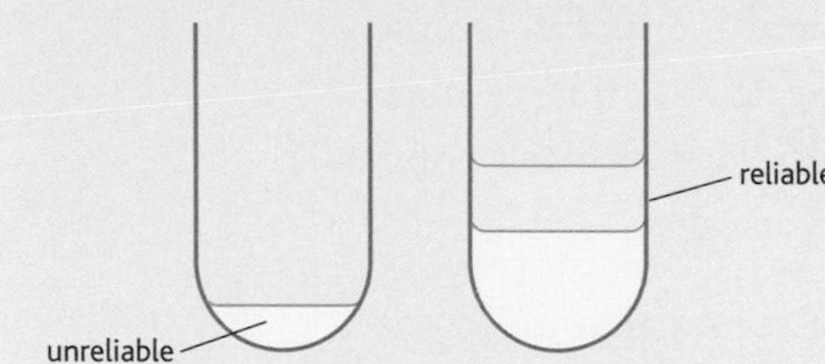

fig. 1.1.26 The depth of precipitate must be measured at the same place in each tube for reliable results.

HSW Measuring the active ingredients in medicines

The labels of medicines show the amounts of the different ingredients, eg iron tablets may contain 200 mg of iron, or a painkiller may contain 500 mg of paracetamol.
How accurate are they? You can check the iron content of iron tablets by grinding up a couple of tablets, dissolving them in dilute sulfuric acid and then titrating the solution against very dilute potassium manganate(VII) solution. The end point of the experiment is when the first permanent purple colour appears.

Comparing class results after such an investigation, you will probably see considerable variation in the amount of iron found in the tablets. This raises several questions.

- How much of this variation is due to experimental error, and how much to variation in the tablets? Think of possible experimental errors.
- How much variation do you think is acceptable in the tablets? In the pharmaceutical industry the amount of variation tolerated changes from drug to drug, from 99–101% of the stated value of active ingredient present to 95–105%. This latter figure gives quite a wide range. How can you have 101% – or 105% – of the stated dose?

Most quality control in the pharmaceutical industry does not involve titrations or other simple laboratory techniques. They are simply not reliable enough – particularly when working with medicines that are needed in micrograms or where an overdose might be harmful. Quality control is ensured in a number of ways. The precise quantities of ingredients for any batch of medicine are measured and recorded. Then random samples from each batch – 10–20 tablets for example – are selected and tested. The tablets or other medicines are crushed if necessary, and dissolved in a known solvent (often methanol). This is filtered and then diluted. Standard solutions are made accurately using pure active drug to a precise concentration. Both the random samples and the standard solution are passed through a high-performance liquid chromatograph (HPLC) and the quantities compared. HPLC is widely used to check the quantities of active ingredients in medicines. The results of this process show if the average content of the medicine lies within the accepted limits of variation.

Questions

1 Using **table 1.1.7**, draw a graph of depth of precipitate against volume of lead nitrate solution added and explain how your graph can help you write a balanced equation for the reaction.

2 Why might a wider level of variation be acceptable in over-the-counter medicines than in prescribed drugs such as cancer treatments?

Yields and atom economy

The yield of a reaction

Suppose you wanted to make 50 g of copper by reacting black copper(II) oxide with hydrogen:

$$CuO(s) + H_2(g) \rightarrow Cu(s) + H_2O(l)$$

Looking up molar masses and the molar volume of a gas, you should be able to calculate that to make 50 g of copper you would react 62.6 g of black copper(II) oxide with 18.9 dm^3 of hydrogen.

However, this calculation assumes that *all* of the copper(II) oxide will be turned into copper – that the **yield** of the reaction will be 100%. Yields approaching 100% are very rare in practical chemical reactions for various reasons. The reactants may not be pure. The reaction may not continue until all the reactants are used up because it is an equilibrium reaction (you will look at this later in the book). Some product is always left behind on the apparatus. Volatile products may evaporate. And human error is often a factor!

The yield is very important. In this example, if the practical yield was likely to be 50% (only half of the copper(II) oxide was converted into copper), you would need twice as much copper(II) oxide as you have calculated to make 50 g of copper.

You can find the yield of a chemical reaction very simply – eg the formation of copper sulfate from this reaction:

$$CuO(s) + H_2SO_4(aq) \rightarrow CuSO_4(aq) + H_2O(l)$$

As black solid copper(II) oxide is added to the sulfuric acid, it reacts to form a solution of copper sulfate, which is blue. Eventually all of the sulfuric acid is used up, and any further copper oxide added simply remains as a black solid, which is insoluble in water. The solid copper(II) oxide can then be filtered out, leaving a solution of copper sulfate. The water can be removed from this solution by evaporation, leaving blue crystals of copper sulfate, $CuSO_4.5H_2O$.

Calculating the yield

A student reacted 6.4 g of copper(II) oxide with excess sulfuric acid and obtained 14.7 g of blue copper sulfate crystals. Calculate the yield.

The balanced equation shows that 1 mol of copper(II) oxide should produce 1 mol of copper sulfate crystals.

$M(CuO) = 63.5 + 16.0\ g\ mol^{-1} = 79.5\ g\ mol^{-1}$

$M(CuSO_4.5H_2O) = 63.5 + 32.1 + (4 \times 16.0) + (5 \times 18.0)\ g\ mol^{-1}$

$= 249.6\ g\ mol^{-1}$

So 79.5 g of copper(II) oxide would produce 249.6 g of blue copper sulfate crystals if the yield of the reaction was 100%.

The number of moles in 6.4 g of copper(II) oxide is:

$$\frac{6.4\ g}{79.5\ g\ mol^{-1}} = 0.08\ mol$$

For a yield of 100%, this would produce 0.08 mol of copper sulfate crystals.

$0.08\ mol \times 249.6\ g\ mol^{-1} = 20.0\ g$ (to 1 decimal place)

But the actual mass of blue copper sulfate crystals produced was 14.7 g, a yield of:

$$\frac{14.7\ g}{20\ g} \times 100\% = \mathbf{73.5\%}$$

Double salts

Double salts are crystals that contain two different salts in a 1:1 ratio. The alums such as $KAl(SO_4)_2.12H_2O$ containing hydrated potassium and aluminium sulphates are one example. Ammonium iron(II) sulphate, $(NH_4)_2SO_4.FeSO_4.6H_2O$, is another example of a double salt.

Making a double salt

You can make ammonium iron(II) sulphate in the laboratory from first principles using iron filings, sulfuric acid and and dilute ammonia. You can calculate the theoretical and actual yield of product.

The idea of atom economy

The concept of **atom economy** was developed by Professor Barry Trost (1941–) at Stanford University in the USA. He argued that what really mattered in a reaction was not the overall yield, but the number of reacting atoms that ended up in the desired product. Any other compounds produced are by-products (often waste) which need to be disposed of and are therefore bad for the environment. A reaction with good atom economy is very efficient at turning reactants into desired products with little waste.

$$\text{Atom economy (\%)} = \frac{\text{mass of atoms in desired product}}{\text{mass of atoms in reactants}} \times 100\%$$

In practice, to compare these masses of atoms we use the relative formula masses of the chemicals involved. Professor Trost was awarded the Presidential Green Chemistry Challenge Award in 1998 and his ideas are increasingly applied to industrial processes around the world.

fig. 1.1.28 Barry Trost first developed the idea of atom economy.

Questions

1 A chemist mixed 12.0 g of phosphorus with 35.5 g of chlorine gas to synthesise phosphorus (III) chloride (phosphorus trichloride). The yield was 42.4 g of PCl_3. Calculate the percentage yield for this reaction. The equation for the reaction is:

$$2P(s) + 3Cl_2(g) \rightarrow 2PCl_3(l)$$

2 2-bromo-2-methylpropane, C_4H_9Br can be used to produce methylpropene, C_4H_8. 2-bromo-2-methylpropane reacts with sodium ethoxide, $NaOC_2H_5$ to produce ethanol, C_2H_5OH and sodium bromide, NaBr along with methylpropene. Using the relative formula masses of the reactants and products, work out the atom economy for this reaction.

HSW Atom economy and the production of ibuprofen

Ibuprofen is a widely used painkiller. For many years it was made by a six-step method developed in the UK in the 1960s. However, the atom economy of the process was only 40% – which meant 60% of the reactants ended up as waste products.

Relative formula mass of reactants = 514.5

Relative formula mass of ibuprofen = 206

Relative formula mass of waste products = 308.5

$$\text{Atom economy} = \frac{206}{514.5} \times 100\% = 40\%$$

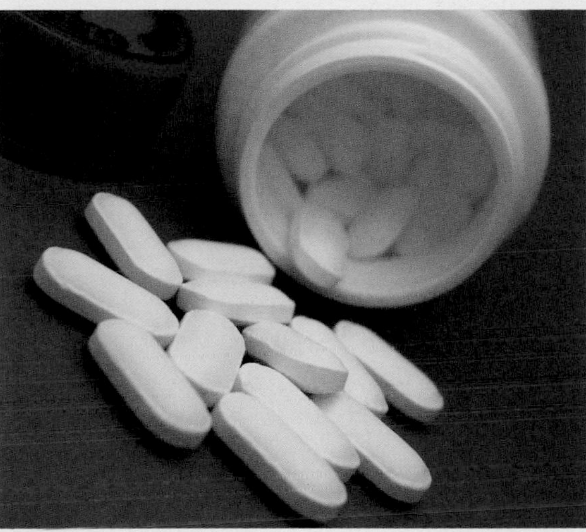

fig. 1.1.29 Around 13 million kilograms of ibuprofen are made each year. The old production method made almost 20 million kilograms of waste products – a high environmental price to cure a headache!

In the mid-1980s scientists worked hard to develop a new, greener way of manufacturing this useful drug. By the mid-1990s they had a three-step process in production with an atom economy of 77%. When you also consider that one of the by-products in the first step is ethanoic acid which can be recovered and is itself a useful end-product, the atom economy becomes an impressive 99%. This new synthesis substantially reduces the environmental impact as well as introducing financial economies into the process.

Relative formula mass of reactants = 266

Relative formula mass of ibuprofen = 206

Relative formula mass of waste products = 60

$$\text{Atom economy} = \frac{206}{266} \times 100 = 77\%$$

2 Energetics and enthalpy changes

Energy and energetics

In this chapter you are going to look at one of the most important areas of chemistry – the energy changes in reactions, and how these energy changes reflect changes at a molecular level.

Energetics is the study of energy transfers between reacting chemicals and their surroundings. In chemical systems, the most common way of transferring energy is by heating. Remember that this means the increased jostling of molecules that causes other molecules to vibrate more. It is not the movement of something called 'heat' – heat is a form of energy, not a substance. The study of these energy transfers is also known as **thermochemistry**.

Exothermic and endothermic reactions

For some chemical reactions to happen, they need energy from heating. Other reactions release energy as heat. A reaction that needs energy from heating to take place is **endothermic** ('endo' means within), and one that releases energy as heat is **exothermic** ('exo' means out).

Whether a reaction takes in or releases energy depends on what is happening to the chemical bonds in the reacting particles. Separating atoms or oppositely charged ions requires energy, while joining oppositely charged ions or atoms together releases energy – in other words:

- Bond breaking requires energy while bond making releases energy.

Both bond breaking and bond making happen in any chemical reaction. Whether the reaction is endothermic or exothermic depends on the balance between the energy required to break bonds and the energy released when new bonds are made. Different bonds have different energy requirements for making and breaking, as you will see.

Energy changes in exothermic reactions

In an exothermic reaction the energy released by bond formation in the products is greater than the energy needed to break the bonds in the reactants. In the exothermic reaction of the combustion of methane:

$$CH_4 + 2O_2(g) \rightarrow CO_2(g) + 2H_2O(g)$$

heat is given out. More energy is released when the new bonds are formed in the products (carbon dioxide and water) than is required to break the bonds in the reactants (methane and oxygen) and the reaction products store less chemical energy than the reactants.

Neutralisation reactions between acids and alkalis are also exothermic, and the rise in temperature produced during the reaction can easily be measured in the lab.

Energy changes in endothermic reactions

In endothermic reactions, the energy required to break the bonds in the reactants is greater than that released when new bonds are formed in the products, so the difference is absorbed from the surroundings. Endothermic reactions are less common than exothermic reactions. For example, the thermal decomposition of calcium carbonate to calcium oxide and carbon dioxide is an endothermic reaction.

Photosynthesis – the process by which plants make glucose and oxygen from carbon dioxide and water using energy from the Sun – is also an endothermic reaction.

fig. 1.2.1 Photosynthesis is perhaps the best known and most important endothermic reaction on Earth!

Enthalpy changes

An exothermic reaction releases energy to the surroundings in the form of heat. This energy is known as the **enthalpy change of reaction** (previously called the heat of reaction). The reaction in which the changes are happening is known as the **system**. Outside the system, everything else is known as the **surroundings**. In the example of the reaction of hydrochloric acid and sodium hydroxide, the acid, the alkali and the products of their reaction are the system. The test tube containing them, the person holding the test tube and the entire rest of the universe are the surroundings.

We think of the system as having a **boundary** which separates it from its surroundings. Sometimes this is a physical boundary – if it prevents particles from entering or leaving the system, the system is said to be **closed**. If the boundary also prevents energy from entering or leaving the system, the system is said to be **isolated**. However, in most cases the boundary does not physically exist but is just a theoretical distinction between a system and its surroundings.

fig. 1.2.2 **In thermochemistry we study the energy transfers between a system and its surroundings.**

All of the energy that leaves a system is transferred to the surroundings (or vice versa). The first law of thermodynamics states that:

- Energy cannot be created or destroyed.

The energy that is transferred to the surroundings is not lost to the universe – it is **dissipated**, or spread through the surroundings. This is an important principle in the behaviour of the universe which is usually called the **principle of conservation of energy**:

- The total energy content of the universe is constant.

Questions

1 Describe at the particle level what happens when energy is transferred as heat from a reaction to the surroundings.

2 What is the difference between an endothermic and an exothermic reaction? Explain how this difference arises.

3 Use fig. 1.2.3 to help you answer the following questions:

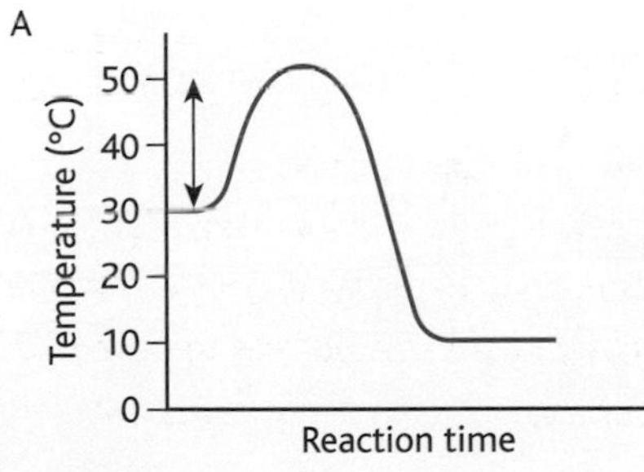

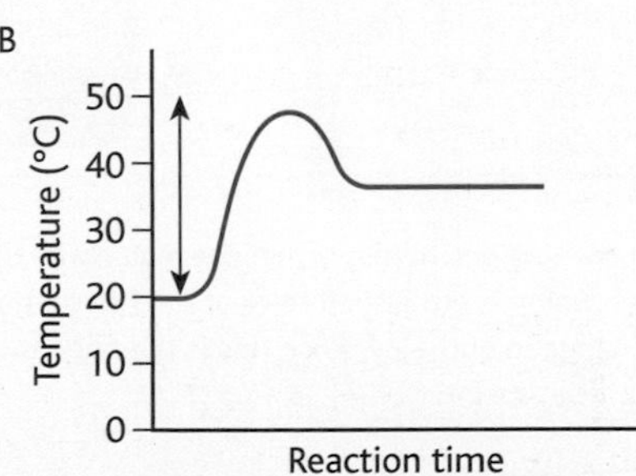

fig. 1.2.3 **The temperature changes during two chemical reactions.**

a Which of these reactions is exothermic and which is endothermic? Explain your reasoning.

b What was the temperature change in (i) the exothermic reaction and (ii) the endothermic reaction?

Measuring energy changes – enthalpy change

How much energy is transferred?

The amount of energy transferred in a given chemical reaction depends to some extent on the conditions under which the reaction occurs. Most chemical changes take place at constant pressure, eg in an open test tube. Chemists define the energy content of a system held at constant pressure as its **enthalpy**, represented by the letter ***H***.

You cannot measure the enthalpy of a system – but fortunately this does not matter! What you really want to know is the **enthalpy change** during a reaction. This is a measure of what the reaction can do for you (or even in some circumstances, what it can do to you!) under conditions of constant pressure. When a system reacts at constant pressure and gives out or takes in energy, we say that it undergoes an enthalpy change, represented as $\Delta \boldsymbol{H}$:

$$\Delta H = H_{\text{products}} - H_{\text{reactants}}$$

fig. 1.2.4 When burning petrol in a car engine, you really do not care what the total enthalpy is either before or after the reaction. All that matters is the change in enthalpy, since this is the only energy that is available to you.

Enthalpy level diagrams

For exothermic reactions, the enthalpy of reaction is *negative*, showing that the energy content of the system has *decreased* – some energy has left the system and entered the surroundings as heat.

$$H_{\text{products}} < H_{\text{reactants}}$$

so ΔH is negative.

During an exothermic reaction energy is transferred to the surroundings and there is a net decrease in the potential energy of the system.

For endothermic reactions, the enthalpy of reaction is *positive*, showing that the energy content of the system has *increased* as it has gained energy from its surroundings.

$$H_{\text{products}} > H_{\text{reactants}}$$

so ΔH is positive.

During an endothermic reaction, energy is transferred from the surroundings and there is a net increase in the potential energy of the system.

The enthalpy changes involved in reactions can be represented in an **enthalpy level diagram**.

In an exothermic reaction, the temperature of the products is higher than the temperature of the surroundings. Energy as heat is transferred to the surroundings as the products cool.

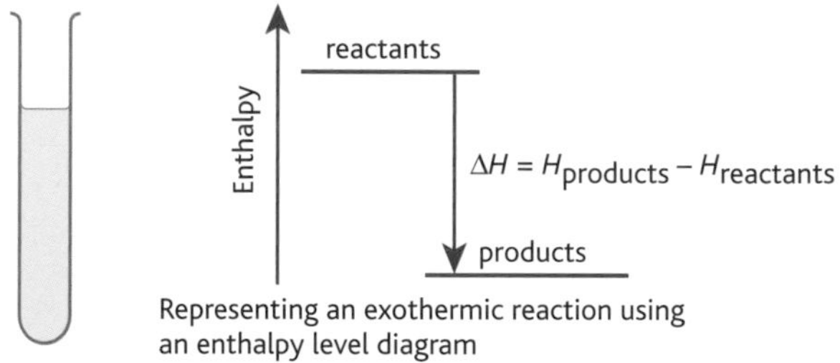

Representing an exothermic reaction using an enthalpy level diagram

In an endothermic reaction, the temperature of the products is lower than the temperature of the surroundings. Energy as heat is transferred from the surroundings as the products warm up.

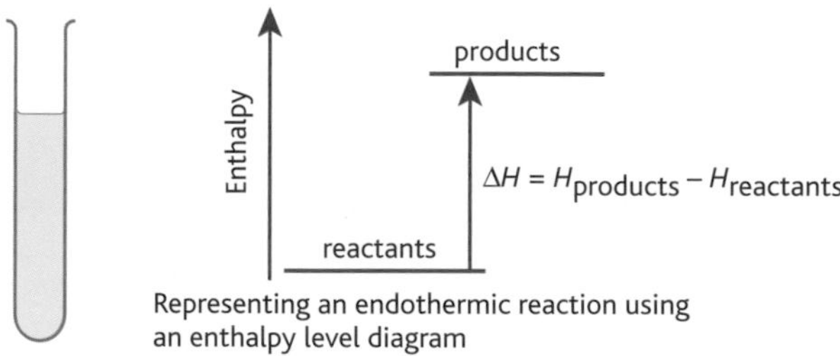

Representing an endothermic reaction using an enthalpy level diagram

fig. 1.2.5 Enthalpy level diagrams show the enthalpy changes in reactions.

HSW Heating and cooling packs

Heat packs

One common example of an exothermic reaction is the heat packs used by sportspeople to relax their muscles.

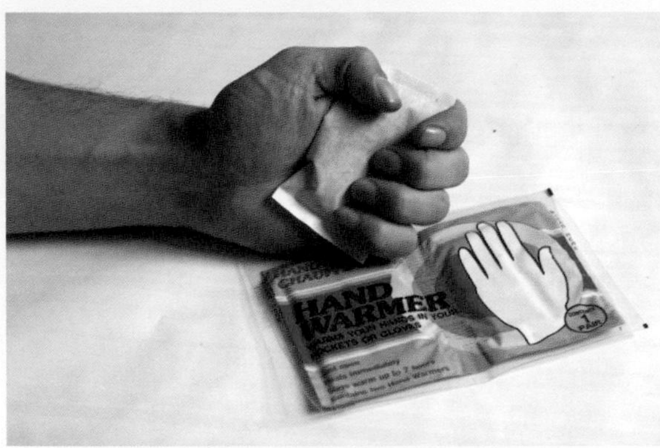

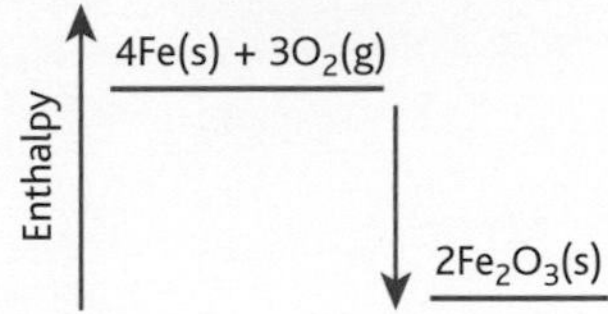

fig. 1.2.6 **Once the exothermic reaction is triggered, the pack gives out heat energy which can be used to relax the muscles before exercise or after injury. They can be used to keep your hands warm in winter as well!**

The heat in one such pack comes from the oxidation of iron, an exothermic reaction. A damp mixture of iron filings with salt and charcoal is contained in a perforated bag. This is in an initially sealed bag which keeps the oxygen out. The pack is activated by either squeezing or breaking a seal, allowing the oxidation reaction to take place, speeded up by the presence of the salt and charcoal.

$$4Fe(s) + 3O_2(g) \rightarrow 2Fe_2O_3(s)$$

Cold packs

Cold packs may be used to treat injury. One type uses the endothermic reaction that happens when ammonium nitrate and water are mixed.

$$NH_4NO_3(s) \rightarrow NH_4^+(aq) + NO_3^-(aq)$$

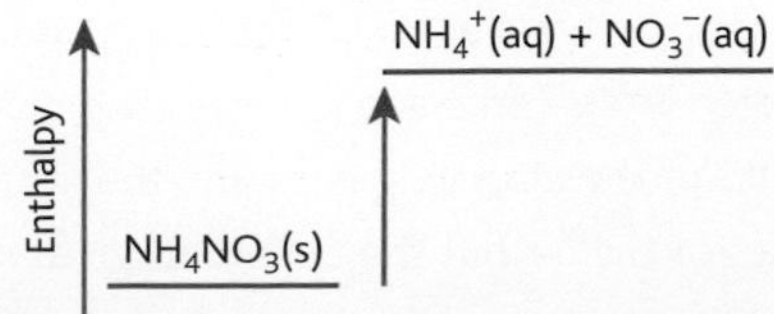

fig. 1.2.7 **Cold packs can be used to help reduce swelling and help recovery in sports injuries. They depend on endothermic reactions.**

HSW Endothermic reactions to save lives?

After a heart attack, brain cells die within minutes because they lack oxygen for respiration. The odds on surviving a heart attack after becoming comatose are not very good – 75% die then or soon afterwards – but occasionally, when people 'drown' in very cold water, they survive for more than half an hour without breathing. Their cooled brain tissues need a lot less oxygen.

At the Dandenong Hospital, Australia, doctors recently tested to see if mild hypothermia (cooling the body) would help heart attack patients. On odd-numbered days, comatose heart attack patients were cooled with ice packs on resuscitation and their core temperature kept low (32–34 °C) for the next 12 hours. On even-numbered days, patients were kept warm after resuscitation. Of the 43 patients cooled, 21 (49%) went back home or for rehabilitation compared with only 9 of the 34 patients (26%) who were kept warm.

A similar study at Bonn University followed 273 patients with an average age of 59. Within 6 months of their heart attack, 41% of those who received cooling treatment died, compared with 55% of those who were kept at normal body temperature. That represents a 14% increase in survival rate for the cooled patients. Among the patients that were cooled, 55% recovered well enough to live independently and work again, compared with only 39% of those who had not been cooled.

As a result of these studies and others, there is a move to design hoods and blankets using cool-pack chemicals to cool the body in emergency situations.

Questions

1 What is the difference between enthalpy and enthalpy change?

2 What is an enthalpy level diagram and why is it useful?

3 How does a cool pack work?

4 Present the data on the impact of cooling on heart attack patients in the two studies graphically. How reliable is the evidence collected? Evaluate the methodology of the trials described.

Heat capacities and calorimeters

You will know by experience that different materials need different amounts of heat to change their temperatures by similar amounts. The same applies to different volumes of the same material – it takes more energy to heat a bathful of water to 35 °C than to heat a cupful of water to the same temperature. The **heat capacity, *C*** of an object (or a body of liquid or air) is the amount of heat required to raise its temperature by 1 K. The SI unit of heat capacity is the joule per kelvin, J K^{-1}.

HSW Kelvin not Celsius

When measuring heat capacities you should use the Kelvin temperature scale rather than the Celsius scale. The Kelvin scale relates to the average kinetic energy of the particles and so reflects their movement at a given temperature.

At absolute zero, 0 K, the kinetic energy of the particles is zero and all particle movements cease. As the temperature rises, the average kinetic energy gets higher and particle movements increase. So a temperature change can be understood as a change in the average kinetic energy of the particles. Using the Kelvin scale helps build up an effective model of what is happening when substances are heated.

Whilst heat capacity is useful to know in certain circumstances, it applies only to a single object. **Specific heat capacity, *c*** gives the heat capacity per unit mass of a particular substance, and is obviously more generally useful. Its SI unit is the joule per kilogram per kelvin, J kg^{-1} K^{-1}. So:

- Specific heat capacity is the amount of energy in joules (J) needed to raise the temperature of one kilogram (kg) of a particular substance by 1 kelvin (K).

This allows you to calculate the amount of energy needed for a reaction. For example, for an endothermic reaction you can use the specific heat capacities of the substances involved and the enthalpy change of the reaction to find out how much energy you need to transfer to the reaction. So if you represent the amount of heat transferred as E and the temperature change as ΔT:

Energy transferred (J) = mass (kg) × specific heat capacity (J kg^{-1} K^{-1}) × temperature change (K)

$E = mc\Delta T$ for a mass m of a substance with a specific heat capacity c

In **table 1.2.1** you can see some examples of the specific heat capacity of a number of substances.

Substance	Specific heat capacity (J kg^{-1} K^{-1})
Carbon (graphite)	0.710×10^3
Copper	0.385×10^3
Ethanol	2.440×10^3
Gold	0.129×10^3
Water (liquid)	4.181×10^3
Iron	0.127×10^3
Lead	0.128×10^3

table 1.2.1 Specific heat capacities.

Measuring energy changes

Cup calorimetry

How do you measure the enthalpy change of a reaction? One widely used method involves carrying out the chemical change in an insulated container called a **calorimeter**, and measuring the temperature change that results. A simple form of this apparatus is a 'coffee-cup calorimeter', which is just two expanded polystyrene cups, one inside the other, with a lid. Because expanded polystyrene is an excellent insulator and also has a very low specific heat capacity, any temperature changes in the calorimeter can easily be measured before the heat finds its way out of the calorimeter. Errors can creep in depending on the uniformity of the insulating cups and the insulating properties of the lid, but the apparatus gives accurate enough results for most situations. **Figure 1.2.8** shows such a calorimeter, and the box on the right shows how it is used to make measurements.

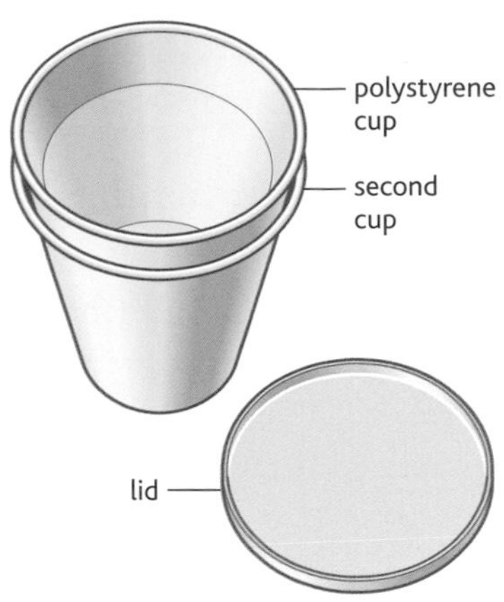

fig. 1.2.8 A coffee-cup calorimeter.

Measurements with a coffee-cup calorimeter

A student placed 25.0 cm^3 of 1.0 M HCl in a coffee-cup calorimeter, and measured its temperature as 22.5 °C. 25.0 cm^3 of 1.0 M NaOH, also at 22.5 °C, was added quickly to the acid. The mixture was stirred and the final temperature was recorded as 29.2 °C. If the specific heat capacity of the solutions is 4.2 J g^{-1} K^{-1} and if their densities can be taken as 1.00 g cm^{-3}, calculate the enthalpy change for this reaction per mole of HCl, assuming that no heat was lost to either the cup or the surroundings.

In the calorimeter, 50 cm^3 of solution increased in temperature from 22.5 °C to 29.2 °C, a rise of 6.7 K (remember that 1 °C is approximately equivalent to 1 K). Taking the total mass of the solutions as 50 g, the energy required for this temperature rise is given by:

$$E = mc\Delta T$$
$$= 50\ \text{g} \times 4.2\ \text{J g}^{-1}\ \text{K}^{-1} \times 6.7\ \text{K}$$
$$= 1407\ \text{J}$$

The student used 25 cm^3 of a 1.0 M solution of HCl, so the number of moles of HCl present is:

$$\frac{25\ \text{cm}^3}{1000\ \text{cm}^3} \times 1.0\ \text{mol}$$
$$= 0.025\ \text{mol}$$

So the enthalpy change, ΔH per mole of HCl is given by:

$$\Delta H = \frac{1407\ \text{J}}{0.025\ \text{mol}}$$
$$= 56.3\ \text{kJ mol}^{-1}\ \text{(to 3 significant figures)}$$

The enthalpy change for this reaction is **–56.3 kJ mol^{-1} of HCl used.**

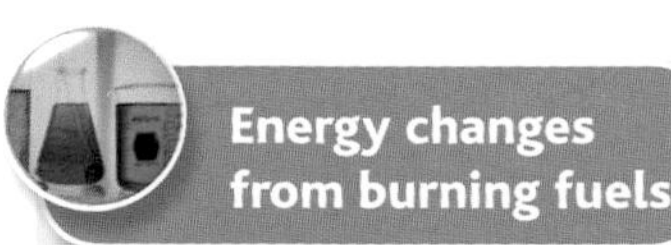

Energy changes from burning fuels

Another simple way of measuring energy changes that can be carried out in the school laboratory is shown in fig. 1.2.9. The heat given out by the burning fuel is used to warm a known mass of water. If you measure the mass of fuel burned and the temperature rise of the water, you can work out the approximate energy change that happens as the fuel burns (the heat of combustion).

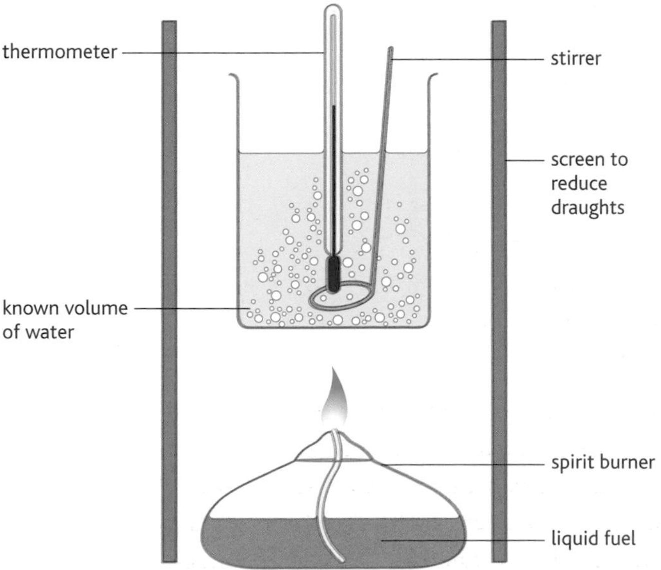

fig. 1.2.9 This is another direct method of calorimetry which is particularly useful for measuring the energy changes when fuels burn.

More accurate measurements – the bomb calorimeter

The calorimeter used to determine the energy change during a reaction accurately is known as a **bomb calorimeter** (so called because the sealed vessel in which the reactions are carried out looks something like a bomb). This apparatus is particularly useful for studying the enthalpy changes when a fuel burns – and it is also used to find out the 'calorific value' (that is, the energy content) of food. Figure 1.2.11 shows a simplified diagram of a bomb calorimeter and explains how it works.

In any calorimeter, however good the insulation, some heat is transferred to the surroundings, which adds an element of inaccuracy and unreliability to the results. This heat loss can be estimated and corrections can be calculated to compensate for it. However, the accuracy can be improved in a more sophisticated calorimeter. For example, for a combustion reaction, the substance is burned in a bomb calorimeter and the temperature change is recorded and graphed. Then an electrical heater is used to create exactly the same temperature change in the calorimeter. In this way the electrical energy needed to bring about the same temperature change can be calculated. This tells you more accurately the energy change that actually occurred during the reaction, because the measured energy change duplicates any heat losses from the calorimeter that occurred during the combustion reaction. So this method avoids the need for any heat loss calculations.

fig. 1.2.10 As you saw in chapter 1.1, information on food packaging usually includes the energy content of the food in kJ. This energy content is calculated by burning the food in a bomb calorimeter. The process of respiration is a slower version of the same reaction.

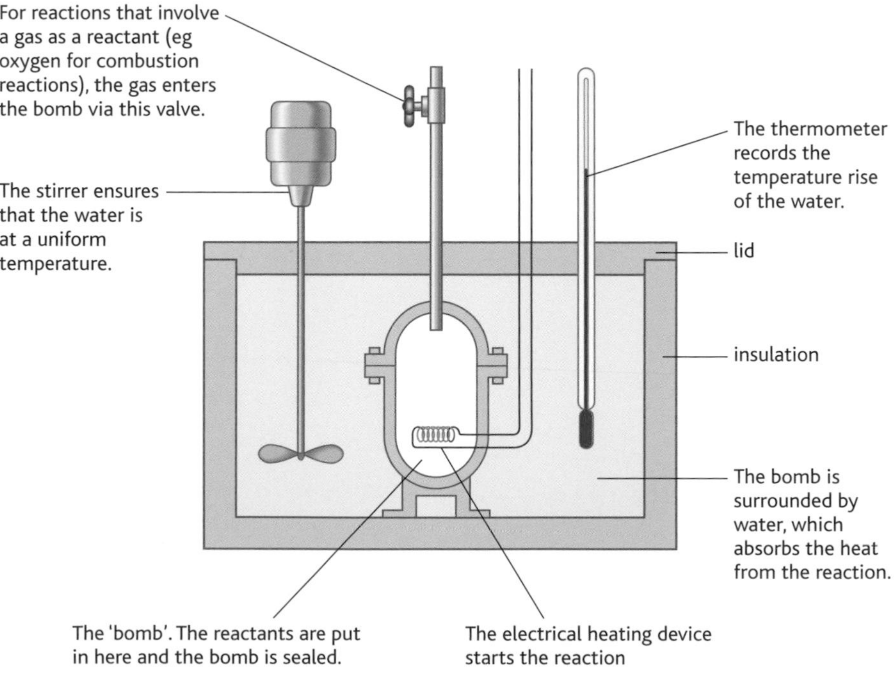

fig. 1.2.11 The energy released by the chemical reaction in the bomb can be calculated from the temperature rise of the water, the specific heat capacity of the water and the heat capacity of the bomb and its contents.

HSW Conditions for measuring enthalpy changes

The enthalpy change of a reaction depends on the physical conditions of temperature and pressure. So chemists specify a set of standard conditions under which enthalpy changes are measured. The standard conditions for thermochemical measurements are agreed as:

- 1 atm pressure and 298 K (25 °C).

An enthalpy change measured under these conditions for a reaction is referred to as a **standard enthalpy change of reaction**, and is represented by the symbol $\Delta H^{\ominus}$ or sometimes $\Delta H^{\ominus}(298\ K)$.

The enthalpy change for a reaction obviously also depends on the quantities of materials reacting – eg if you are burning more fuel, the enthalpy change for the reaction will be greater. Unlike specific heat capacity, enthalpy changes are expressed per mole of reaction as you saw in the worked example earlier, not per kilogram.

As well as temperature and pressure, you need to remember two other points about the standard conditions for a chemical reaction:

- The reacting substances must be in their normal physical states under the standard conditions. Where two or more forms of a substance exist, the most stable one at 1 atm and 298 K is used. So, for example, standard enthalpy changes involving sulfur relate to rhombic sulfur, which is stable at room temperature, not monoclinic sulfur, which is stable above 95.5 °C.
- Where solutions are used, they should have a concentration of 1 mol dm^{-3}.

So when you are presented with data from thermochemical experiments you can assume that the reaction took place under standard conditions of temperature and pressure, that the physical state that is most stable under those conditions was used, and that the quantities of reacting substances are known. You can also assume that the measurements have been taken in a calorimeter that is perfectly insulated so no heat energy has been transferred elsewhere. Other possible sources of inaccuracy, eg error in making up solutions or cooling air currents that might affect readings, are ignored. So when you consider the validity of any results, you need to bear these assumptions in mind.

Questions

1 Why is specific heat capacity a much more useful measure than heat capacity?

2 A gold ring with a mass of 5.5 g is warmed up from 25.0 °C to 28.0 °C. How much energy has been transferred to the gold? (specific heat capacity of gold is 0.129 J g^{-1} K^{-1})

3 A coffee-cup calorimeter contains 55.0 cm^3 of a dilute solution of copper(II) sulfate at a temperature of 22.8 °C. A small amount of zinc powder also at 22.8 °C is added to the solution. Copper metal is formed and the temperature of the solution rises to 32.3 °C. The copper is collected, dried and weighed, when it is found to have a mass of 0.324 g.

  a Calculate the total amount of energy released in this reaction, ignoring the heat capacity of the zinc and the calorimeter. Take the specific heat capacity of the solution as 4.2 J g^{-1} K^{-1}.

  b Calculate the enthalpy change for the reaction per mole of copper formed.

4 How might you increase the reliability of the data collected using:

  a a coffee-cup calorimeter

  b the direct calorimeter shown in fig. 1.2.9?

Important enthalpy changes

Representing standard enthalpy changes

You have seen that standard enthalpy changes of reaction are useful measurements that allow you to compare how much energy is given out or taken in at constant pressure by different reactions. There are a number of useful enthalpy changes that we will be looking at here. Some of these can be found experimentally, eg using a bomb calorimeter to determine the standard enthalpy of combustion; others need to be calculated using data measured from other related reactions.

A standard enthalpy change may be presented like this:
$\Delta H^{\ominus}_{c}$[C(graphite)] = –393.5 kJ mol^{-1}

This is what the symbols mean:

- ΔH: enthalpy change
- $^{\ominus}$: this superscript indicates standard conditions 1 atm pressure, 298 K (25 °C)
- $_{c}$: the subscript indicates the type of enthalpy change – c = combustion, f = formation, a = atomisation, etc.
- negative numbers: exothermic reactions – energy is given out to the surroundings
- positive numbers: endothermic reactions – energy is taken in from the surroundings.

Fractions of moles

Sometimes when you write thermochemical equations you may need to use fractions such as '½O_2'. This is so that you have the correct number of moles for the reaction. For example, if you are quoting the standard enthalpy change when carbon burns to form carbon monoxide (see below) you need to quote the enthalpy change for burning one mole of carbon. To get a balanced equation for the reaction forming carbon monoxide, you need half a mole of oxygen molecules (which provides one mole of oxygen atoms).

Standard enthalpy change of combustion

Chemists define the **standard enthalpy change of combustion, $\Delta H^{\ominus}_{c}$** of a substance as the enthalpy change that happens when one mole of the substance is completely burnt in oxygen under standard conditions. The term 'completely burnt' is important, because many elements form more than one oxide. For example, carbon can burn to form two oxides:

$$C(graphite) + \tfrac{1}{2}O_2(g) \rightarrow CO(g)$$
$$C(graphite) + O_2(g) \rightarrow CO_2(g)$$

Complete combustion of carbon produces carbon dioxide – so it is the enthalpy change in the second process that is the standard enthalpy change of combustion of carbon. In 'chemist's shorthand' this is:

$\Delta H^{\ominus}_{c}$[C(graphite)] = –393.5 kJ mol^{-1}

fig. 1.2.12 Combustion is usually a highly exothermic process!

Different fuels have different standard enthalpy changes of combustion (see table 1.2.2).

Fuel	Methane	Hexane	Methanol	Ethanol	Carbon	Hydrogen
$\Delta H^{\ominus}_{c}$ (kJ mol^{-1})	–890.3	–4163.0	–726.0	–1367.3	–393.5	–285.8

table 1.2.2 The standard enthalpy changes of combustion for some fuels.

Why do the standard enthalpy changes of combustion vary so much? When these fuels release energy, they combine with oxygen. The amount of energy involved in this depends on two factors:

- *The number of bonds that need to be broken and made* – this depends on the size of the molecule. For example, in methane there are just 4 carbon–hydrogen bonds, compared with hexane that has 5 carbon–carbon bonds and 14 carbon–hydrogen bonds, a total of 19 bonds. There is a lot more bond-breaking and bond-forming per mole when hexane is burnt in oxygen than when methane is burnt. Another example compares the combustion of methane and methanol. The same products are formed (CO_2 + H_2O), but methanol already contains an O—H bond. The energy released during combustion comes from the making of bonds with oxygen. If methanol already has one O—H bond made, it will give out less energy when it burns. As a general rule, the more oxygen a fuel has in its molecule, the less energy it will give out when it burns.

- *The type of bonds involved* – bonds between different elements involve different **bond energies**. For example, it takes more energy to break a carbon–hydrogen bond than a carbon–carbon bond. You will be looking at this in more detail later in this chapter. It means that the chemical make-up of the fuel and the bonds that need to be broken or made affect the enthalpy of combustion.

Standard enthalpy change of formation

Another useful enthalpy change is the energy released when a compound is formed. The **standard enthalpy change of formation, $\Delta H^{\ominus}_{f}$** of a compound is the enthalpy change when one mole of the compound is formed from its elements under standard conditions. For example, the standard enthalpy change of formation of methane (CH_4) is for the change:

$$C(graphite) + 2H_2(g) \rightarrow CH_4(g)$$

The accepted value for the enthalpy change of this reaction is –74.8 kJ mol^{-1}. Once again, we can use 'chemist's shorthand':

$$\Delta H^{\ominus}_{f}[CH_4(g)] = -74.8 \text{ kJ mol}^{-1}$$

The units of enthalpy of formation are kJ mol^{-1}, as we would expect. Note that 'per mole' refers to the *formation of one mole of the compound*, not to the quantity of elements reacting. If we say that 'the standard enthalpy change of formation of sodium carbonate, Na_2CO_3 is –1130.7 kJ mol^{-1}', this means that for every mole of sodium carbonate formed from its elements, the change in enthalpy is –1130.7 kJ.

For an element under standard conditions its enthalpy of formation is zero. This is sometimes useful in calculations.

Reaction	Standard enthalpy change of formation (kJ mol^{-1})
C(graphite) + $2H_2(g) \rightarrow CH_4(g)$	–74.8
C(graphite) + $O_2(g) \rightarrow CO_2(g)$	–394
2C(graphite) + $2H_2(g) \rightarrow C_2H_4(g)$	+52.2
C(graphite) + $2H_2(g)$ + ½$O_2(g) \rightarrow CH_3OH(l)$	–239

table 1.2.3 The standard enthalpy changes of formation for a number of compounds.

Standard enthalpy change of atomisation

The standard enthalpy change of atomisation, $\Delta H^{\ominus}_{at}$ of an element is the enthalpy change when one mole of its atoms in the gaseous state is formed from the element under standard conditions. The standard enthalpy change of atomisation of carbon (as graphite) is +716.7 kJ mol^{-1} – this is the enthalpy change for:

$$C(graphite) \rightarrow C(g)$$
$$\Delta H^{\ominus}_{at}[C(graphite)] = +716.7 \text{ kJ mol}^{-1}$$

Note that the standard enthalpy change of atomisation refers to the *formation of one mole of atoms of the element* – it is *not* the enthalpy change when one mole of the element is atomised. Atomisation is always endothermic, since it involves increasing the separation between atoms, which requires energy.
Standard enthalpy changes of atomisation may be calculated from other measurements (in the case of a solid, from the enthalpies of fusion and vaporisation together with its specific heat capacity), or they may be measured by spectroscopic means in the case of gases like oxygen.

Standard enthalpy change of neutralisation

The **standard enthalpy change of neutralisation, $\Delta H^\ominus_n$** is the energy change when an acid and a base react to form 1 mole of water under standard conditions.

The enthalpy change of neutralisation for dilute solutions of strong acids with strong bases is always close to –58 kJ mol^{-1}. The reason is that these acids and alkalis are fully ionised, so the reaction is the same in every case.

Enthalpies of neutralisation can be measured approximately by mixing solutions of acids and alkalis in a coffee-cup calorimeter.

$$H^+(aq) + OH^-(aq) \rightarrow H_2O(l) \quad \Delta H^\ominus = -57.9 \text{ kJ mol}^{-1}$$

For example:

$$HCl(aq) + NaOH(aq) \rightarrow NaCl(aq) + H_2O(l) \quad \Delta H^\ominus_n = -57.9 \text{ kJ mol}^{-1}$$

Questions

1 For each of the following, write an equation for the reaction in which the enthalpy change is the standard enthalpy change of combustion.

a $C_2H_6(g)$ b $CH_3OH(l)$

2 For each of the following, write an equation for the reaction in which the enthalpy change is the standard enthalpy change of formation.

a $K_2CO_3(s)$ b $CH_3CH_2OH(l)$

3 The standard enthalpy change of formation is defined using the term *standard conditions*. What does this mean and why is it necessary to have a standard set of conditions?

4 Which of the combustion reactions shown in table 1.2.2 is:

a the most exothermic

b the least exothermic?

c How would you explain the difference?

5 a In table 1.2.3, which reaction is endothermic?

b Why are all standard enthalpy changes of atomisation endothermic, when only a minority of standard enthalpy changes of formation are endothermic?

Working with enthalpy changes of reaction

Enthalpy changes for 'difficult' reactions

You have seen that the standard enthalpy change of combustion of carbon is –393.5 kJ mol^{-1}:

$$C(graphite) + O_2(g) \rightarrow CO_2(g) \qquad \Delta H_c^{\ominus} = -393.5 \text{ kJ mol}^{-1}$$

Suppose you wanted to know the enthalpy change for the decomposition of carbon dioxide to its elements:

$$CO_2(g) \rightarrow C(graphite) + O_2(g) \qquad \Delta H^{\ominus} = ?$$

This reaction would be very difficult to carry out. Fortunately the law of conservation of energy comes to our rescue – because the first reaction, to make carbon dioxide, has a standard enthalpy change of –393.5 kJ mol^{-1}, the enthalpy change for the reverse reaction must be +393.5 kJ mol^{-1}. If it were different the implication would be that energy had either disappeared somewhere or appeared out of nowhere. So the law of conservation of energy tells you that the enthalpy change for a reverse reaction has the same numerical value as the enthalpy change for the forward reaction, but the opposite sign. You cannot get energy out of nowhere – so the enthalpy changes for forward and reverse reactions must add together to give zero.

Taking this idea a bit further, think about the possible reactions of carbon with oxygen – carbon forms two oxides. You can draw a 'triangle of reactions' in which carbon reacts with oxygen to form carbon monoxide, which reacts with more oxygen to form carbon dioxide, which then decomposes to form carbon and oxygen again. This is shown in fig. 1.2.13(a).

Can you work out the energy changes involved here? The standard enthalpy of formation of carbon dioxide is –393.5 kJ mol^{-1}, while the standard enthalpy of combustion for carbon monoxide and oxygen forming carbon dioxide is –283.0 kJ mol^{-1}. On the basis of these figures, the law of conservation of energy tells you that the standard enthalpy of formation of carbon monoxide is:

$$[-393.5 \text{ kJ mol}^{-1} - (-283.0 \text{ kJ mol}^{-1})] = -110.5 \text{ kJ mol}^{-1}$$

The enthalpy level diagram in fig. 1.2.13(b) makes this clearer. This is exactly the figure you get if you measure the enthalpy change for the formation of carbon monoxide experimentally under standard conditions. Notice how the enthalpy change is zero if we go round a closed loop in this cycle, eg from $C + O_2 \rightarrow CO + ½O_2 \rightarrow CO_2 \rightarrow C + O_2$ again.

There are many cases where the law of conservation of energy can be used to combine enthalpy changes like this.

(a)

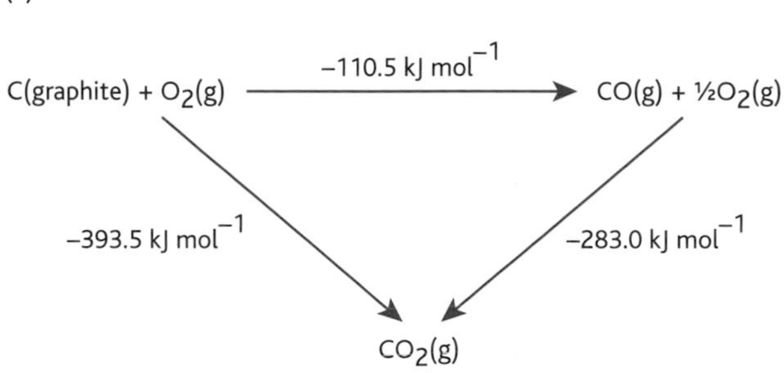

(b)

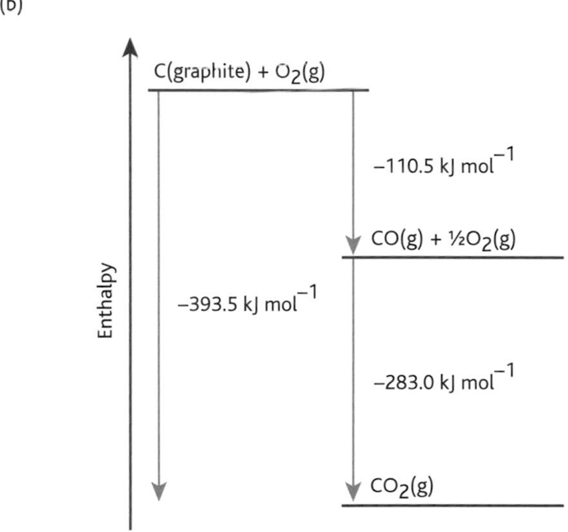

fig. 1.2.13 The energy changes in this cyclical process are governed by the law of conservation of energy, just like all other energy changes.

Hess's law

The chemist and doctor Germain Henri Hess used the law of conservation of energy to find a way of working out the energy changes in a reaction. He developed **Hess's law** which states that:

- The total enthalpy change for a reaction is independent of the route taken.

In other words, if we have a reaction:

A + B → C + D

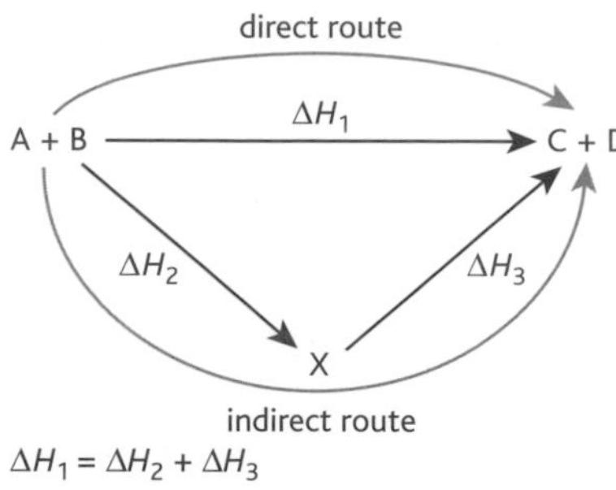

fig. 1.2.14 The law which chemists use when combining enthalpy changes for different reactions was named for the Swiss–Russian chemist, Germain Henri Hess (1802–50).

it does not matter whether you go directly from A + B to C + D, or whether you go via some intermediate compound X – the standard enthalpy change for the overall reaction will always be the same.

HSW Using Hess's law and energy cycles to calculate enthalpy changes

Hess's law is an example of a mathematical model that enables you to calculate standard enthalpy changes of formation for substances that might otherwise be difficult to measure. This might seem unnecessary – but in fact it is very useful for determining how stable compounds will be. This type of information is particularly useful for substances such as fuels.

One example of using Hess's law is for the formation of sucrose (table sugar) from carbon, hydrogen and oxygen:

$$12C(graphite) + 11H_2(g) + 5\tfrac{1}{2}O_2(g) \rightarrow C_{12}H_{22}O_{11}(s)$$

No chemist has ever been able to get this reaction to happen, so you cannot measure the enthalpy change directly. However, you *can* measure the standard enthalpy change of combustion of sucrose, and this can lead you to the standard enthalpy change for this reaction, as fig. 1.2.15 shows.

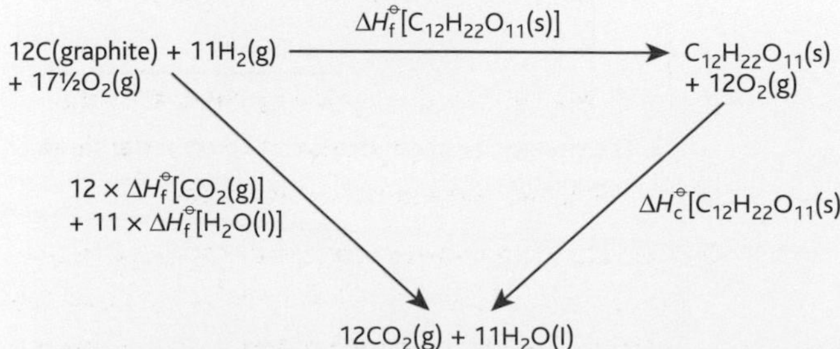

fig. 1.2.15 By measuring the standard enthalpy change of formation of carbon dioxide and water, and the standard enthalpy change of combustion of sucrose you can calculate the enthalpy change you want to know for the formation of sucrose. The extra oxygen is needed to balance the cycle.

Looking at the cycle of reactions in fig. 1.2.15, you can see that the standard enthalpy change of formation of sucrose can be calculated by applying Hess's law:

$$\Delta H_f^\ominus[C_{12}H_{22}O_{11}(s)] = 12 \times \Delta H_f^\ominus[CO_2(g)] + 11 \times \Delta H_f^\ominus[H_2O(l)] - \Delta H_c^\ominus[C_{12}H_{22}O_{11}(s)]$$

The figures for the enthalpy changes on the right-hand side of this equation are:

$$\Delta H_f^\ominus[CO_2(g)] = -393.5 \text{ kJ mol}^{-1}$$

$$\Delta H_f^\ominus[H_2O(l)] = -285.8 \text{ kJ mol}^{-1}$$

$$\Delta H_c^\ominus[C_{12}H_{22}O_{11}(s)] = -5639.7 \text{ kJ mol}^{-1}$$

Substituting these into the equation we get:

$$\Delta H_f^\ominus[C_{12}H_{22}O_{11}(s)] = 12 \times (-393.5) + 11 \times (-285.8) - (-5639.7) \text{ kJ mol}^{-1}$$

$$= -4722 + -3143.8 + 5639.7 \text{ kJ mol}^{-1}$$

$$= -2226.1 \text{ kJ mol}^{-1}$$

So the standard enthalpy of formation of sucrose (to 4 significant figures) is -2226 kJ mol^{-1}.

In this way Hess's law allows you to combine known data to calculate enthalpy changes that cannot be measured directly, eg the standard enthalpies of formation of hydrocarbon fuels.

Hydrogen peroxide is a powerful oxidising agent which is used to bleach paper and cloth, and also to turn hair blonde. Hydrogen peroxide decomposes slowly to form water and oxygen:

$$H_2O_2(l) \rightarrow H_2O(l) + \tfrac{1}{2}O_2(g)$$

Figure 1.2.16 shows the cycle of reactions that can be used to calculate $\Delta H^\ominus$ for this reaction.

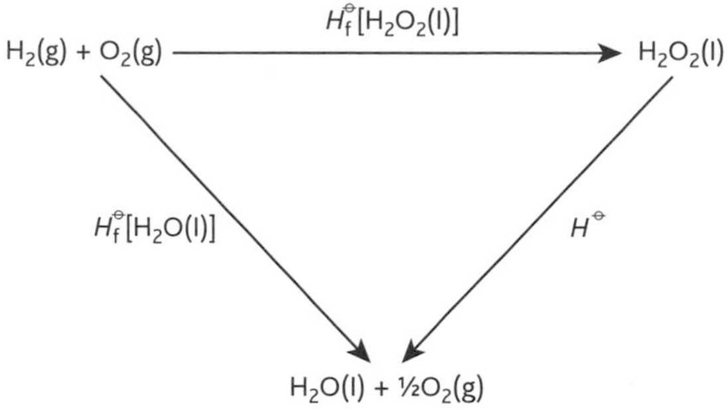

fig. 1.2.16 **You can use these reactions to calculate $\Delta H^\ominus$ for the decomposition of hydrogen peroxide.**

The standard enthalpy change of formation of hydrogen peroxide from its elements is –187.8 kJ mol^{-1}, and that for water from its elements is –285.8 kJ mol^{-1}. Remember that the enthalpy change of formation of an element in its standard state is zero, by definition. Applying Hess's law, the value of $\Delta H^\ominus$ for the decomposition of $H_2O_2(l)$ is given by:

$$\Delta H^\ominus = -\Delta H_f^\ominus[H_2O_2(l)] + \Delta H_f^\ominus[H_2O(l)] + \tfrac{1}{2}\Delta H_f^\ominus[O_2(g)]$$

$$= -(-187.8) + (-285.8) + 0 \text{ kJ mol}^{-1}$$

$$= -98.0 \text{ kJ mol}^{-1}$$

So the standard enthalpy change for the decomposition of hydrogen peroxide to water and oxygen is –98 kJ mol^{-1}.

Using Hess's law in this way, you can find $\Delta H^\ominus$ for any process using standard enthalpies of formation. In general, for any chemical process the standard enthalpy change $\Delta H^\ominus$ is given by:

$\Delta H^\ominus$ = sum of standard enthalpy changes of formation of products – sum of standard enthalpy changes of formation of reactants

Calculating $\Delta H^\ominus$

If the standard enthalpy changes of formation of nitrogen(II) oxide, NO and nitrogen(IV) oxide, NO_2 are +90.2 kJ mol^{-1} and +33.2 kJ mol^{-1} respectively, calculate $\Delta H^\ominus$ for the change:

$$NO(g) + \tfrac{1}{2}O_2(g) \rightarrow NO_2(g)$$

The standard enthalpy change for this process is given by:

$$\Delta H^\ominus = \Delta H_f^\ominus[NO_2(g)] - \Delta H_f^\ominus[NO(g)]$$

(Note that the standard enthalpy change of formation of oxygen is zero, by definition.) Substituting the data gives:

$$\Delta H^\ominus = +33.2 - (+90.2) \text{ kJ mol}^{-1}$$

$$= -57.0 \text{ kJ mol}^{-1}$$

For the reaction shown, **$\Delta H^\ominus$ is –57.0 kJ mol^{-1}.**

Questions

1 Define the term 'standard enthalpy change of combustion'. Explain how the standard enthalpy change of combustion may be used to calculate the standard enthalpy change of formation of a given compound indirectly.

2 Hydrogen chloride, HCl can be made by heating potassium chloride with concentrated sulfuric acid:

$$H_2SO_4(l) + 2KCl(s) \rightarrow 2HCl(g) + K_2SO_4(s)$$

Given the information:

$H_2SO_4(l) + 2KOH(s) \rightarrow K_2SO_4(s) + 2H_2O(l)$
$\Delta H^\ominus = -342$ kJ

$HCl(g) + KOH(s) \rightarrow KCl(s) + H_2O(l)$
$\Delta H^\ominus = -204$ kJ

calculate the standard enthalpy change for the formation of hydrogen chloride from potassium chloride and concentrated sulfuric acid.

3 The standard enthalpy change of combustion of propanoic acid is –1527.2 kJ mol^{-1}. Given that the standard enthalpy change of formation of water is –285.8 kJ mol^{-1} and that of carbon dioxide is –393.5 kJ mol^{-1}, calculate the standard enthalpy change of formation of propanoic acid.

Bond enthalpies and mean bond enthalpies

The idea of **bond enthalpy** is useful in calculating the energy change in a reaction involving covalent bonds. You have already seen that breaking bonds is endothermic and making bonds is exothermic. Bond enthalpy data can tell you which bond will break first in a reaction, how easy or difficult this bond breaking is and therefore how rapidly a reaction will take place at room temperature.

The energy needed to break a particular covalent bond, or the energy released when the bond is formed, is called the **bond dissociation enthalpy** (sometimes called bond energy).

H—H(g) → H(g) + H(g) +436 kJ mol^{-1}

Scientists have found that the bond enthalpy for a given bond does not vary much in different compounds. For instance, the C—C bond is very similar in ethane (C_2H_6) and butane (C_4H_{10}). These compounds behave in similar ways, so it is not surprising that their bonds are similar. However, the same type of bond in different compounds can have slightly different bond enthalpies depending on the combinations of other atoms in the molecule and their effect on the bonds.

Mean bond enthalpy

Chemists use the idea of **mean bond enthalpy** – the mean (average) value of the bond dissociation enthalpy of a particular type of bond over a wide range of different compounds. Mean bond enthalpies can be used to work out the enthalpy change in a reaction using Hess's law. Because they use average values taken from many different compounds, enthalpy changes calculated in this way are usually slightly different from those obtained experimentally.

Bond enthalpies apply to substances in the gaseous state. Many reactions involve solids, liquids or aqueous solutions, so this is another possible source of inaccuracy in your calculations.

As an example, consider methane, CH_4 which contains four C—H bonds. The bond enthalpy of the C—H bond in methane can be found from the change:

$CH_4(g) \rightarrow C(g) + 4H(g)$

which is sometimes called **atomisation** – in this case, the atomisation of methane.

The enthalpy change for the atomisation of methane can also be found by applying Hess's law to the series of changes shown in fig. 1.2.17.

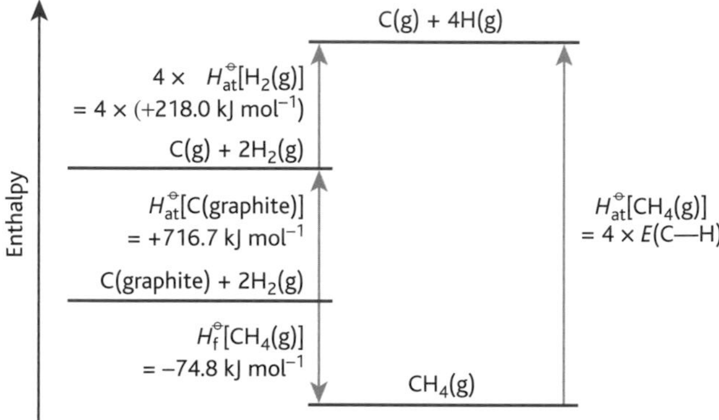

fig. 1.2.17 **Calculating the bond enthalpy of the C—H bond in methane.**

Figure 1.2.17 shows that $\Delta H^{\ominus}$ for the atomisation of methane is:

$$-(-74.8) + 716.7 + 4 \times 218.0 \text{ kJ mol}^{-1}$$
$$= 1663.5 \text{ kJ mol}^{-1}$$

This process involves the breaking of four C—H bonds, so the C—H bond enthalpy in methane, written as E(C—H) or ΔH_d(C—H), is:

$$E(\text{C—H}) = \frac{1663.5 \text{ kJ mol}^{-1}}{4}$$
$$= +415.9 \text{ kJ mol}^{-1}$$

This is the energy released when one mole of carbon atoms and one mole of hydrogen atoms combine to form one mole of C—H bonds, or it is the energy required to break one mole of C—H bonds. The calculation assumes that all four C—H bonds in methane are the same, so we have calculated the *average* C—H bond enthalpy. This is why we write E(C—H), meaning 'the average C—H bond enthalpy'. Bond enthalpy values will vary slightly depending on where they are found, eg the C—H bond enthalpy in methane will be slightly different from the value of the C—H bonds in ethanol, but these variations are overcome by using the *average* bond enthalpy.

Bond	ΔH_d (kJ mol^{-1})	Bond	ΔH_d (kJ mol^{-1})	Bond	ΔH_d (kJ mol^{-1})	Bond	ΔH_d (kJ mol^{-1})
C—H	413	C—Cl	346	C=C	612	N—N	163
C—O	358	H—H	436	C≡C	838	N≡N	945
C=O	743	H—O	463	C—N	286	H—F	565
C—C	347	H—N	388	C—F	467	F—F	158

table 1.2.4 These bond enthalpies are the mean values calculated for a number of polyatomic (many-atom) molecules.

The value for the C—H bond enthalpy calculated opposite for methane is very close to the C—H bond enthalpy average over a number of compounds in table 1.2.4.

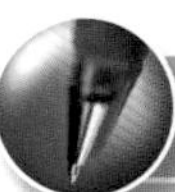

Enthalpy change of formation from bond enthalpies

Using the mean bond enthalpies in table 1.2.4, estimate the enthalpy change of formation of methanol vapour from its elements.

To carry out this calculation you also need to know the enthalpy changes of atomisation of the elements in methanol. These are as follows:

$\Delta H^{\ominus}_{at}$[C(graphite)] = +716.7 kJ mol^{-1}
$\Delta H^{\ominus}_{at}$[H_2(g)] = +218.0 kJ mol^{-1}
$\Delta H^{\ominus}_{at}$[O_2(g)] = +249.2 kJ mol^{-1}

The enthalpy level diagram in fig. 1.2.18 shows the enthalpy changes in the formation of methanol.

From the enthalpy diagram, the enthalpy of formation of methanol can be calculated from:

$\Delta H^{\ominus}_f$ = +716.7 + (4 × 218.0) + 249.2 – [(3 × 413) + 358 + 463] kJ mol^{-1} = –222 kJ mol^{-1}

The enthalpy change of formation of methanol vapour from its elements is **–222 kJ mol^{-1}.**

(Note that this is not a *standard* enthalpy change, since we are forming methanol in the vapour state, whereas methanol is actually a liquid under standard conditions.)

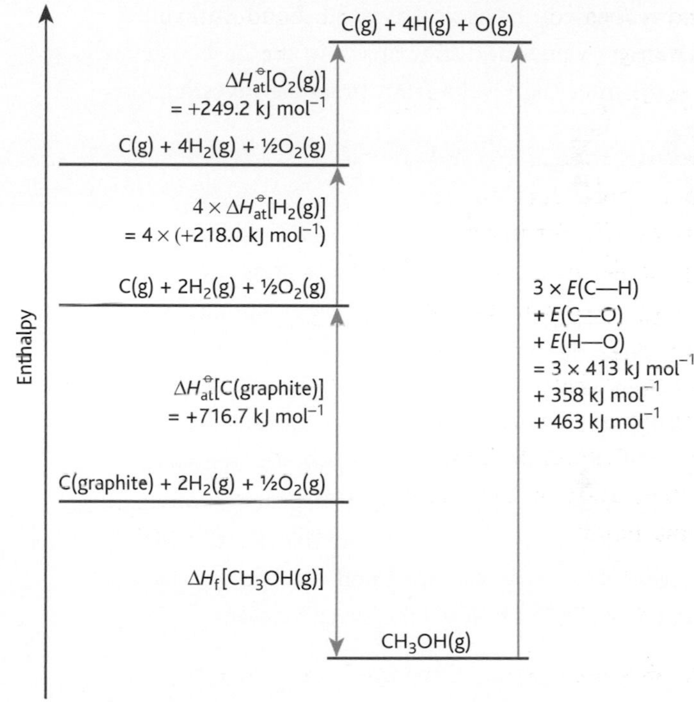

fig. 1.2.18 Enthalpy diagram for the formation of methanol.

Questions

1 a $C_2H_4(g) + H_2(g) \rightarrow C_2H_6(g)$
The enthalpy change of this reaction can be found by experiment to be –136 kJ mol^{-1}. Calculate the enthalpy change using the mean bond enthalpies given in table 1.2.4.

b The values will be different. Why is this?

c Sketch a fully labelled enthalpy level diagram for this reaction.

Using bond enthalpies

Although bond enthalpies are a theoretical concept, they are a valuable tool for chemists:

- They allow you to estimate the enthalpy changes in reactions.
- They can be used to compare the strengths of bonds between different atoms (see below).
- They help in building up an understanding of the structure and bonding of compounds.
- They also help in developing an understanding of the mechanisms of chemical reactions.

However, as you have already seen, bond enthalpies are average values and they apply to the gaseous state. So sometimes the results from practical investigations differ from the expected theoretical bond enthalpies. Practical work can show differences in the bond enthalpies of some bonds in different molecules – eg the carbon–carbon single bond ranges from 330 to 346 kJ mol^{-1}. This difference, although small, would be significant if you were calculating the enthalpy of formation of a compound containing a large number of carbon–carbon bonds, such as a large organic molecule. So if you use bond enthalpies in Hess cycle calculations, bear in mind that there may be some small discrepancies in the answers that you reach. On the other hand, sometimes the differences between theoretical values and experimental values can drive forward the development of new chemical ideas.

HSW Bond enthalpies and the changing benzene model

Benzene (C_6H_6) was first isolated in 1825 by Michael Faraday from whale oil. Over the years much was discovered about this unusual compound, but no one could decide on the structure. Then Friedrich August Kekulé, a German chemist, dozed off by the fire and in his dream saw snake-like molecules which held their own tails to make a ring structure. Kekulé held onto the image and in 1865 proposed his famous benzene ring model, with alternating single and double bonds. This structure was accepted for many years, until work on bond enthalpies raised doubts and changed the model for ever.

This simplified structure, leaving out the hydrogens, is often used to represent benzene.

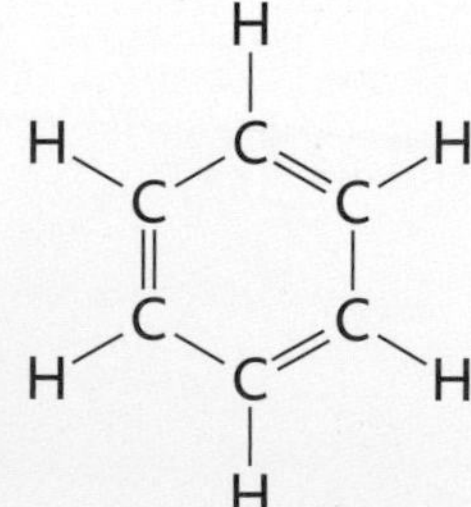

fig. 1.2.19 **Kekulé and his model of benzene with its different length single and double bonds.**

Cyclohexene contains one carbon–carbon double bond. The enthalpy change when this molecule undergoes an addition reaction with hydrogen is –120 kJ mol^{-1}.

If benzene has three C=C double bonds you would expect Kekulé's benzene structure to have an enthalpy change of –360 kJ mol^{-1} when reacting with hydrogen.

In fact, in experiments it was shown that the enthalpy change when benzene is hydrogenated is –208 kJ mol^{-1}. The molecule is more stable than the Kekulé model would lead you to believe.

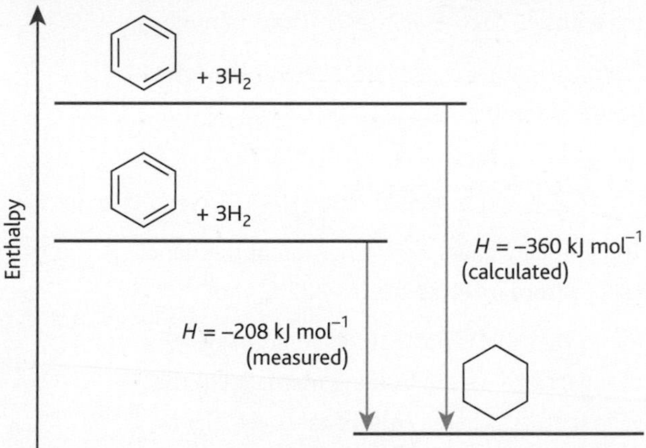

fig. 1.2.20 **The difference between the theoretical enthalpy change when benzene reacts with hydrogen and the changes seen experimentally show that benzene is more stable than expected.**

This difference was too big to ignore, so in the 1920s and 1930s chemists had to look again at the structure of benzene. Largely as a result of these bond enthalpy discoveries, Linus Pauling (an American Nobel prize-winning chemist) came up with a new model of benzene with six carbon–carbon bonds all of the same length and strength, all somewhere between a single and a double bond, which explained the properties of the molecule and is still used today.

Bond enthalpies can also help you to predict which bonds will break first in a reaction, and how easy it is to break the bond. The higher the bond enthalpy, the more energy is needed to break the bond, and the less likely it is to break in a reaction. Bonds with relatively low bond enthalpies are the easiest and therefore usually the first bonds to break. A reaction that involves breaking bonds with relatively low bond enthalpies is more likely to take place at room temperature than a reaction between molecules held by bonds with high bond enthalpies.

Other ways of using enthalpy changes

As you have just seen, bond enthalpies can be very useful to chemists. There are a number of other ways in which enthalpy changes of various sorts can be useful.

For example, if you know the enthalpy change of combustion of a range of fuels, you can work out which will be the most efficient in terms of energy released per gram of fuel burned. You can also work out which fuel produces most or – more important – least carbon dioxide for the energy produced.

Fuel	Formula	Enthalpy change of combustion (kJ mol^{-1})	Molar mass (g mol^{-1})
Carbon	C	–393.5	12
Methane	CH_4	–890.3	16
Methanol	CH_3OH	–726.0	32
Propane	C_3H_8	–2219.2	44
Butane	C_4H_{10}	–2876.5	58

table 1.2.5 The enthalpy change of combustion of some fuels.

So, for example, to discover which fuel in table 1.2.5 is the most efficient, divide the enthalpy of combustion by the molar mass. Methane is the clear winner (–55.6 kJ g^{-1}) with propane and butane close behind. Carbon (coal or charcoal) and methanol are much less efficient (–32.8 kJ g^{-1} and –22.7 kJ g^{-1} respectively).

Stability

Chemists often talk of the **stability** of a compound in relation to enthalpy changes – but what do they mean?

If you think about the example of hydrogen peroxide again, you saw that its standard enthalpy of formation is –187.8 kJ mol^{-1}. This tells you that one mole of hydrogen peroxide stores 187.8 kJ less energy than one mole of hydrogen and one mole of oxygen. Your everyday experience tells you that a decrease in energy is associated with an increase in stability (see fig. 1.2.21), and you can say that:

- hydrogen peroxide is more stable than its elements

or:

- hydrogen peroxide is stable with respect to its elements.

It is important to *compare* the stability of hydrogen peroxide with something (in this case, its elements). As you know, hydrogen peroxide decomposes to water and oxygen which suggests that it is unstable with respect to these two substances. The decomposition of hydrogen peroxide to water and oxygen is an exothermic change, which confirms this suggestion, and so we can say:

- hydrogen peroxide is stable with respect to its elements but unstable with respect to water and oxygen.

Nitrogen(II) oxide, NO is another interesting example of stability. The figures in the box at the top of page 45 show that nitrogen(II) oxide is unstable with respect to its elements, since its standard enthalpy of formation is positive (the reaction that forms it is endothermic). Yet nitrogen(II) oxide can be stored for long periods at room temperature, as can ethyne ($\Delta H_f^\ominus$ = +228.0 kJ mol^{-1}). The reason for this apparent contradiction is that the stability we have been considering so far is a particular type of stability, called **energetic stability** or **thermodynamic stability**.

Although nitrogen(II) oxide and ethyne are energetically unstable, they both have a different kind of stability called **kinetic stability**. If an energetically unstable substance has kinetic stability, its situation is very similar to the children's bricks in fig. 1.2.21. The tower of bricks is unstable, but it requires a small push to make it fall over so it can be stood on end indefinitely – at least until someone bumps into the the table! It is just the same with nitrogen(II) oxide and ethyne – although they are energetically unstable, they require a small 'nudge' to push them to decompose.

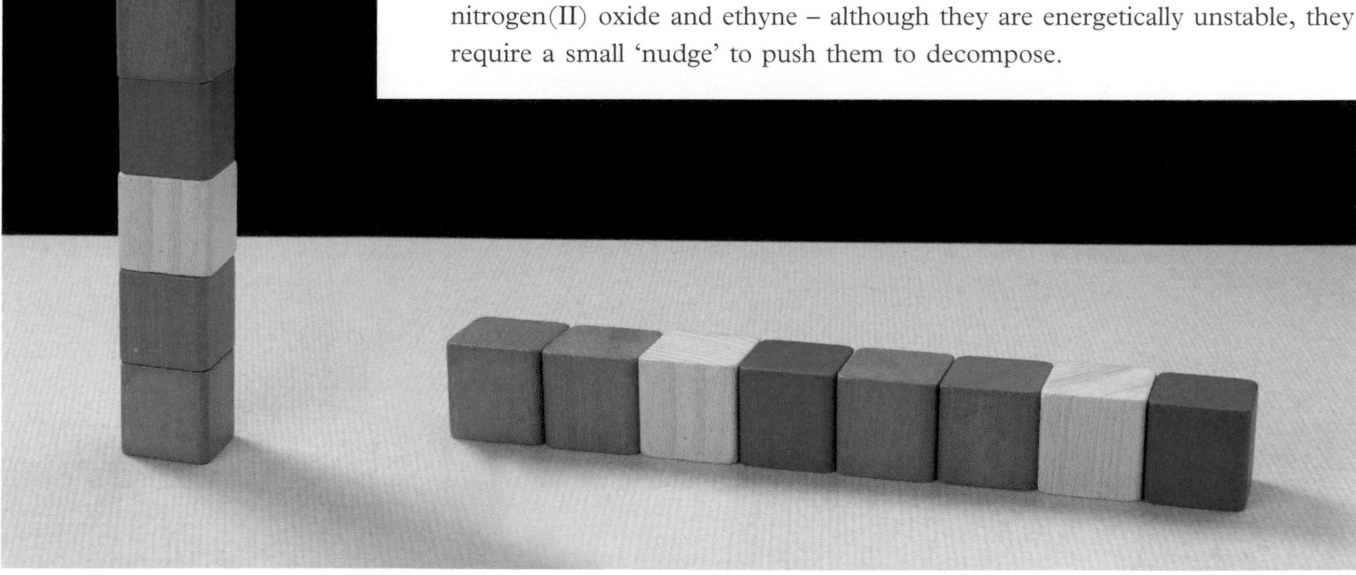

fig. 1.2.21 These bricks are less stable when they are built into a tower than when they are arranged flat on the bench because they have less potential energy lying flat than standing in a tower. The same principle can be applied to chemical atoms when they are arranged and bonded in different ways in molecules.

fig. 1.2.22 $\Delta H^{\ominus}$ for the reaction C(diamond) to C(graphite) is –1.9 kJ mol^{-1}, so energetically these diamonds are destined to become graphite. However, the kinetic stability of diamond means that this is likely to take a very long time.

HSW Will it or won't it?

The example of the reaction of nitrogen(II) oxide demonstrates how careful you need to be when talking about the stability of substances. Although nitrogen(II) oxide is kinetically stable with respect to its elements, it is kinetically and energetically unstable with respect to nitrogen(IV) oxide in the presence of oxygen. Exposed to the air, colourless nitrogen(II) oxide therefore rapidly forms brown fumes of nitrogen(IV) oxide.

Chemists often use the value of $\Delta H^{\ominus}$ to predict whether a particular reaction is likely to happen or not. Exothermic reactions form products that are more energetically stable than the reactants, suggesting that reactions with a large negative value of $\Delta H^{\ominus}$ are very likely to happen. Although many spontaneous reactions *are* highly exothermic, the value of $\Delta H^{\ominus}$ alone is not sufficient to make predictions about whether a particular reaction is likely to 'go' or not. There are several reasons for this:

1 $\Delta H^{\ominus}$ tells us about the *energetic* changes in a reaction, but tells us nothing about the *kinetic* stability of the reactants. The enthalpy change of combustion of petrol is enormous, yet the safety of much of our system of road transport relies on the fact that a mixture of petrol and air is kinetically stable. So petrol does not burn until it has the 'push' of heat being applied.

2 $\Delta H^{\ominus}$ tells us about enthalpy changes under *standard conditions*. Actual enthalpy changes under the very different conditions of temperature, pressure and concentrations found in most laboratories are often very different from the carefully controlled standard changes.

3 Other factors apart from enthalpy changes often affect chemical changes. These factors are concerned with the way in which the system and its surroundings are organised. You will look at some of these factors later in the course.

Questions

1 Using the data provided in table 1.2.5, which fuel would you recommend using to produce the least carbon emission for the most energy?

2 Explain how inaccuracies can occur in the measurement and calculation of bond enthalpies.

3 In benzene a discrepancy of 182 kJ between the experimentally measured and calculated bond enthalpies led to Pauling's new model of benzene. A similar discrepancy for a 20-carbon carbohydrate molecule would be ignored. Explain the difference in the response of scientists in these two cases.

4 The rocket fuel hydrazine (N_2H_4) reacts with fluorine to form gaseous nitrogen and hydrogen fluoride. Write a balanced equation for this reaction. Using data from table 1.2.4 and any other data you need from a data book, calculate:

a the energy needed to break the bonds in this reaction

b the energy needed to form the bonds in the products (remember that nitrogen gas involves a triple N≡N bond)

c the enthalpy change involved in the combustion of hydrazine with fluorine.

3 Atomic structure and the periodic table

In chapter 1.1 you saw how calculations can help chemists work out what happens in reactions, using the following quantities:

- The **relative atomic mass** of an element is the average mass of its isotopes compared with the mass of an atom of carbon-12.
- A single **atomic mass unit** is the mass of a carbon-12 atom divided by 12.
- For a compound, the **relative formula mass** is the sum of the relative atomic masses of all the atoms in the chemical formula. For a molecular compound this is referred to as the **relative molecular mass**.

How can you measure these quantities?

Measuring the mass of an atom

To find the relative atomic mass of an atom, you need to measure its mass and compare it with the mass of an atom of carbon-12 (^{12}C). But atoms are far too small for you to measure their mass directly by weighing. An instrument called a **mass spectrometer** provides the answer (see **fig. 1.3.1**).

1 The sample being measured must be in the gaseous state for its particles to move through the machine. The sample is injected into the mass spectrometer and is first vaporised.

2 The vapour is bombarded with high-energy electrons, which collide with atoms of the sample. They knock one or more electrons out of the atoms within the sample particles to form positive ions. Losing electrons makes no significant difference to their mass, but because the particles are now charged they can be accelerated in an electric field.

3 The ions are accelerated by an electric field.

4 They pass through a velocity selector, which makes sure they are all travelling at the same velocity. This means any differences in the effect of the magnetic field in the next section will be due to different masses or charges of the ions, not to any difference in their speeds.

5 The ions enter a uniform magnetic field, which deflects them. The amount they are deflected depends on both the mass of the ion and the charge on it. Heavier ions are deflected less than lighter ions, and ions with a small positive charge are deflected less than ions with a bigger positive charge. The strength of the magnetic field is steadily increased. At any particular setting, only ions of one particular mass:charge ratio will pass through and be detected – any other ions will be deflected too much or too little to pass through (see **fig. 1.3.1**).

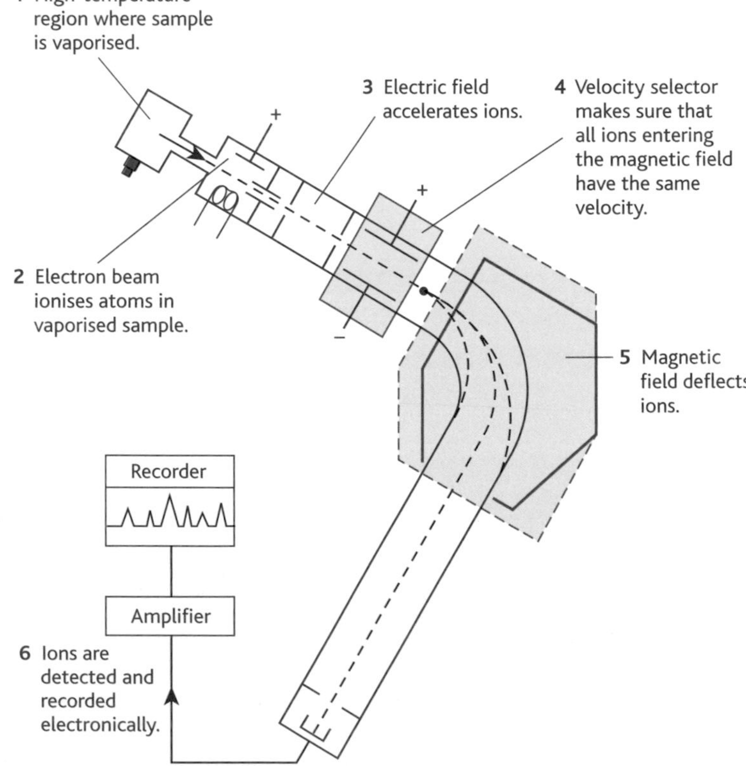

fig. 1.3.1 **The mass spectrometer is a vital tool for chemists working in many different fields.**

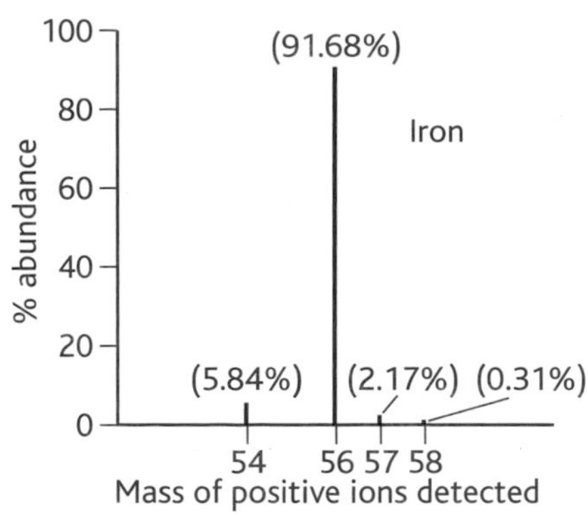

fig. 1.3.2 There are four peaks on the mass spectrum of iron, representing the four isotopes of iron.

6 The detector detects how many ions pass through the machine at each magnetic field setting and each setting of the velocity selector. It shows how many ions of each different mass:charge ratio ($^m/_z$ value) there are in the sample.

A **mass spectrum** is obtained as a result (see **fig. 1.3.2**). The relative heights of the peaks show the relative abundance of the different ions present. This gives you a picture of the relative abundance of the different isotopes of each atom. You will soon find out how to interpret this type of data yourself.

HSW The mass spectrometer in the pharmaceutical industry

Over the last 15 years or so, mass spectrometers have been used increasingly by pharmaceutical industries. The development of a new drug costs millions of pounds (around US$800 million) and it takes around 12–15 years before a new drug reaches the market. The mass spectrometer is a very useful tool for several reasons. It is very sensitive, so small samples can be measured very accurately. It can also differentiate accurately between very similar compounds, so it can detect isomers (different molecules that have the same atoms, but arranged differently in space) and other molecules that have only small differences between them. Isomers are important in drug development because the human body is very sensitive to them. You will find out more about this when you study organic chemistry.

In the pharmaceutical industry the mass spectrometer is usually used in combination with high-performance liquid chromatography (HPLC). This is very similar to the paper chromatography you will be familiar with. In HPLC a mixture of chemicals is injected into one end of a steel tube which is packed with silica, SiO_2. The compounds in the mixture are separated as solvent is pumped through the tube at high pressure. Once the compounds have been separated they are passed through the mass spectrometer for detailed analysis.

Modern techniques allow hundreds of thousands of compounds to be synthesised at the same time. Using HPLC along with mass spectrometry, chemists can quickly identify compounds that might have potential biological activity by measuring the relative molecular mass very sensitively. As a result possible drug candidates are identified early and no time is wasted on useless compounds.

Only one compound in every 4000–10 000 makes it past the early stages of research to be tested for biological activity in the lab, and thousands more fail during *in vitro*, animal and human testing. Very few indeed go on to make an effective drug ready for use. Mass spectrometry with HPLC is used throughout the development process, eg to identify the breakdown products of a drug in the body or to assess the purity of a sample.

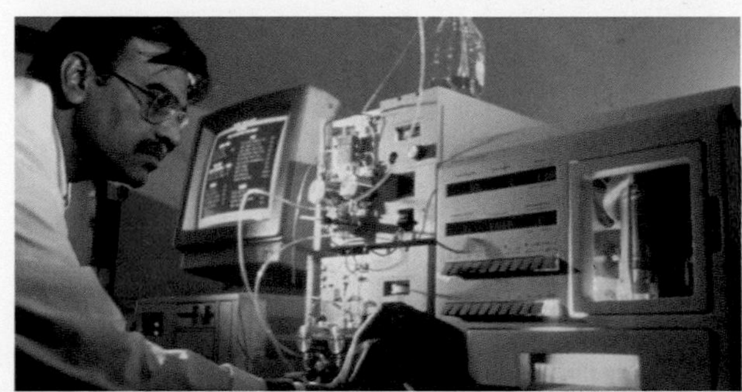

fig. 1.3.3 Using HPLC

Questions

1 Draw a flow chart to show the journey of a chemical through a mass spectrometer.

2 List at least four advantages of using mass spectrometry in drug development.

Using data from a mass spectrometer

A mass spectrum can look daunting, but in fact it is relatively straightforward to interpret and use the data. The peaks on the mass spectrum of an element show you the different isotopes of the element.

Finding the relative atomic mass from a mass spectrum

(a)

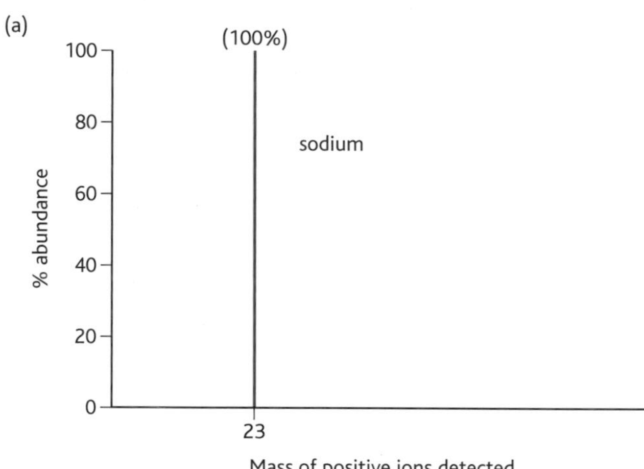

(b)

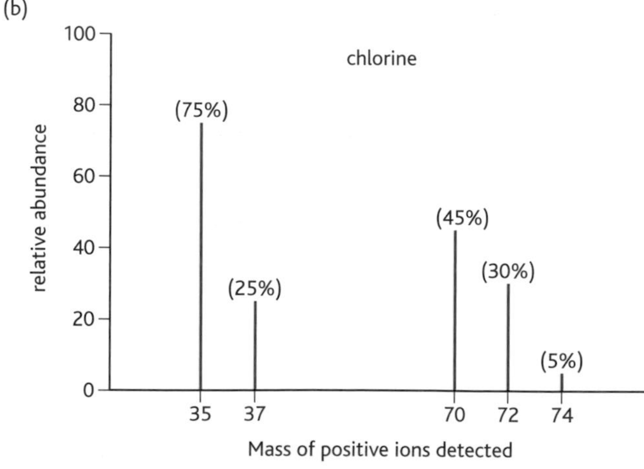

fig. 1.3.4 **(a) The mass spectrum of sodium. (b) The mass spectrum of chlorine.**

As **fig. 1.3.4** shows, the mass spectrum of sodium consists of a single line. This tells us that all the atoms in a sample of sodium have the same mass. Sodium is a very simple case – it has a relative atomic mass of 23. Sodium has an atomic number of 11, so sodium atoms contain 11 protons and 12 neutrons.

The mass spectrum of chlorine contains two regions, one corresponding to the Cl atom with two lines, and the other corresponding to the Cl_2 molecule with three lines. The first region tells you that a chlorine atom may have one of two possible masses, each of which has a different relative abundance. These are the different isotopes of chlorine, and the mass spectrometer gives you their relative masses.

Relative atomic mass

Using fig. 1.3.4(b), work out the relative atomic mass of chlorine.

The mass spectrum shows that chlorine atoms have two peaks, at 35 and 37, telling you there are two isotopes of chlorine, ^{35}Cl and ^{37}Cl. Chlorine has atomic number 17, so ^{35}Cl has 17 protons and 18 neutrons while ^{37}Cl has 17 protons and 20 neutrons.

As you saw in chapter 1.1, you need to calculate the weighted mean to take into account the relative abundances of the different isotopes:

$A_r(Cl) = (35 \times 75\%) + (37 \times 25\%)$

$A_r(Cl) = \mathbf{35.5}$

Figure 1.3.4(b) also tells you about the relative abundances of the different molecules of chlorine. A chlorine molecule is made up of two chlorine atoms, each of which could be ^{35}Cl or ^{37}Cl.

- The peak at 70 corresponds to ions produced from chlorine molecules made up of two ^{35}Cl atoms.
- The peak at 72 represents molecules made up of one ^{35}Cl and one ^{37}Cl atom.
- The final peak at 74 represents molecules made up of two ^{37}Cl isotopes.

This technique is very useful for calculating the exact relative atomic and molecular masses of substances. For example, in natural ores silicon is found in the proportions shown in **table 1.3.1**.

Isotope	Relative atomic mass	Percentage abundance (%)	Mass abundance
Silicon-28	28	92	25.76
Silicon-29	29	5	1.45
Silicon-30	30	3	0.9

table 1.3.1 The relative abundance of silicon isotopes in a sample. $\frac{A_r}{M_a} = 25.76 + 1.45 + 0.9 = 28.11$.

HSW Mass spectrometry and chickens!

In 2007 a problem arose in the UK food chain. Many supermarkets and butchers sell 'corn-fed' chickens at a premium price. People often choose to buy them because the chickens are supposed to be free range, grown slowly and fed a diet which is at least 50% maize (sweetcorn). This is supposed to give tastier meat with a rich, yellow tinge from well-managed chickens. However, it was suspected that some of these chickens were actually cheaply reared birds that had been given food containing yellow dye instead of a corn-rich diet. This suspicion was confirmed using mass spectrometry.

fig. 1.3.5 It costs a lot more to rear chickens in conditions like these, feeding them maize and allowing the birds to grow and mature slowly, than to rear them intensively in barns.

Almost 99% of the stable carbon isotopes in carbon dioxide is ^{12}C and about 1% is ^{13}C. The proportion of these stable isotopes that occurs in plant material depends on the way in which the plants fix carbon dioxide during photosynthesis. Maize plants use a different biochemical pathway from most UK plants – and as a result maize plants have a different ^{12}C:^{13}C ratio. This in turn means that animals that have eaten a lot of maize have a different ^{12}C: ^{13}C ratio from animals that have eaten non-maize food. This obvious difference was detected when samples of the chicken were analysed and passed through a mass spectrometer. As a result, customers concerned about chicken welfare and the quality of their meat could be more confident that in future they would be getting the meat they had paid for!

Questions

1. Look back at the mass spectrum of iron in fig. 1.3.2. Write the data from that spectrum in a table and use them to calculate the relative atomic mass of iron.
2. Table 1.3.2 shows the isotopes of krypton based on the mass spectrum in fig. 1.3.6. Use these figures to calculate:
   a. the mass abundance of each krypton isotope
   b. the relative atomic mass of krypton.

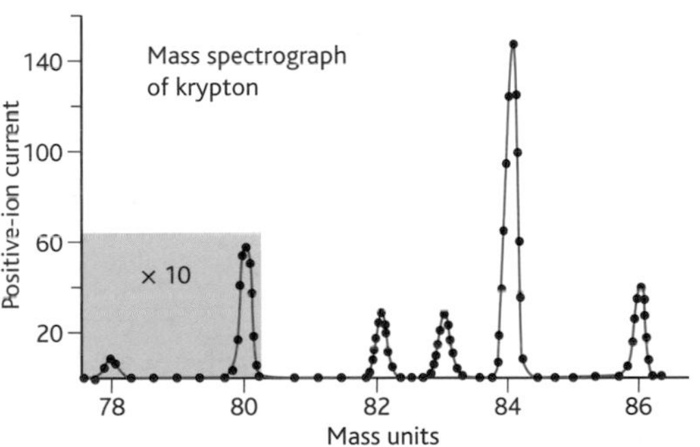

fig. 1.3.6 The mass spectrum of krypton.

Isotope of krypton	Percentage abundance %	Mass abundance
^{78}Kr	0.12	
^{80}Kr	2.0	
^{82}Kr	12	
^{83}Kr	12	
^{84}Kr	57	
^{86}Kr	17	

table 1.3.2 The isotopes of krypton.

HSW Using the mass spectrometer

You have met some of the applications of mass spectrometers – here are some more.

Radioactive elements

There are a number of elements that undergo **radioactive decay**. Radioactive decay takes place when an unstable nucleus breaks apart to become more stable, and emits **alpha**, **beta** or **gamma radiation** in the process.

- Alpha particles consist of two protons and two neutrons (they are helium nuclei).
- Beta particles are high-energy electrons.
- Gamma rays are electromagnetic radiation.

During radioactive decay, the radioactive isotope of one element becomes a different isotope, often of a different element. For example, thorium-228 emits an alpha particle to become radium-224.

Radioactive decay occurs spontaneously and randomly – you can't predict precisely when an atom will break up and emit radiation – and it is not affected by external conditions such as chemical reactions, temperature or pressure. However, in a sample of radioactive material there is one factor that is constant. Experiments have shown that the time taken for half of the atoms in a sample of a radioisotope to decay is constant for that particular isotope. This time is called the **half-life**, $t_{1/2}$ of the isotope. For example, the half-life of radon-222 is 3.82 days. Radon-222 decays to form polonium-218 and emits an alpha particle. If you start with 1 g of radon-222, after 3.82 days there will be 0.5 g of radon-222 left. In another 3.82 days you would have 0.25 g of radon-222 left, and so on (see **fig. 1.3.7** and **table 1.3.3**.)

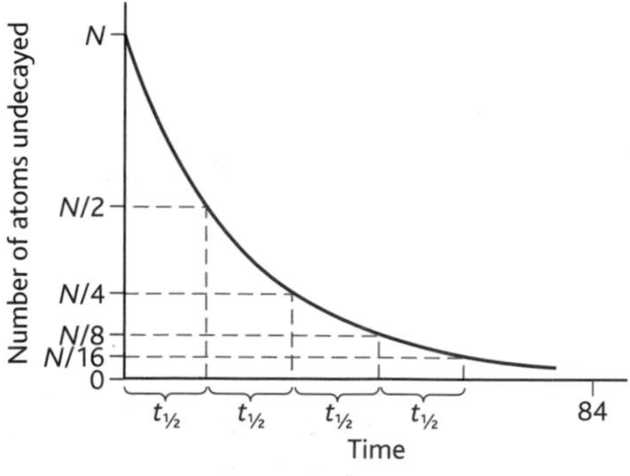

fig. 1.3.7 **The number of atoms of a radioactive isotope decreases with time, and this decay has a constant half-life which is specific for different isotopes.**

Radioisotope	Half-life
Polonium-212, $^{212}_{84}Po$	3×10^{-7} s
Bismuth-212, $^{212}_{83}Bi$	60.6 min
Sulfur-35, $^{35}_{16}S$	87 days
Carbon-14, $^{14}_{6}C$	5730 years
Plutonium-240, $^{240}_{94}Pu$	6580 years
Uranium-238, $^{238}_{92}U$	4.5×10^{9} years

table 1.3.3 **The half-lives of some radioactive elements.**

Some radioactive isotopes have very short half-lives, and decay so rapidly that they do not contribute to the relative atomic mass of the element. For example, polonium has 25 isotopes, but only three have a long enough half-life to contribute to the relative atomic mass. However, polonium-210 has a half-life of 138 days – and in 2006 this became the radioactive isotope that everyone had heard about when it was used to poison the Russian Alexander Litvinenko who was living in London. The presence of this isotope was confirmed in many different places, and technology such as the mass spectrometer giving a detailed analysis of all the different isotopes present, helped scientists identify the most probable source of the material used – the Russian nuclear power industry.

Another well-known radioactive material is uranium, which is used both in the nuclear power industry and in the production of nuclear weapons. The mass spectrometer shows you that in a sample of natural uranium the three isotopes found are uranium-234 (0.0057%), uranium-235 (0.7196%) and uranium-238 (99.276%). The half-lives of uranium isotopes are long – about 704 million years for ^{235}U and around 4.47 billion years for ^{238}U – and because of this they can be used to calculate the ages of rocks in the Earth.

Radioactive dating

Historians and archaeologists use mass spectrometers to help them date ancient remains. One of the best-known dating techniques, known as **radiocarbon dating**, depends on the radioactive isotope carbon-14. Molecules of carbon dioxide containing an atom of carbon-14 are produced in the upper atmosphere by cosmic rays, which are high-energy subatomic particles from space. The ^{14}C isotope makes up 0.000 000 000 10% of the total carbon in the

CO_2 in the atmosphere. The rest largely contains the non-radioactive (stable) ^{12}C isotope, along with some ^{13}C. The level of ^{14}C remaining can be plotted to give a curve which can be used to calculate time intervals.

Plants take up carbon dioxide containing carbon-14 during photosynthesis in exactly the same way as they take up carbon dioxide containing carbon-12, because carbon-12 and carbon-14 are chemically identical. As a result the proportion of carbon-14 and carbon-12 in living plant matter is the same as the proportion in the air. This ratio is also found in animal matter, since animals ultimately get the energy they need from eating plants. However, when an organism dies, it stops taking in carbon-14. The unstable carbon-14 decays very slowly over time. So in once-living material, the ratio of carbon-14 to carbon-12 falls over time, providing a way of measuring the age of the organism. The proportion of the carbon-14 isotope present in once-living tissues halves every 5730 years. This fact makes it possible to calculate the age of a once-living sample of biological material from the percentage of carbon-14 left in the sample.

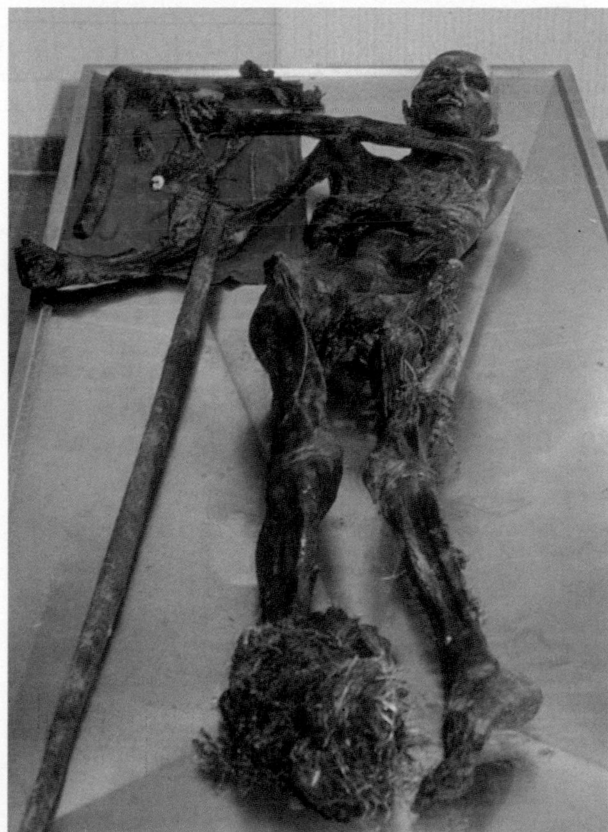

fig. 1.3.8 **This frozen body was discovered in 1991, surprisingly well preserved. The use of mass spectrometry to date the body has shown it to be around 5300 years old!**

The Dead Sea Scrolls (a collection of documents written between the second century BC and the first century AD) and the Iceman (a frozen body discovered in Italy in 1991 and found to be 5300 years old) are just two well-known examples of the use of radiocarbon dating. The accuracy of the technique has been tested by using it to date artefacts whose age is already known and confirmed by other methods. Mass spectrometry has also been used to expose possible fakes. For example, many people thought that the Shroud of Turin was the burial cloth in which Christ was wrapped following his crucifixion. Carbon-14 dating of the cloth of the shroud suggests that it is considerably less than 2000 years old and may therefore be a forgery.

Carbon-14 is not the only isotope used in dating processes. For example, rocks may be dated using the isotopes potassium-40, argon-39 and argon-40. All of these isotopes have very long half-lives, which make them suitable for dating very old material. The proportions of different isotopes of rubidium and strontium, and also uranium and lead, can be used to date rocks. For example, rubidium-87 decays to form strontium-87. The ratios of the different isotopes can be used to calculate the age of a sample (**fig. 1.3.9**) – and if the graph line is straight, the results are reliable! All the data from the different dating methods lead to the same result – that the Earth is around 4.5 billion years old.

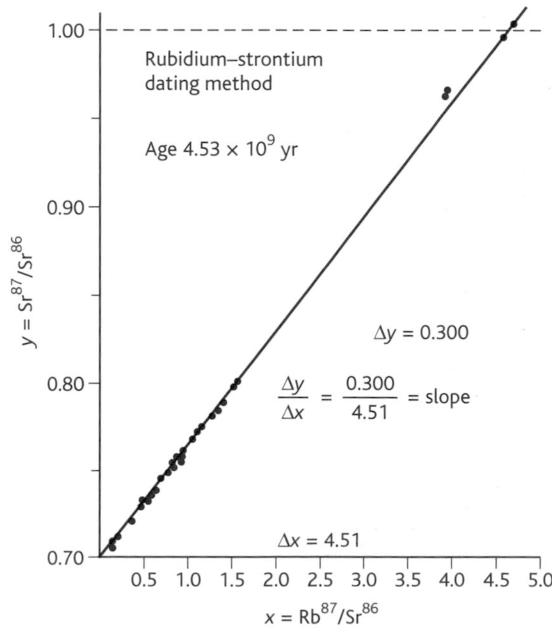

fig. 1.3.9 **Plot of ^{87}Sr versus ^{87}Rb standardised against ^{86}Sr** ***(Source: Kane 1987).***

HSW Using the mass spectrometer (continued)

Drug testing – detecting the cheats

Sporting excellence is achieved by relatively few people. To be great at any sport you need lots of talent as well as the ability to stick to a punishing training regime. Some people try to take a short cut, and use performance-enhancing drugs. For example, **anabolic steroids** can be used to build bigger muscles and so improve athletic performance. These drugs are an artificial form of chemicals made naturally in the body – in particular, the male sex hormone **testosterone**. Everyone, men and women alike, makes testosterone and excretes it from their body in urine. Another similar compound known as **epitestosterone** is also excreted. In a normal person, the ratio of testosterone to epitestosterone (the T:E ratio) is almost never greater than 4:1. If someone has been taking anabolic steroids, this ratio can be greatly raised. A ratio of 6:1 or more starts alarm bells ringing.

fig. 1.3.10 **In 2006 the American Floyd Landis won the Tour de France. However, mass spectrometry showed evidence of anabolic steroids in his urine – he was stripped of his title and banned from competing for two years.**

There is a constant battle between the sporting authorities and athletes who are trying to cheat. The mass spectrometer is an important tool in this battle, used to analyse the T:E ratio in a urine sample. Unfortunately the system is not foolproof. For example, some people have naturally high T:E ratios, so can appear to have cheated when they haven't. Other people have very low ratios, so even when they use performance-enhancing anabolic steroids their readings are normal. One promising new development is the use of even more sensitive mass spectrometry which detects the ratio of carbon-13 to carbon-12. In synthetic testosterone this ratio is always different from the ratio in natural hormones and their breakdown products. This makes it possible to distinguish between testosterone produced in the body and synthetic testosterone taken as a performance enhancer.

Mass spectrometers in space

Space is still the final frontier for human exploration. We know relatively little about our nearest neighbours in space, although over the last few decades we have begun to find out rather more. The big question that many people ask is whether there is life on other planets. Only by discovering what the atmosphere of the planets is like, what the body of the planet is made of, what the temperature range is and what the weather is like will we have any chance of discovering if there ever has been – or ever could be – life on our neighbours in the universe.

Sending a mass spectrometer into space to identify the different elements in the atmosphere of a planet has been a major step forward. In the 1980s, the Pioneer Venus Mission set up by NASA (the US National Aeronautics and Space Administration) sent a mass spectrometer into the upper atmosphere of Venus. The spectrometer identified gases which included helium, carbon dioxide and molecular nitrogen, along with carbon monoxide, atomic oxygen and atomic nitrogen. In the 1990s there was another expedition – the Galileo probe to Jupiter. Some of the images sent back by Galileo were quite amazing, and information from the mass spectrometer combined with other instruments gave a far clearer picture of the planet than had ever been possible before.

fig. 1.3.11 **Jupiter photographed by the Galileo space probe.**

Jupiter has an atmosphere made up largely of hydrogen and helium along with a number of other molecules, some organic and some inorganic. These include methane, ammonia, water vapour and hydrogen sulfide. The atmosphere of Jupiter also contains other noble gases in addition to helium – these include neon, argon, krypton and xenon.

But the planet with the greatest fascination for people on Earth is Mars. In the 1990s excitement rose to fever pitch when a meteorite found in Alaska was shown to have come from Mars. It contained evidence that convinced a team of American scientists that life might exist on the planet. The material was analysed in a mass spectrometer and was found to contain hydrocarbons, and in particular a group of chemicals known as polycyclic aromatic hydrocarbons, which the scientists felt were very similar to the results they would expect from the simple decay of organic material. In addition, the structure of the rock showed tiny structures which the American team thought might be fossilised microbacteria. Controversy still rages over this evidence.

Mass spectrometers also play an important role in human exploration of space. The US space shuttles and the International Space Station rely on mass spectrometers to analyse gases and warn the astronauts of any problems. NASA has even developed a tiny mass spectrometer for these missions, known as the Trace Gas Analyser. It weighs only around 2 kg and fits into a shoebox-sized case. One of these devices can be fitted onto the suit of an astronaut to detect any traces of leaking gases such as hydrazine during a spacewalk outside the spacecraft.

fig. 1.3.12 A tiny mass spectrometer makes space exploration safer for everyone who uses the International Space Station.

Questions

Percentage of original ^{14}C left in sample (%)	Time since organism died (years)
100	0
95	420
90	870
85	1 350
80	1 850
75	2 400
70	2 950
65	3 550
60	4 200
55	4 950
50	5 750
45	6 600
40	7 600
35	8 700
30	9 950
25	11 450
20	13 300
15	15 700
10	19 050
5	24 750
1	38 100

table 1.3.4 Carbon left in a sample versus age.

1 Use the data in table 1.3.4 to plot a graph showing the relationship between the percentage of carbon-14 left in a sample and its age.

2 Calculate the approximate ages of specimens found with these amounts of their total ^{14}C left.
   a 3% b 43% c 89%.

3 In 1988, fragments from the Shroud of Turin underwent radiocarbon dating in Switzerland, the UK, the US and France. All of the laboratories dated the fragments between 1260 and 1390 AD. What range of carbon-14 left in the fabric of the Shroud would give these readings? What percentage would have given a reading dating the shroud as genuinely coming from around 2000 years ago?

The arrangement of electrons in atoms

Although we can identify atoms by the mass of their nucleons, it is their **electronic structure** which determines the way atoms behave and their chemical properties.

HSW Deciphering the electronic structure of the atom

Thomson's plum pudding model

By the end of the nineteenth century it had been established that atoms contained electrons with a negative charge. Given that atoms themselves have no charge, this negative charge must be cancelled out by positively charged particles within the atom. Joseph John Thomson was a gifted physicist and teacher – not only did he win a Nobel prize for his work on cathode rays and electrons, but seven of his pupils and his own son George all won Nobel prizes as well! He established that the mass of an electron was about 1/2000 that of a hydrogen atom. Then in 1898, as a result of his work on matter, Thomson proposed his 'plum pudding' model for the atom, with negative electrons embedded in a sphere of positive charge. This seemed to explain the behaviour of matter as he observed it.

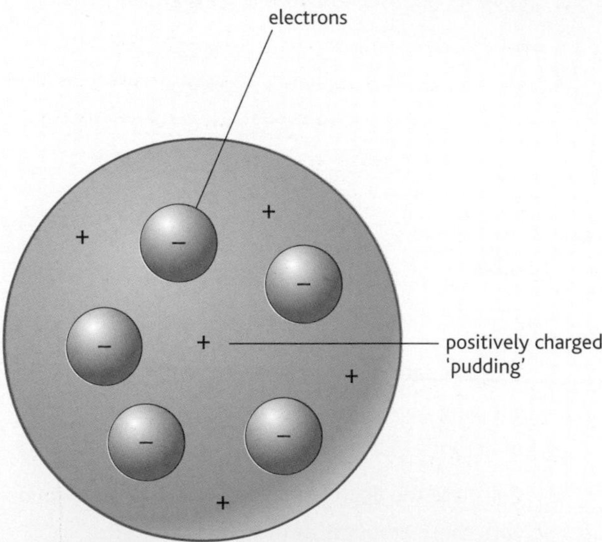

fig. 1.3.13 **The plum pudding model of the atom developed by Joseph Thomson, one of the great and most inspirational chemists.**

At this stage scientists thought the mass of an atom was contained in its electrons. If the mass of one electron was about 1/2000 that of a hydrogen atom, a hydrogen atom must contain around 2000 electrons. Further work showed that the particles with positive charge in atoms had far greater mass than electrons. These positively charged particles were called **protons**.

Rutherford's atom with a nucleus

The model of the nuclear atom was first proposed by Ernest Rutherford and his colleagues Hans Geiger and Ernest Marsden in 1911. They directed a beam of alpha particles (positively charged radioactive particles) at thin metal foils. Most alpha particles passed straight through the foils and were detected on the other side. But in a very small percentage of cases the alpha particles were deflected, sometimes through an angle greater than 90°. Rutherford suggested that most of the mass of the atom was concentrated in a tiny positively charged region, with electrons circulating around it. When an alpha particle came close to a positive nucleus, the electrostatic repulsion between the two would deflect the alpha particle. Because the nucleus was very small, only a few particles approached a metal nucleus closely enough to be repulsed. The rest passed through the large areas of the metal atoms containing the negatively charged electrons – and empty space!

Using his model, Rutherford calculated that the nucleus of an atom is of the order of 10^{-14} m across. Atoms themselves are of the order of 10^{-10} m – so the diameter of the nucleus is about 1/10 000 of the diameter of the atom.

fig. 1.3.14 **If the goalkeeper represents the nucleus of an atom, and the football represents an alpha particle, the goalposts would need to be 20 km either side of the keeper to represent the rest of the atom!**

Rutherford's model pictured the atom as a miniature solar system, in which electrons orbited the nucleus, much as planets orbit the sun. The positive charge on the nucleus of each atom was balanced by the negative charge on the orbiting electrons, which moved in a space far larger than the nucleus.

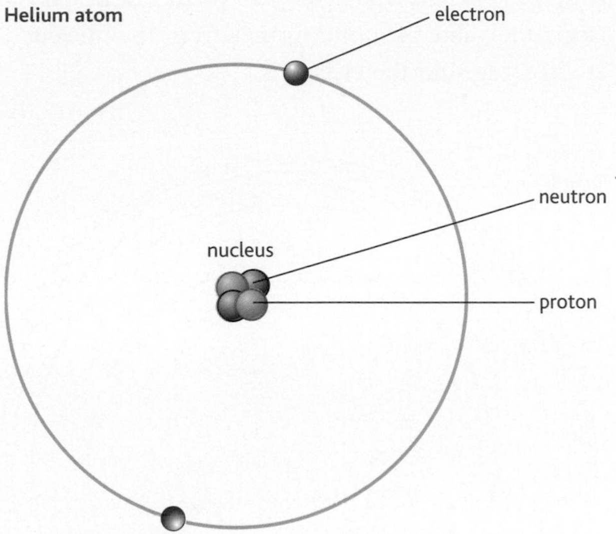

fig. 1.3.15 **Rutherford's model of the atom. In this model the mass of the electron is 1/2000 the mass of the proton, and the mass of the proton equals the mass of the neutron.**

Bohr's electron shells

In the early twentieth century the Danish physicist Niels Bohr worked with both Thomson and Rutherford before he developed a ground-breaking model of the hydrogen atom.

If a gas is heated, or if an electrical charge is passed through it, the gas gives out light. If you pass this light through a prism or diffraction grating (a very fine grid) the light is split to form a spectrum. The spectrum is made up of a series of separate lines or bands and is known as a **line spectrum** or **emission spectrum**. This line spectrum is always identical for any given element. Bohr set out to relate the line spectrum of hydrogen to the structure of the hydrogen atom.

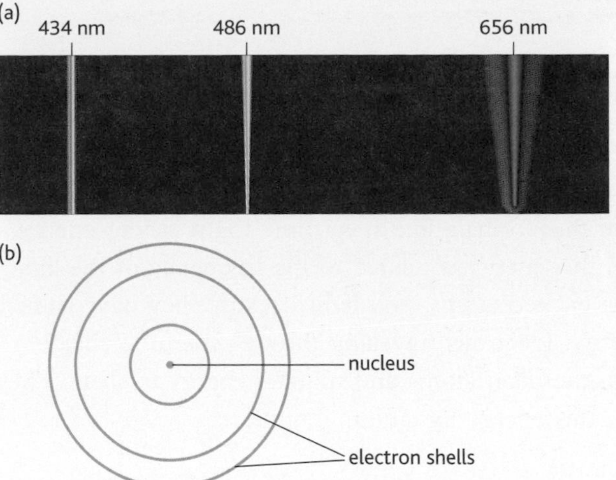

fig. 1.3.16 **Bohr used the hydrogen spectrum to develop a model for the hydrogen atom with possible electron shells arranged like this.**

Bohr's idea was that electrons are arranged in shells around the positive nucleus. Each shell represents a different energy level. Moving out from the centre of the atom, successive shells become closer together, in the same way as the energy levels in the line spectrum become closer. The spectrum is produced as atoms are **excited** (electrons are raised from one energy level to a higher one). As they fall back to their original level, light energy is emitted.

Bohr's new model of the atom fitted exactly the experimental results for the line spectrum of hydrogen and in 1922 he won a Nobel prize for his work. The equations derived from this model could be used to calculate the radius of the hydrogen atom and its energy levels, particularly its ionisation energy. However, the model could not successfully explain the behaviour of atoms with more than one electron (even helium). So Bohr, along with his pupils Heisenberg and Schrödinger amongst others, set to work on new ideas which would explain the atom further. This was the beginning of **quantum mechanics**, which still underpins our present model of atomic structure. You will be looking at this model in more detail in the rest of this chapter.

Questions

1 Use the information given here and your own research to develop a timeline showing the progression of ideas about the structure of the atom.

2 Explain the role of experimental data in the development of both Rutherford's and Bohr's theories of the structure of the atom.

Electrons in atoms – the modern story

More about atomic spectra

Why do atoms that have been excited, such as neon gas in a neon light (discharge tube, **fig. 1.3.24**), emit light that forms a line spectrum? Light carries energy, and this energy is related to the frequency of the light. The excited atoms emit light because they have gained energy. Electrons travelling through the tube collide with the neon atoms and transfer energy to them. They lose this energy by emitting light.

An electron may transfer some of its kinetic energy to a neon atom in a collision. E is the energy transferred from the electron to the atom. This energy then leaves the atom in the form of a photon of light. The frequency of the emitted light is determined by the value of E.

The fact that this light consists of only a limited number of frequencies tells you that only a limited number of energy changes or **transitions** can take place within the atom (**fig. 1.3.17**).

A ball on a staircase can have only certain fixed levels of potential energy, since it cannot come to rest between steps. An atom is similar to this – it can exist in a number of fixed energy states, and can move between these states.

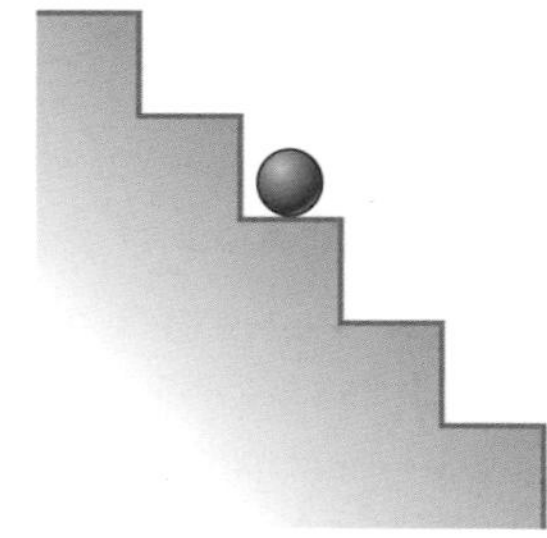

fig. 1.3.17 This model shows the way electrons occupy specific energy levels.

Using the wavelengths of light in the line spectrum of the hydrogen atom (**fig. 1.3.16(a)**), scientists have obtained a series of energy differences, ΔE for the hydrogen atom. These values of ΔE can be used in turn to build up a series of energy levels in the hydrogen atom.

Energy levels and electron shells

The absorption and emission of light by an atom can be explained in terms of electron transitions between different fixed energy levels in the atom. These energy levels correspond to how electrons are arranged within atoms.

Quantum mechanics says that the electrons in atoms are arranged in a series of **shells**, which resemble the layers in an onion. Each shell is described by a number known as the **principal quantum number,** $\boldsymbol{n}$ which tells you about the size of the shell. The larger the value of n, the further from the nucleus you are likely to find the electron.

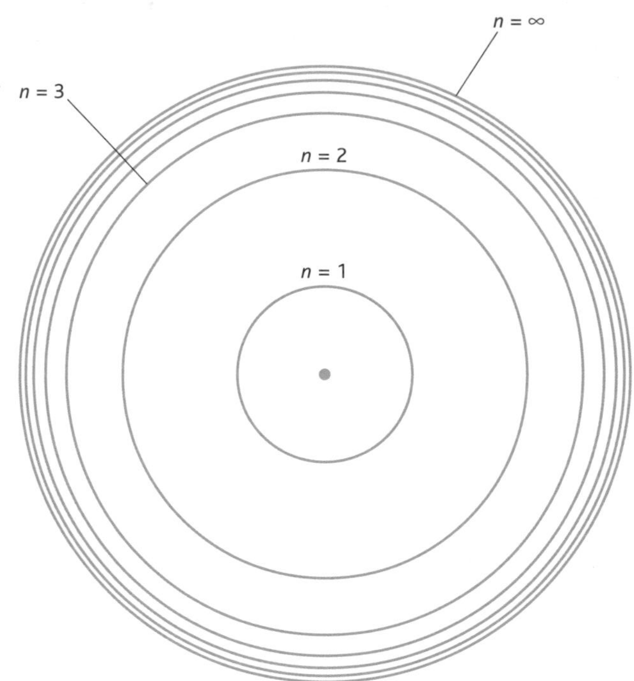

fig. 1.3.18 A model of a hydrogen atom to show the possible electron shells. As you move out from the centre of the atom, successive shells get closer together, as do the energy levels.

Ionisation energies

Ionisation is the complete removal of an electron from an atom. Ionisation is an **endothermic** process, since work must be done on an electron in order to overcome the attractive force between it and the nucleus. The amount of energy needed to remove an electron from an atom is known as the **ionisation energy**. This can be measured by gradually increasing the voltage applied to a gas until it conducts electricity and emits light, which tells you an electron has been freed.

By looking in detail at the ionisation energies of different electrons in an atom, you can build up a detailed picture of the electronic structure of that atom. You will be looking at ionisation energies in more detail later in this chapter.

An atom in its **ground state** is at its lowest energy level. The energy needed to remove the electron from the hydrogen atom in its ground state is the ionisation energy of hydrogen, normally quoted for 1 mole of hydrogen atoms:

Ionisation energy of hydrogen = 1312 kJ mol^{-1}

For an atom with more than one electron, the energy to remove the first electron is called the **first ionisation energy**, the energy to remove a second electron is called the **second ionisation energy**, and so on. The first ionisation energy is a measure of how tightly – or loosely – the outer electron is attracted to the positive nucleus. The less tightly it is bound (the more easily it is removed) the more reactive the element will be. To calculate the *total* energy required to remove the first and second electrons from an atom, you have to add the first and second ionisation energies together. But the different energies required to remove the first and subsequent electrons from an atom help to confirm that the electrons are to be found at different energy levels.

Subshells

Quantum mechanics also tells you that each shell may contain a number of **subshells**. These subshells are described by letters: s, p, d, f, g. etc. The following subshells are available in each shell:

Shell	Subshells
1	1s
2	2s, 2p
3	3s, 3p, 3d
4	4s, 4p, 4d, 4f

Shell 1 is closest to the nucleus so it takes the most energy to remove electrons in this shell. Within a shell the subshells have different energies, with electrons in the lowest energy subshells being closest to the nucleus:

s (lowest energy) < p < d

Each type of subshell contains one or more **orbitals** (see **table 1.3.5**). An orbital is the region where the electrons are most likely to be found.

Figure 1.3.19 shows an energy level diagram for atoms. Notice that all the orbitals in a particular subshell are at the same energy level. As *n* increases, the energy gap between successive shells gets smaller (eg compare the gap between shells 2 and 3 with the gap between shells 6 and 7). As a result of this, the orbitals in neighbouring shells may overlap, eg the 3d orbital has an energy level above that of the 4s orbital but below that of the 4p orbital.

	First shell (n=1)	Second shell (n=2)	Third shell (n=3)	Fourth shell (n=4)
Subshells	s	s p	s p d	s p d f
Number of orbitals	1	1 3	1 3 5	1 3 5 7

table 1.3.5 Shells, subshells and orbitals

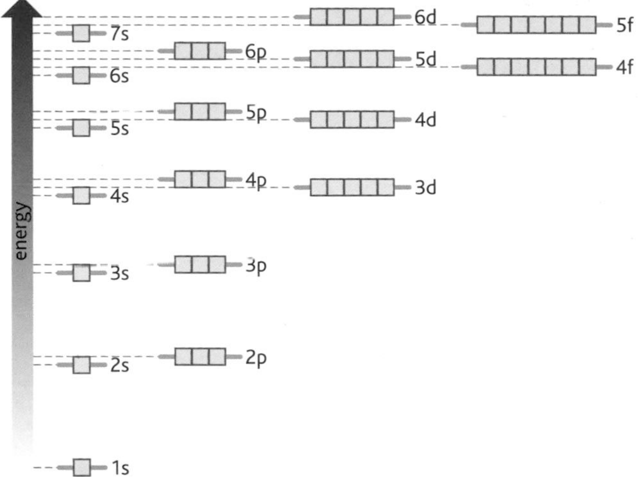

fig. 1.3.19 An energy level diagram for an atom. Each orbital is indicated by a single square box, .

Questions

1. What is ionisation energy and why is it so important in the development of a useful model of the electronic structure of an atom?
2. Rank the following orbitals in order of energy level, starting with the lowest (the one from which it would take most energy to remove an electron): 3p, 1s, 2s, 3d, 4s.

Filling orbitals and shells

Electron spin

An atom is in its lowest state of energy (its ground state) when its electrons are in the orbitals with the lowest possible energy levels. There are some straightforward rules about how electrons are arranged in orbitals in elements in their ground states. One of the factors influencing the filling of orbitals is **electron spin**.

An electron in an atom behaves like a tiny magnet. Imagine that an electron spins on its axis, just as the Earth does (see **fig. 1.3.20**). A moving charge creates a magnetic field. An electron can spin either clockwise or anticlockwise, and can be represented by a small arrow, pointing up (↑) to represent spin in one direction and down (↓) to represent spin in the opposite direction.

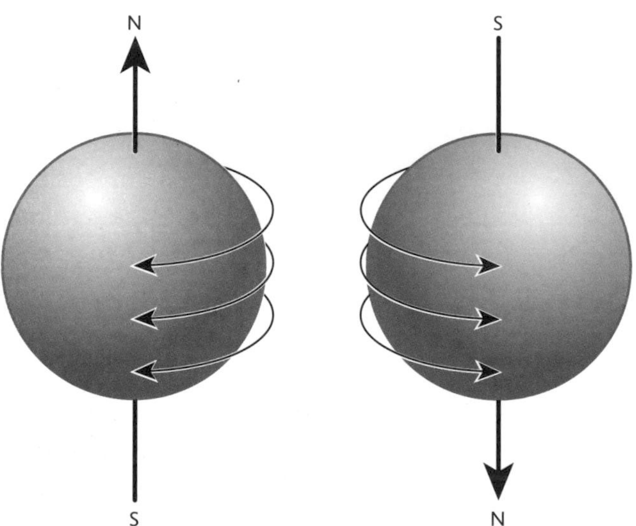

fig. 1.3.20 The magnetic properties of the electron can be explained by imagining that the electron spins on its axis.

Two electrons in the same orbital cannot have the same spin. This means each orbital can contain a maximum of two electrons, having opposite spin. This can be summarised as follows:

Subshell	Number of orbitals	Maximum number of electrons in subshell	
s	1	[↑↓]	= 2
p	3	[↑↓] [↑↓] [↑↓]	= 6
d	5	[↑↓] [↑↓] [↑↓] [↑↓] [↑↓]	= 10
f	7	[↑↓] [↑↓] [↑↓] [↑↓] [↑↓] [↑↓] [↑↓]	= 14

Filling the orbitals

Hydrogen to nitrogen

You know that the hydrogen atom in its ground state has a single electron in its 1s shell. You can represent the **electronic configuration** of hydrogen as $1s^1$. Electronic configurations are often shown with each orbital represented by a square box:

	1s
H	[↑]

The atomic number of helium is 2 ($Z = 2$), which means the helium atom has two protons in its nucleus and so must also have two electrons. These both occupy the 1s orbital, provided that they have opposite spins, so the electronic configuration of helium is $1s^2$:

	1s
He	[↑↓]

The next two elements are lithium ($Z = 3$) and beryllium ($Z = 4$). The 1s shell is full, so the next two electrons go into the next lowest energy orbital, the 2s orbital:

	1s	2s
Li $1s^2 2s^1$	[↑↓]	[↑]
Be $1s^2 2s^2$	[↑↓]	[↑↓]

Boron, carbon and nitrogen are the next elements, with $Z = 5$, 6 and 7 respectively. The 2s orbital is full, so the fifth, sixth and seventh electrons go into the next lowest energy orbital, the 2p orbital:

	1s	2s	2p
B $1s^2 2s^2 2p^1$	[↑↓]	[↑↓]	[↑] [ ] [ ]
C $1s^2 2s^2 2p^2$	[↑↓]	[↑↓]	[↑] [↑] [ ]
N $1s^2 2s^2 2p^3$	[↑↓]	[↑↓]	[↑] [↑] [↑]

Notice that all three of the 2p orbitals are always shown (by empty boxes) even if only one or two of them contains an electron. It doesn't matter which of the 2p orbitals is filled first, because they all have the same energy.

Two important principles

In the electronic configuration for carbon and nitrogen, the electrons in the 2p subshells are placed in different orbitals, rather than placing them in the same orbital with opposite spins. This is **Hund's rule**, which says that when electrons are placed in a set of orbitals with equal energy, they 'spread out' to maximise the number of unpaired electrons. So in the nitrogen atom the three 2p orbitals each contain a single unpaired electron.

Electrons will fill the lowest-energy orbitals first, then the remaining orbitals in order of increasing energy. This means that a shell is not always completely filled before electrons start filling the next shell. So a 4s orbital will fill before a 3d orbital, because the 4s orbital is at a lower energy level than a 3d orbital. The order of filling orbitals based on their energy level is 1s, 2s, 2p, 3s, 3p, 4s, 3d, 4p, 5s, 4d, 5p, 6s, 4f, 5d, 6p, ….

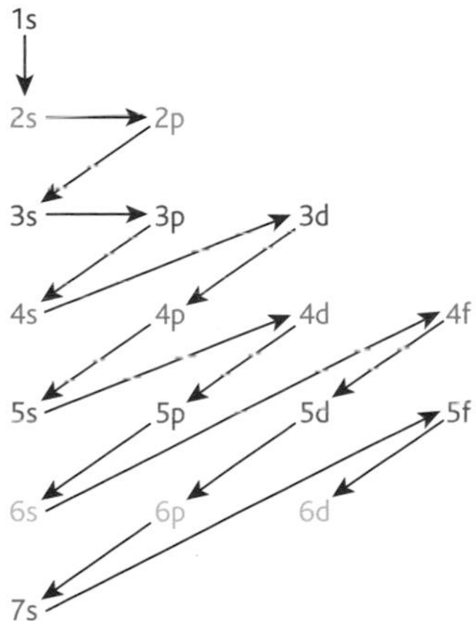

fig. 1.3.21 **The order of filling orbitals. Electrons fill the orbital with the lowest energy first, which is why the order is not strictly numerical!**

Oxygen to neon

In the elements oxygen, fluorine and neon the remainder of the second shell is completed by adding three more electrons:

	1s	2s	2p
O $1s^2 2s^2 2p^4$	[↑↓]	[↑↓]	[↑↓] [↑] [↑]
F $1s^2 2s^2 2p^5$	[↑↓]	[↑↓]	[↑↓] [↑↓] [↑]
Ne $1s^2 2s^2 2p^6$	[↑↓]	[↑↓]	[↑↓] [↑↓] [↑↓]

Electronic configurations

Write down the electronic configuration of a vanadium atom (Z = 23), and represent this diagrammatically, showing how electron spins are paired.

Vanadium has 23 protons in its nucleus, so the atom will have 23 electrons. Following fig. 1.3.21 (and remembering Hund's rule too) we can fill subshells in order of increasing energy until we run out of electrons. This gives the configuration:

V $1s^2 2s^2 2p^6 3s^2 3p^6 3d^3 4s^2$

The arrangement of electrons in orbitals can be represented diagrammatically as:

1s	2s	2p	3s	3p	3d	4s
[↑↓]	[↑↓]	[↑↓] [↑↓] [↑↓]	[↑↓]	[↑↓] [↑↓] [↑↓]	[↑] [↑] [↑] [ ] [ ]	[↑↓]

The 4s orbital is at a lower energy level than the 3d orbitals, so the 4s orbital fills first. However, when you write this, the orbitals are grouped together in their shells: …$3d^3 4s^2$. The three electrons in the 3d subshell go into separate orbitals so as to maximise the number of unpaired electrons.

Using the rules outlined above, the electronic configuration of a polonium atom (Z = 84) is written as follows:

Po $1s^2 2s^2 2p^6 3s^2 3p^6 3d^{10} 4s^2 4p^6 4d^{10} 4f^{14} 5s^2 5p^6 5d^{10} 6s^2 6p^4$

Writing electronic configurations for ions

For an ion, you simply add or subtract the right number of electrons from the outer shell – taking electrons for a positive ion, adding them for a negative ion. Remember Hund's rule when removing electrons: one electron comes out of each completely filled orbital in the outer shell before any unpaired electrons and removed. For example, the oxygen ion, O^{2-} is $1s^2 2s^2 2p^6$, and the calcium ion Ca^{2+} is $1s^2 2s^2 2p^6 3s^2 3p^6$.

In the next chapter you will be looking at how you can use electronic configurations to help you predict how elements react.

Questions

1 Use the periodic table at the back of the book to write down the electronic configurations of the following:
a Li atom **b** O atom **c** Mg atom **d** P atom
e Cl atom **f** Ga atom **g** Mg^{2+} ion **h** Cl^- ion.

2 An element has the electronic configuration $1s^2 2s^2 2p^6 3s^2 3p^6 4s^2 3d^{10} 4p^2$. Write down its atomic number.

Representing electrons and orbitals

HSW Shorthand notation

Chemical changes concern only the outer electrons in the atom, not those in the inner shells. The inner electrons are strongly attracted to the nucleus and they are not involved in forming chemical bonds. So it is helpful to simplify electronic configurations to focus on the outer electrons.

For example, the electronic configurations for potassium and calcium are:

K $1s^2 2s^2 2p^6 3s^2 3p^6 4s^1$

Ca $1s^2 2s^2 2p^6 3s^2 3p^6 4s^2$

The core of electrons in the first three shells of both atoms are identical – it is only the fourth shell that differs. The inner electrons have the configuration of the noble gas argon. The 'shorthand notation' for these elements uses the symbol of the noble gas followed by the outer electrons:

K [Ar]$4s^1$

Ca [Ar]$4s^2$

Electron density maps

You have seen that electrons can undergo transitions from one shell to another, and that these transitions are associated with changes in the energy of the atom. But exactly how are electrons arranged within the atom?

The model of the behaviour of electrons in atoms that is widely accepted by scientists today considers electrons as waves rather than as particles. The mathematical theory behind this idea is very complex, but fortunately the results of the theory – explained below – are quite straightforward.

You may never before have thought about a particle behaving as a wave, so the idea might seem very strange. It becomes more straightforward if we talk in terms of probabilities. An electron behaves as though it is spread out around the nucleus as a sort of cloud – in fact, it is sometimes referred to as an **electron cloud**. Calculations from quantum mechanics provide us with a map of the **electron density** of this electron cloud, showing the shapes of the areas in space where the electron is likely to be.

How does the cloud fit with the idea of shells and orbitals? An orbital is a region where the *probability* of finding the electron is greatest – although it does not rule out the possibility that it may be somewhere else altogether. It's rather like a pet cat. The most likely place to find it is in or around the house, but that doesn't rule out the possibility that it could be somewhere else altogether – in a field, over the road or under the neighbour's bed!

Figure 1.3.22(a) shows the **electron density map** (or **plot**) calculated for the 1s orbital. There are several ways of interpreting a diagram like this. One is to say that an electron in a 1s orbital spends most of its time in a sphere close to the nucleus. Another is to say that the most likely place to find an electron in a 1s orbital is in a sphere close to the nucleus. You can also interpret the diagrams in terms of electron density, showing how much of an electron's charge is likely to be found in a given volume – in other words, the space in which the electron is most likely to be travelling. These interpretations are essentially similar, and are all attempts to picture what happens to electrons in a way that relates to your own experience of the world – which is very different from the subatomic world of the electron!

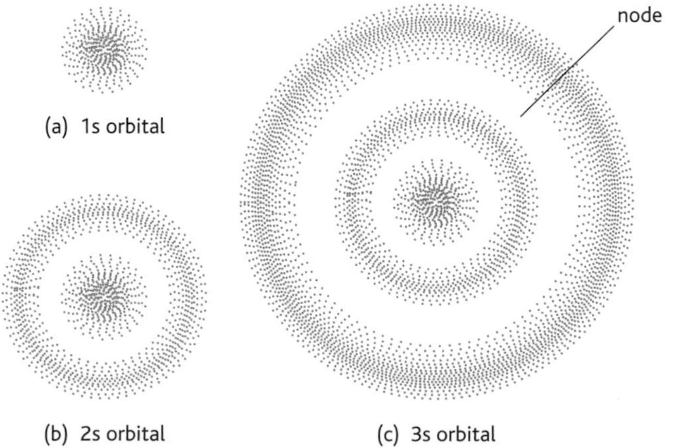

fig. 1.3.22 The electron clouds for the s orbitals in the first, second and third shells.

The electron density maps for the s orbitals in the second and third shells are shown in fig. 1.3.22(b) and (c). The distribution of electron density in these orbitals is similar to that in the 1s orbital (they are spherically symmetrical around the nucleus), but there are regions called nodes where the electron wave has

zero amplitude and where the electron density is zero as a result – so in theory you won't find the electron there! This is also the case for the p and d orbitals, which are shown in fig. 1.3.23.

(a) The shapes and orientations of the orbitals within a p subshell

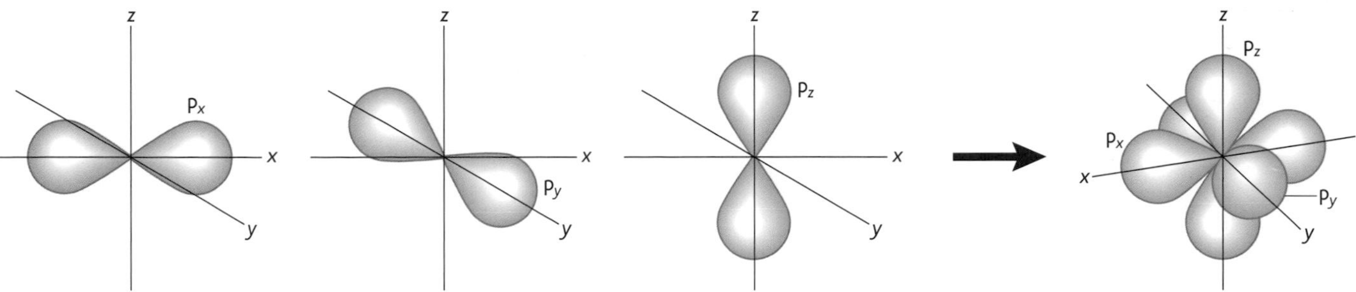

(b) The shapes and orientations of the orbitals within a d subshell

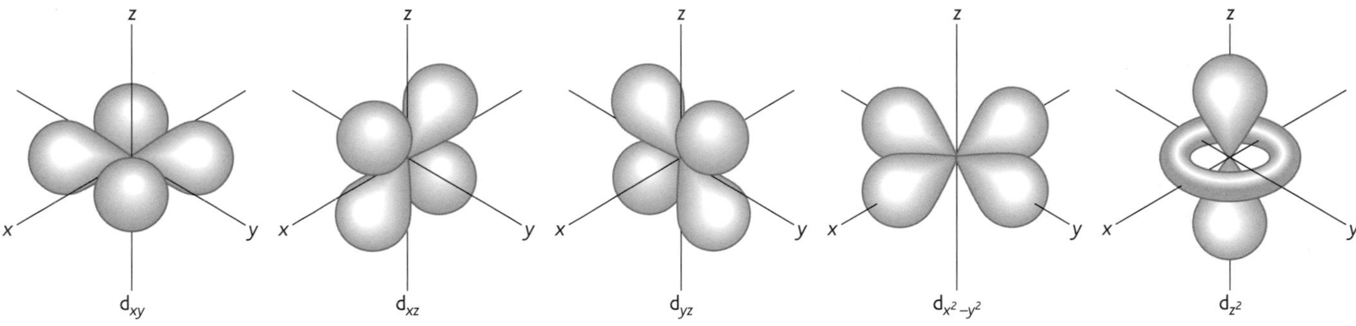

fig. 1.3.23 The shapes and orientations of the three orbitals within a p subshell and the five orbitals within a d subshell.

The p orbitals are approximately dumbbell shaped, and lie at 90° to each other as fig. 1.3.23 shows. As with the s orbitals, the size of p orbitals increases with principal quantum number, so a 3p orbital is larger than a 2p orbital. Each orbital is labelled as if it lay along one of the axes of an *xyz* coordinate system. Notice that each p orbital has a node at the nucleus of the atom.

Questions

1 Write shorthand electronic configurations for the following elements and explain why these are useful:
a Be **b** N **c** S **d** Ti **e** Ge **f** Kr.

2 Explain what is represented by the orbital diagrams in figs. 1.3.22 and 1.3.23.

3 The gold atom, Au has 79 electrons. On this basis you might expect the electronic configuration of the gold atom in its ground state to be:

[Xe] $4f^{14}5d^{9}6s^{2}$

However the actual configuration is:

[Xe] $4f^{14}5d^{10}6s^{1}$

Explain this.

Electronic configuration and chemical properties

Chemistry is a potentially explosive science. When chemicals are mixed together the results may range from no apparent reaction at all to a massively exothermic reaction in which energy is released as light, heat and sound. Over the years, many chemists have suffered from their inability to predict the outcome of mixing chemicals. Almost two centuries ago the young Michael Faraday got his first chance to work at the Royal Institution when Sir Humphry Davy injured his eyes. He was carrying out an experiment on the dangerously unstable substance nitrogen(III) chloride, and could neither read or write as a result. He called on the young Faraday to make notes of his work for him. Years later Faraday himself was to complain in his journal of eye injuries from explosions, feeling fortunate that none had caused permanent damage!

Modern chemists are much more aware of safety. They also have the tools to predict with great accuracy how different chemicals will react with one another. Central to this understanding is the **periodic table**.

Periods, groups and electronic configurations

The way an element reacts with others is determined by a combination of factors, particularly the electronic configuration of its atoms. The most important thing is the arrangement of the electrons in the outer shell, because this determines how reactive the element is and how it combines with other elements.

The modern periodic table arranges the elements in order of their atomic number. Vertical columns in the periodic table are called **groups** and the horizontal rows are **periods**. Hydrogen occupies a unique position at the top of the periodic table and does not fit naturally into any group. All the elements in a group have the same number of electrons in their outer shell. In some groups, such as group 1, this leads to a remarkable similarity between the elements. Other groups show less similarity, eg group 4, but the elements in each group have characteristics in common.

All the elements in a period have the same number of electron shells – the elements in period 1 have one shell (principal quantum number 1), those in period 2 have two shells and so on. The elements in each group or period show particular characteristics and trends in their chemical and physical behaviours that can be explained in terms of their atomic numbers. This provides chemists with valuable information about what is likely to happen when particular chemicals react. Modern **periodic law** states that:

- The properties of the elements are a function of their atomic numbers.

Look at the periodic table in fig. 1.3.25. You can see it has regions called **blocks** – elements in the s block have outer electrons in s subshells, those in the **p block** have outer electrons in p subshells, and so on for the **d block** and **f block**.

Using this table you can see, for example, that the gases helium, neon and argon all have electronic structures with full subshells. Measurements show that they have exceptionally high ionisation energies (see fig. 1.3.33 later in the chapter) – a very large amount of energy is needed to remove an electron from these atoms. This confirms that this electronic configuration is a particularly stable one, and helps to explain why they rarely react with other chemicals.

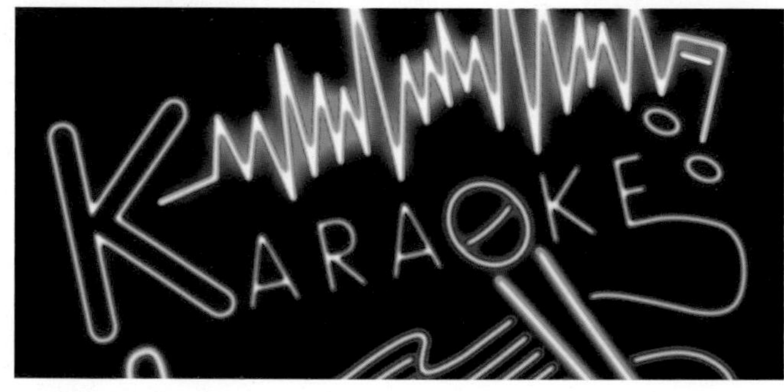

fig. 1.3.24 These spectacular lights are provided by neon and other noble gases. Electrical energy causes the atoms to become excited as electrons in the outer shells jump to a higher energy level. As they return to their ground state they give out light energy of varying wavelengths, which is seen as different colours (see page 62).

Elsewhere in the table, you can see that sodium and lithium each have a single electron outside a full shell of inner electrons. Similarly fluorine and chlorine have outer shells that are just one electron short of a full shell. You can see that the position of an element in the periodic table is determined by its electronic structure, which directly influences its physical and chemical properties. The periodic table provides chemists with valuable information about what is likely to happen when particular chemicals react.

	1	2													3	4	5	6	7	0
1s									1 H											2 He
2s	3 Li	4 Be												2p	5 B	6 C	7 N	8 O	9 F	10 Ne
3s	11 Na	12 Mg												3p	13 Al	14 Si	15 P	16 S	17 Cl	18 Ar
4s	19 K	20 Ca	3d	21 Sc	22 Ti	23 V	24 Cr	25 Mn	26 Fe	27 Co	28 Ni	29 Cu	30 Zn	4p	31 Ga	32 Ge	33 As	34 Se	35 Br	36 Kr
5s	37 Rb	38 Sr	4d	39 Y	40 Zr	41 Nb	42 Mo	43 Tc	44 Ru	45 Rh	46 Pd	47 Ag	48 Cd	5p	49 In	50 Sn	51 Sb	52 Te	53 I	54 Xe
6s	55 Cs	56 Ba	5d	57 La	72 Hf	73 Ta	74 W	75 Re	76 Os	77 Ir	78 Pt	79 Au	80 Hg	6p	81 Tl	82 Pb	83 Bi	84 Po	85 At	86 Rn
7s	87 Fr	88 Ra	6d	89 Ac	104 Rf	105 Db	106 Sg	107 Bh	108 Hs	109 Mt	110 Ds	111 Rg								

lanthanides

4f	58 Ce	59 Pr	60 Nd	61 Pm	62 Sm	63 Eu	64 Gd	65 Tb	66 Dy	67 Ho	68 Er	69 Tm	70 Yb	71 Lu

actinides

5f	90 Th	91 Pa	92 U	93 Np	94 Pu	95 Am	96 Cm	97 Bk	98 Cf	99 Es	100 Fm	101 Md	102 No	103 Lr

blocks

- s block
- p block
- d block
- f block

fig. 1.3.25 The modern periodic table, shown here in a relatively simple wide form. The names of the individual elements and their relative atomic masses may also be shown. There are 91 naturally occurring elements, but more (at present 103) are shown on the periodic table. This is because a number of unstable, radioactive elements have been synthesised in recent years by scientists. These elements are also, for completeness, included in the periodic classification.

HSW Whose idea was it anyway?

The history of the development of the periodic table is peppered with good ideas that were ignored. It is worth remembering that the modern model of the atom was a long way off when these scientists were trying to explain the behaviour of the elements they observed, and to predict what reactions would take place.

Early ideas about atomic mass

As far back as 1799, Joseph Proust showed that the proportions by mass of the elements were constant in each particular compound he studied. A year later John Dalton proposed that elements were made up of atoms, and that each element is made up of one particular type of atom which differs from all the others by its mass. Unfortunately Dalton wasn't very good at actually working out what these different atomic masses were. When other, more careful, scientists tried they often came up with different answers, so they were not entirely convinced by his work. In 1828 Jöns Jacob Berzelius published a list of atomic masses which were quite accurate, but after Dalton's earlier inaccurate efforts no one was interested.

Döbereiner's triads – the idea of periodicity

In 1829 Johann Döbereiner showed that many of the known elements could be arranged in groups of three, with all the members of the group showing similar properties. He called these groups **triads**.

This was the first time elements had been grouped in this way. A number of scientists continued working to group the elements. In 1862 Alexandre-Émile Beguyer de Chancourtois developed an early version of the periodic table, showing similarities between every eighth element. Unfortunately, when his article was published the diagram was not printed. Without this diagram the periodicity of the elements was not clear. As a result of this error, his ground-breaking work was largely ignored!

Newlands' law of octaves

One year later John Newlands announced his **law of octaves**. This involved arranging elements in order of their atomic masses in groups of eight. Although this showed a basic 'periodic table' with similar elements grouped together, Newlands' work was dismissed because it contained some major flaws. He assumed all the elements had been discovered, even though new ones were turning up each year. He put two elements in the same place several times to make things fit. He even put dissimilar elements like copper, lithium and sodium together when he couldn't decide where else they fitted in. Yet again, the scientific community was unimpressed with the whole idea of periodicity. It was even suggested that arranging the elements alphabetically might well produce as many circumstantial similarities as Newlands' carefully arranged octaves.

fig. 1.3.26 One example of Döbereiner's triads is the group of chlorine, bromine and iodine – elements we still recognise as related today.

Mendeleev's periodic table

Finally, in 1869/70 Dmitri Mendeleev and Julius Meyer published clear representations of periodicity in the elements. Meyer plotted various physical properties of the known elements against their atomic masses and produced curves that demonstrated the periodic relationships. However, Mendeleev published first, and scored a major publicity coup by producing a table with gaps in it for as yet undiscovered elements. Based on his table he predicted the properties of these missing elements. Then in 1895 and 1896 gallium and germanium were discovered, two of the elements predicted by Mendeleev. Experimental evidence confirmed that they showed exactly the properties he had deduced from his periodic table. This convinced any remaining sceptics of the periodicity of the chemical elements based on their atomic mass, and the name Mendeleev became associated with the modern periodic table.

fig. 1.3.27 Dmitri Mendeleev founded the modern periodic table.

Property	Predicted	Found experimentally
	Eka-aluminium	**Gallium**
Relative atomic mass	68	69.9
Density (g cm^{-3})	6.0	5.9
Melting temperature	low	29.8 °C
	Eka-silicon	**Germanium**
Relative atomic mass	7?	72.3
Oxide	EsO_2, density 4.7 g cm^{-3}	GeO_2, density 4.23 g cm^{-3}
Chloride	$EsCl_4$, boiling temperature < 100 °C, density 1.9 g cm^{-3}	$GeCl_4$, boiling temperature 84 °C, density 1.84 g cm^{-3}

table 1.3.6 Mendeleev's predicted properties for some undiscovered elements (which he called eka-aluminium and eka-silicon), and the experimental evidence which confirmed his ideas when gallium and germanium were discovered.

Questions

1 The gases helium to radon are sometimes known as the inert gases. Why is this?

2 What is the basis of the vertical groups and horizontal periods of the periodic table in terms of the electronic structure of the atoms?

3 In which block (s, p, d or f) in the periodic table would you find the following elements?

a Be **b** Ni **c** Cs **d** S **e** Kr

4 Discuss the role of human error, experimental inaccuracy and creative thinking in the development of the periodic table.

The blocks of the periodic table

One useful aspect of the periodic table is the way it can be divided into blocks determined by the electron orbitals which are being filled. The four main blocks of elements are shown in fig. 1.3.25 by different shading, and examples are shown in figs. 1.3.28–31.

The s-block elements

The s block contains the elements of groups 1 and 2. They are known as the **s-block elements** because the outermost electrons are in the s subshells. These electrons are easily lost to form positive ions. The s block includes the **metals** sodium, potassium, calcium and magnesium which are very reactive, forming stable ionic compounds with non-metals. Indeed, they are often referred to as the **reactive metals**. They have lower melting temperatures and boiling temperatures and lower densities than most other metals. Like all metals, they conduct electricity as a result of metallic bonding with its sea of electrons between the atoms. You will look more closely at bonding in the next chapter, and the properties of group 2 elements in unit 2.

Helium and hydrogen are also s-block elements. Electrons are filling the 1s shell in both, but the characteristics of these two elements – a reactive gas and an unreactive gas, both non-metals – are so different from all the other s-block elements that many scientists treat them as a separate group.

fig. 1.3.28 The s-block metals are very reactive, as the behaviour of this piece of sodium with cold water demonstrates.

The d-block elements

The d block lies between groups 2 and 3 and contains elements with successive electrons being added in d orbitals. The **d-block elements** are often called the **transition metals**, though some of them, such as zinc, do not fit this description well. They include familiar everyday metals such as copper, iron, silver and chromium, and they are much less reactive than the elements in groups 1 and 2 because inner d orbitals are being filled while the outer s subshell is full.

The transition metals all conduct electricity and heat and many are shiny and hard and can be pulled and hammered into shape. Mercury is a striking exception, with such a low melting temperature that it is a liquid at room temperature.

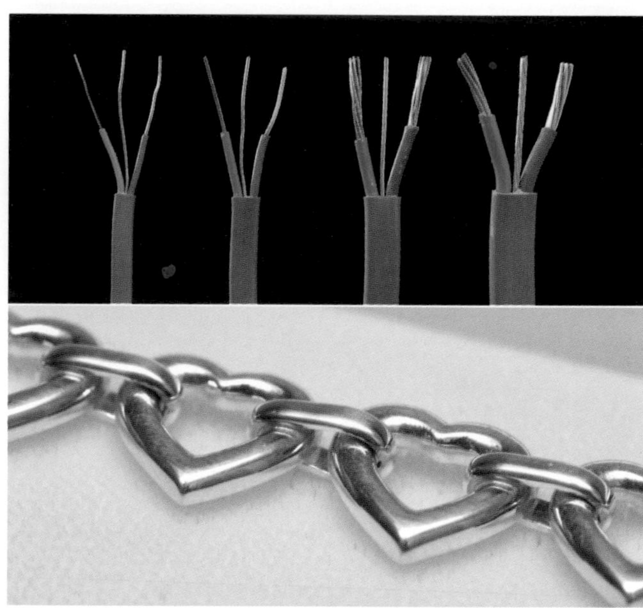

fig. 1.3.29 The d-block elements are much less reactive than the s-block elements and are useful, like this copper and silver, as the uncombined elements.

The f-block elements

In the **f-block elements** successive electrons are being added in the f subshells. The top row of 14 elements (the **lanthanides**) are all very similar metals. The second row (the **actinides**) are all radioactive. Only the actinides up to uranium are naturally occurring. The others have all been synthesised by chemists and are very unstable, often with half-lives of only milliseconds.

The p-block elements

The **p-block elements** are those in groups 3, 4, 5, 6, 7 and 8. For these elements electrons are being added to p orbitals in the outer shell. The p block contains all the **non-metals** (apart from hydrogen and helium) and **metalloids** as well as some metals.

Tin and lead are p-block metals, and they form positive ions and form ionic bonds with non-metals. However, many of the metals in the p block do not have strong metallic characteristics, although they all conduct electricity and heat. These **post-transition metals** are relatively unreactive and many of their reactions resemble those of non-metals.

The metalloids occur in a diagonal block. Although they act like non-metals in most ways, they have one important property which is metallic in nature. They are conductors of electricity, albeit poor ones. In particular, the metalloids silicon and germanium have been responsible for the microchip revolution of the late twentieth century.

The non-metals all form covalent bonds with other non-metals and ionic bonds with metals. Carbon (solid), oxygen (gas) and bromine (liquid) are all examples of non-metals. The majority of them do not conduct electricity, although some forms of carbon do. Some non-metals, eg carbon, form giant covalent structures with atoms linked by strong covalent bonds. However, many non-metals exist as small molecules. You will be looking at group 7 of the p-block elements, the halogens, in unit 2.

fig. 1.3.30 Many pieces of technology which everyone takes for granted rely on microchips to work, and many of these chips are made from the metalloid silicon.

fig. 1.3.31 Although a few non-metals (like the sulfur shown here) are solids, the majority are gases, like the chlorine used as a chemical weapon in the First World War.

The **noble gases** in group 8 are extremely unreactive as a result of their completely filled shells.

HSW Compounds of the noble gases?

It was originally thought that the noble gases were completely chemically unreactive or **inert** (they used to be called the inert gases). As the model of atoms with shells of electrons developed, it seemed obvious that the noble gases, with their complete outer shells of electrons, would not bond ionically or covalently with other atoms.

In 1933 the Nobel prize-winning chemist Linus Pauling predicted that the heavier noble gases would make compounds, but it wasn't until the 1960s that the theory became reality. In 1962 Neil Bartlett, a British chemist working in the US, managed to produce xenon hexafluoroplatinate, using a very powerful oxidising agent. Since then a number of noble gas compounds have been made, but few of them have proved very useful. Xenon makes the most compounds – the lightest noble gases remain completely inert, though chemists are still trying to persuade them to react!

Questions

1 Why are the transition metals so different in chemical character from the s-block elements?

2 The s-block elements and noble gases have one crucial difference in their electronic structure. Explain what this difference is and how it affects the reactivity of the elements.

Patterns in the periodic table

You have seen how the periodic table was built up from observations relating the atomic numbers and the properties of the elements. The electronic structure of the elements and the way they react allow us to group them in blocks. Now you are going to look at how repeating trends of properties appear in the elements of the periodic table. These repeating patterns are known as the **periodicity** of the elements and the properties are known as **periodic properties**. The readiness with which an atom loses or gains electrons determines the way it bonds with other atoms, and that in turn has a big effect on many of its physical properties.

Trends in the periodic table

Moving across the periodic table, elements gain electrons. Moving down a group, elements gain electron shells. This changes the size of the atoms, which in turn obviously affects both their physical and chemical properties. You are going to look at the way the atoms change in diameter, and then at how this affects some of the patterns in properties of the elements, using periods 2 and 3 as an example.

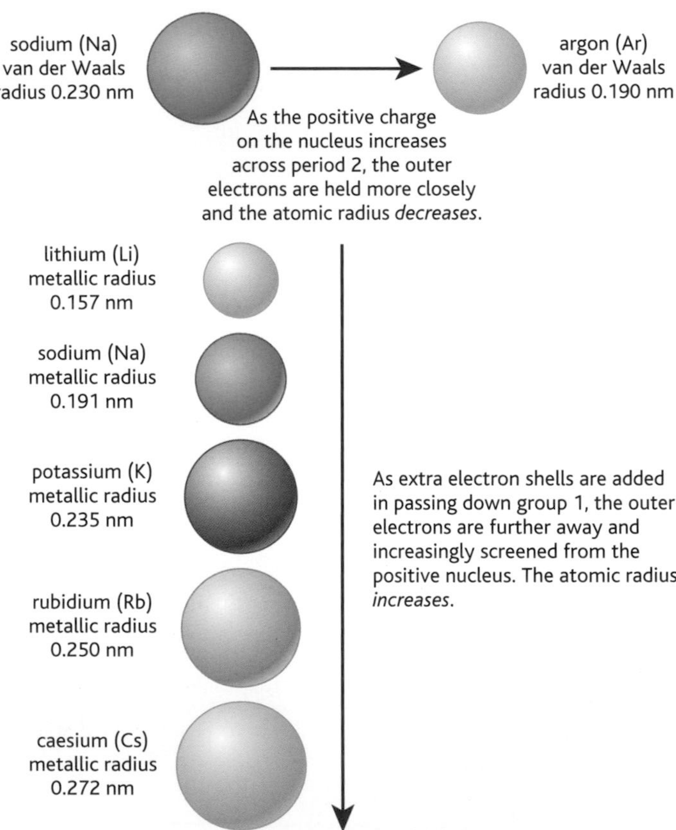

fig. 1.3.32 **The trends in the atomic radii of the elements are directly related to their electronic structure.**

- *The atomic radius generally decreases across a period* – as you can see from **fig. 1.3.32**, as you move across a period the nuclear charge becomes increasingly positive as the number of protons in the nucleus increases. Although the number of electrons also increases, the outer electrons are all in the same shell. This means that they are attracted more strongly to the nucleus, thus reducing the atomic radius across a period.
- *The atomic radius generally increases down a group* – the outer electrons enter new energy levels passing down a group, so although the nucleus also gains positive protons, the electrons are both further away and screened by more electron shells. As a result they are not held so tightly and the atomic radius increases (see **fig. 1.3.32**).

The atomic radius changes when atoms form ions. A positive ion always has a smaller ionic radius than the original atom. This is because the loss of electron/s means that the remaining electrons each have a greater share of the positive charge of the nucleus so are more tightly bound, and when the ion is formed a whole electron shell is usually lost. Conversely a negative ion has a larger radius than that of the parent atom. Even though the additional electrons are in the same shell as existing electrons, the addition of extra negative charge means that the electrons are less tightly bound to the nucleus and so the radius is larger. The term **ionic radius** is used to describe the size of ions, and you will read about trends in ionic radii in chapter 1.4.

Periodic trends in ionisation energy

One of the most important periodic properties is **ionisation energy**. This measures the energy change associated with, for example:

$$Na(g) \rightarrow Na^{+}(g) + e^{-}$$

As you saw earlier, the first ionisation energy of an element is directly related to the attraction of the nucleus for the most loosely bound of the outer electrons. The more tightly held the outer electrons, the higher the first ionisation energy. There are three main factors that affect the ionisation energy of an atom:

- *The attraction between the nucleus and the outermost electron* – decreases (reducing the ionisation energy) as the distance between them increases.
- *The size of the positive nuclear charge* – a more positive nucleus has a greater attraction for the outer electron, leading to a higher ionisation energy.
- *Inner shells of electrons repel the outer electron, screening or shielding it from the nucleus* – the more electron shells there are between the outer electron and the nucleus, the less firmly held is the outer electron and the lower will be the ionisation energy.

Ionisation energy increases across a period – it becomes harder to remove an electron. This is the result of the increasing positive nuclear charge across the period without the addition of any extra electron shells to screen the outer electrons. The atomic radius gets smaller and the electrons are held more firmly, so it requires more energy to bring about ionisation. **Figure 1.3.33 (a)** shows the first ionisation energy of the first 55 elements. The end of each period is marked by the high ionisation energy of a noble gas. As you have seen, this distinctively high ionisation energy is a result of the stable electronic structure of the noble gases and is indicative of their unreactive natures.

As **fig. 1.3.33 (b)** shows, first ionisation energies do not increase smoothly across a period. This is due to the presence of subshells within each shell. In the period lithium to neon, the first ionisation energy of beryllium is larger than that of boron. The same relationship is seen for magnesium and aluminium in the next period. Removing one electron from an atom of boron or aluminium removes the single electron in the p subshell. However, for beryllium or magnesium an electron must be removed from a full s subshell. Full subshells are particularly stable, so this requires more energy than removing the single p electron from boron or aluminium.

The unexpectedly high first ionisation energies of nitrogen and phosphorus can be explained by the fact that both these elements contain half-full outer p subshells. Half-full subshells also appear to be associated with greater stability. Removing an electron from an atom of oxygen or sulfur removes the fourth electron in the p subshell, leaving a half-full p subshell. On the other hand, removing one electron from a nitrogen or phosphorus atom requires breaking into a half-full subshell, a process that requires more energy. Successive ionisation energies like these provide clear evidence for the existence of energy shells and subshells in the structure of the atom.

Ionisation energy generally decreases down a group – it becomes easier to lose an electron. You will be looking at this in more detail later, when you consider the properties of group 2 elements.

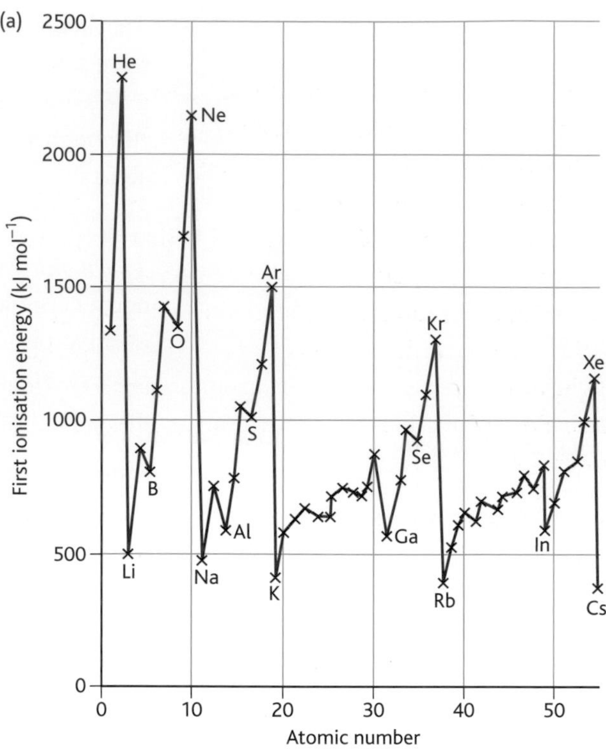

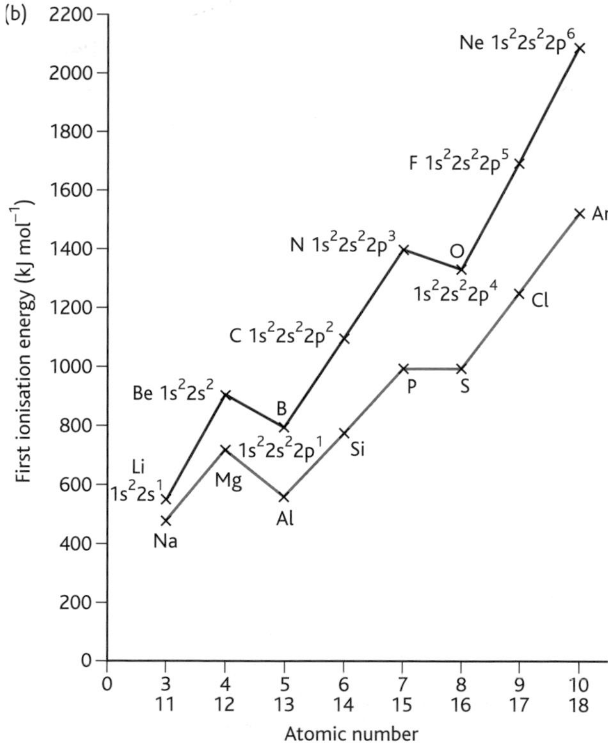

fig. 1.3.33 (a) The first ionisation energy of the first 55 elements of the periodic table. Clear patterns are immediately obvious. (b) A closer look at patterns within individual periods reveals more subtle fluctuations.

Patterns in physical properties

Periodic trends are not limited to the chemical properties of the elements – they are clearly seen in the physical properties as well. Periodicity is seen in the melting temperatures of the elements, their boiling temperatures, their densities and many other properties. These physical properties are closely linked to the structure and bonding of the atoms in the element and the ways in which the atoms are packed together. This in turn depends on the electronic structure of the atoms. You will be looking at bonding in more detail in chapter 1.4.

The melting temperature of a given substance is the temperature at which the pure solid is in equilibrium with the pure liquid, at atmospheric pressure. The melting temperature is affected by both the packing and the bonding of the atoms within a substance. This changes as you move across a period. Many scientists prefer not to analyse trends in melting temperatures of elements across periods because there is considerable debate about the validity of some of the explanations commonly given. However, it is possible to see patterns in some of the periods – fig. 1.3.34 shows the melting temperatures of the elements in periods 2 and 3.

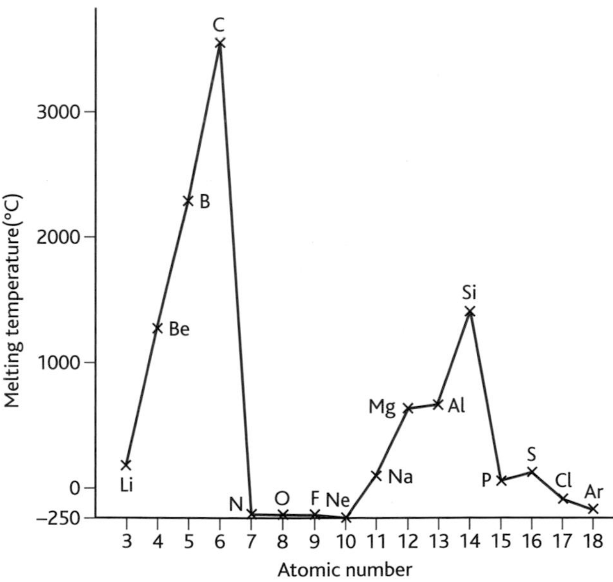

fig. 1.3.34 The melting temperatures of the elements of periods 2 and 3 show clear periodicity.

The relatively high melting temperatures of the metals such as lithium, magnesium and aluminium are due to their metallic structure. The atoms are held tightly together in a sea of electrons. It takes a lot of energy to separate them which gives these elements a relatively high melting temperature.

Moving across the periods you meet giant molecular structures like those found in the metalloids silicon and carbon in the form of diamond. They have strong covalent bonds between the atoms which holds them tightly within a crystal structure. As a result it is very difficult to remove individual atoms and these elements also have extremely high melting temperatures.

In contrast, the simple molecular structures found in most non-metals on the right of the periodic table mean they exist as small, individual molecules. Although the forces holding the atoms within the molecules together are strong covalent forces, the molecules themselves are only held together by weak intermolecular forces (you will learn more about these in chapter 1.4). This means the molecules can be separated relatively easily and most non-metals have low melting temperatures as a result.

At least partly because of these structural differences, dramatic changes are seen in melting temperatures across a period, mirroring the changes of structure and bonding in the elements.

Questions

1 Why does ionisation energy increase across a period?

2 Why does ionisation energy decrease down a group?

3 Why are the first ionisation energies of sulfur and aluminium lower than those of phosphorus and magnesium respectively?

4 How do periodic trends in observations such as atomic radius, ionisation energy and melting temperature confirm the current model of the electronic structure of atoms?

4 Bonding

Bonding in ionic compounds

What is a chemical bond?

The forces holding atoms together are called **chemical bonds**, and understanding them is a central part of chemistry, since changes in bonding underlie all chemical reactions. The physical and chemical properties of both elements and compounds are affected by the type of chemical bonding between the atoms. As you saw in chapter 1.3, the bonds between atoms in elements determine physical properties such as the element's melting temperature. The orientation of bonds is also crucial in determining the shape of a molecule.

In chapter 1.3 you saw that it is the electrons in the outer shells of atoms that are involved in the process of bonding. Atoms can lose, gain or share electrons when they react.

Ionic bonding

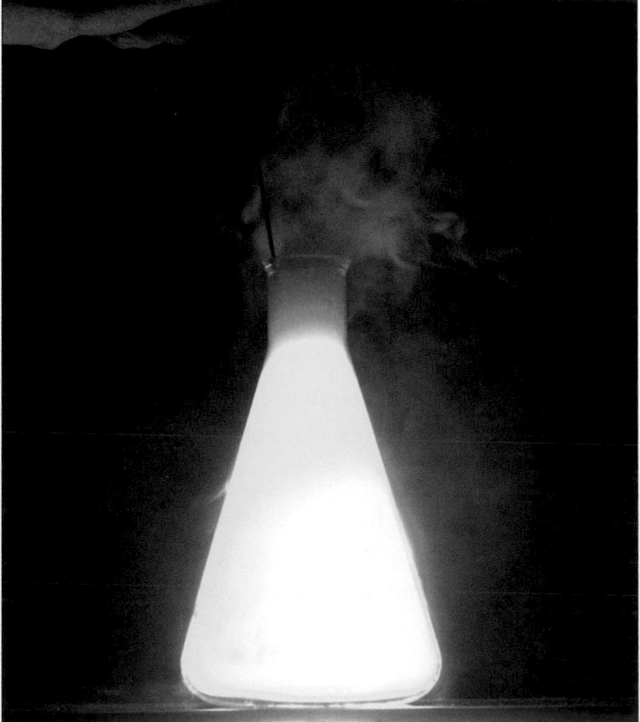

fig 1.4.1 Sodium (a soft metal with a low melting temperature) reacts violently with chlorine (a greenish gas) to produce sodium chloride (a white solid with a high melting temperature).

Ionic compounds are usually formed when metals bond with non-metals. Oppositely charged ions are formed, which are held together by an extremely strong **electrostatic** force of attraction. This attractive force is called the **ionic bond**. The ions are held in an arrangement known as a **giant lattice structure**.

Ionic bonding can be thought of as the net electrostatic attraction between the ions. An ion attracts ions with the opposite charge and repels ions with the same charge. The **lattice** structure of a particular ionic compound is simply the arrangement of ions in a way that maximises the attractive forces between oppositely charged ions and minimises the repulsion between similarly charged ions. The forces exerted by the ions in a giant lattice act equally in all directions, holding the ions together tightly. Giant lattices of ions form **ionic crystals**.

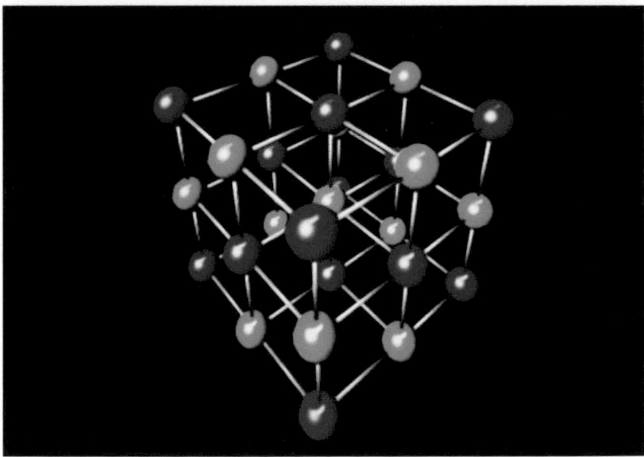

fig 1.4.2 A model of the ionic lattice of sodium chloride, NaCl.

Why does sodium chloride contain Na^+ and Cl^- ions rather than (for example) Na^- and Cl^+ ions, or even Na^{2+} and Cl^{2-} ions? The answer lies in the electronic structure of the sodium and chlorine atoms:

Na $1s^2 2s^2 2p^6 3s^1$
Cl $1s^2 2s^2 2p^6 3s^2 3p^5$

In chapter 1.3 you saw that full (and half-full) subshells of electrons are particularly stable, and the ionisation energies of the elements (see **fig. 1.3.32**) suggest that the electronic configuration of the noble gases, with full outer shells, is particularly stable. Sodium can achieve a full outer shell by losing the electron in its 3s subshell to become an Na^+ ion. This gives it the same electronic configuration as the noble gas neon, and Na^+ and Ne are said to be **isoelectronic** (they have the same number and arrangement of electrons):

Na^+ $1s^2 2s^2 2p^6$
Ne $1s^2 2s^2 2p^6$

Just as sodium achieves a full outer shell of electrons by losing an electron, chlorine can do so by gaining an extra electron in its 3p subshell. The Cl^- ion so formed is isoelectronic with the argon atom:

Cl^- $1s^2 2s^2 2p^6 3s^2 3p^6$
Ar $1s^2 2s^2 2p^6 3s^2 3p^6$

So sodium and chlorine can each achieve the electronic configuration of a noble gas by transferring the electron in sodium's 3s subshell to chlorine's 3p subshell. We can represent this as shown in **fig. 1.4.3**, which also shows the same process between a calcium atom and two fluorine atoms.

This complete transfer of electrons from one atom to another forms an ionic or **electrovalent bond**. In the process of electron transfer, each atom becomes an ion which is isoelectronic with the nearest noble gas, and the substance formed is held together by electrostatic forces between the ions.

The tendency for the ions formed by elements to have a full outer shell of electrons is expressed in the **octet rule**:

- When elements react, they tend to do so in a way that results in an outer shell containing eight electrons.

Like most simple rules this one has exceptions, although it is useful in many situations.

Accounting for electrons – dot and cross diagrams

Counting electrons is helpful to show how atoms may donate or receive electrons in order to obtain a full outer shell. This is usually done using **dot and cross diagrams**, which show the electrons as dots or crosses. **Figure 1.4.3** shows two examples of such 'electronic book-keeping'. Only the electrons in the outer shell are shown, because these are the ones involved in chemical bonds.

Dot and cross diagrams distinguish between the electrons in different atoms using dots and crosses, but remember that all electrons are identical. When a sodium atom loses an electron to a chlorine atom, it is impossible to tell which of the electrons surrounding the Cl^- ion came from the sodium atom.

(a) Na + Cl ⟶ $[Na]^+$ $[Cl]^-$

(b) Ca + F F ⟶ $[Ca]^{2+}$ $[F]^-$ $[F]^-$

fig 1.4.3 The transfer of electrons in the formation of (a) sodium chloride and (b) calcium fluoride. Each atom forms an ion with an outer shell containing eight electrons.

Questions

1 If an ion is isoelectronic with another element, what does this mean? Give an example.

2 Draw dot and cross diagrams to show the bonds in the simplest compounds formed between:
  a chlorine and potassium
  b chlorine and calcium
  c oxygen and sodium
  d oxygen and calcium.

Giant ionic lattices

When sodium and chlorine react to produce sodium chloride, a compound results with properties quite different from those of its two elements, as **table 1.4.1** highlights.

Property	Sodium	Chlorine	Sodium chloride
Appearance and state	Soft metal, melting temperature 98°C	Greenish gas, boiling temperature −35°C	White crystalline solid, melting temperature 801°C
Reaction with water	Reacts vigorously with water	Soluble in water	Soluble in water
Electrical conductivity	Good conductor of electricity in solid and liquid states	Does not conduct electricity	Good conductor of electricity in liquid state and in aqueous solution

table 1.4.1 The properties of sodium, chlorine and sodium chloride.

The explanation for the different properties of sodium chloride and its elements lies in the fact that sodium chloride is an ionic compound, made up of a lattice of Na^+ and Cl^- ions. In contrast, sodium consists of a lattice of metal atoms, and chlorine is a gas made up of Cl_2 molecules. The structure of substances has a major effect on their chemical and physical properties.

Trends in ionic radii

In chapter 1.3 you saw how atomic radii vary across and down the periodic table. The **ionic radius** is the radius of an ion in a crystal. The radius of a positive ion is smaller than the element's atomic radius because the remaining electrons are more strongly attracted to the positive nucleus. In the same way, negative ions are larger than the atom that formed them because the additional negative charge means all the electrons are bound less tightly to the nucleus.

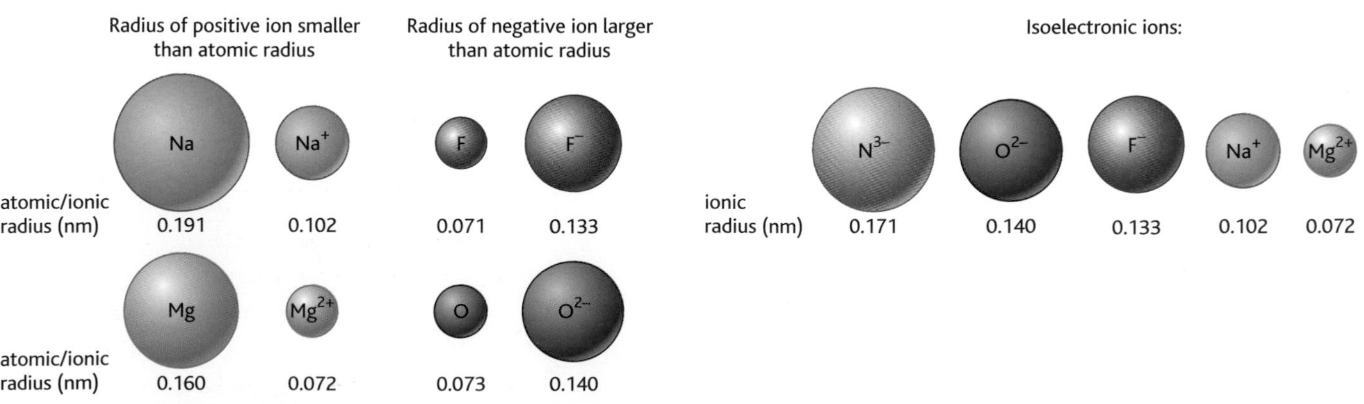

fig. 1.4.4 You can see clear trends in ionic radii both across the periods and down the groups of the periodic table.

Types of lattice structure

If you pass X-rays through a crystal, they are scattered or diffracted by the electrons in the atoms or ions in the structure. This produces a diffraction pattern which can be used to tell you about the arrangement of the atoms or ions in the crystal. These patterns are known as **electron density maps**. You will soon be looking at this in more detail.

Electron density maps show that the exact arrangement of ions in an ionic lattice varies depending on the relative sizes of the different ions present. For example, sodium chloride has the **face-centred cubic structure** shown in fig. 1.4.5(a). Each ion has six nearest neighbours, so we say that the **coordination number** is 6. As you can see from this diagram, the chloride ions are considerably larger than the sodium ions, and so the sodium ions fit into the spaces between the chloride ions. The packing of ions in the NaCl lattice is commonly found in ionic compounds, and is known as the **rock salt structure**.

The structure of caesium chloride, shown in fig. 1.4.5(b), is different because the caesium ion is larger than the sodium ion. More chloride ions can therefore fit around the caesium ion – you can see here that each caesium ion has a coordination number of 8. This crystal structure is known as a **body-centred cubic structure**.

Ionic compounds tend to have very high melting and boiling temperatures because of the large number of ionic bonds between the oppositely charged ions. These need to be broken or weakened in order to melt an ionic solid.

We can relate the crystal structure model to the formula of CsCl. Figure 1.4.5(b) shows that each unit cell contains one caesium atom at its centre. At each corner of the cell is a chloride ion. Each of these ions is shared between seven other unit cells, so effectively one-eighth of each chloride ion 'belongs' to the unit cell. Therefore:

$1 \times Cs^+$ ion $= 1 \times Cs^+$

joins with

$8 \times {}^1/_8 Cl^-$ ion $= 1 \times Cl^-$

Therefore the formula is CsCl.

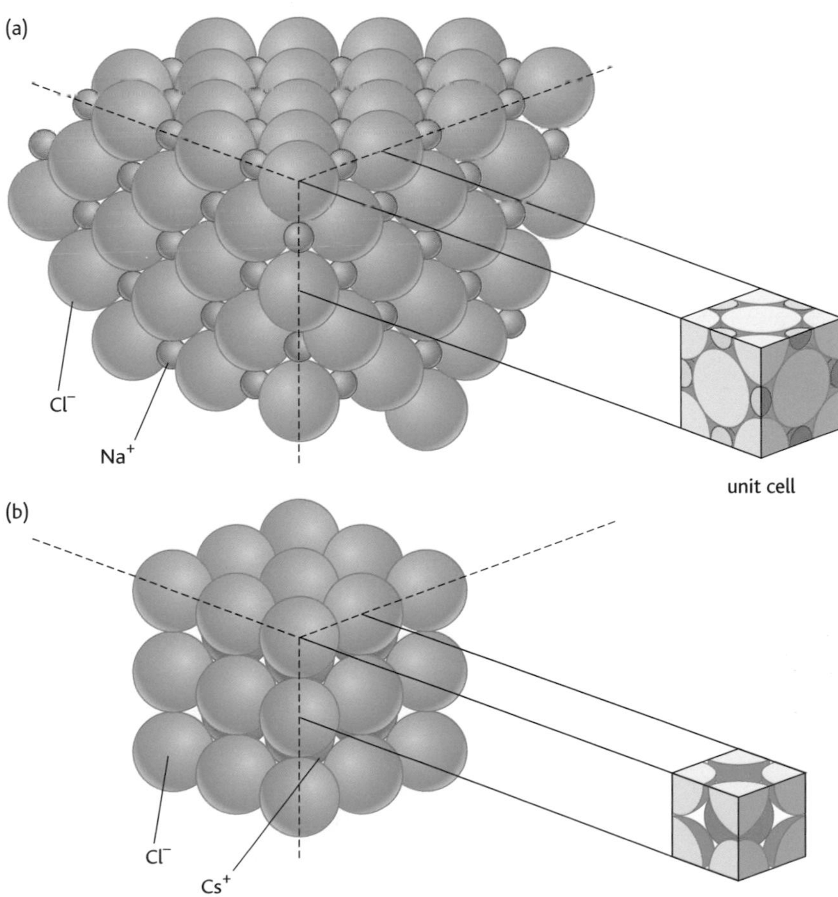

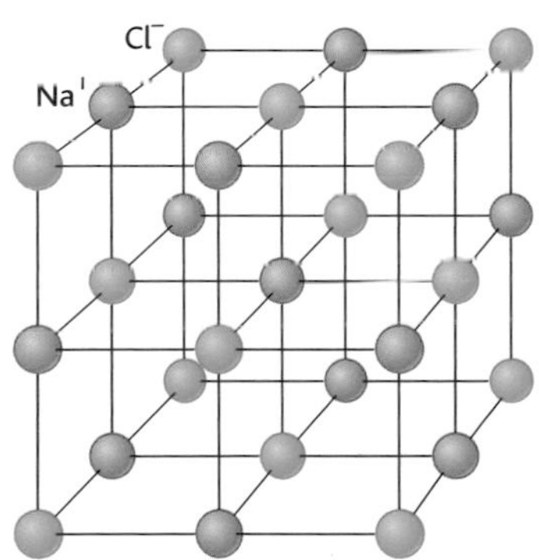

fig. 1.4.5 **(a)** Sodium chloride has a face-centred structure. Each cation is surrounded by 6 anions, and vice versa. The unit cell is shown in two different ways. **(b)** Caesium chloride has a body-centred cubic structure, with a coordination number of 8.

Questions

1 Look at the radii of the isoelectronic ions shown in fig 1.4.4. They all have the same number of electrons. Explain the differences in the radii.

2 What effect does ionic radius have on:

  a the reactivity of an ionic compound

  b the arrangement of the ions in an ionic lattice?

HSW Evidence for the existence of ions

It is impossible to see ions, yet the models used to explain ionic bonding and many other aspects of chemistry depend on their existence. What is the evidence for the existence of ions?

Physical properties

The physical properties of what we know as ionic compounds are explained by the existence of charged ions. In some ways ionic compounds are similar to metals in that they have high melting and boiling temperatures and are quite hard. These facts suggest that the particles that make up ionic substances are held tightly to each other, and the existence of positively and negatively charged ions would explain this property.

As you have seen, the model for a solid ionic compound consists of a lattice containing both positive and negative ions. The properties of ionic solids reflect the strong interactions that exist between ions with opposite charges in this lattice. Such a structure also accounts for the observation that ionic solids conduct electricity when molten but not in the solid state. This fits with the model – in the solid the ions will be held immobile in the lattice, and will be unable to move to conduct electricity. However, when the ionic substance melts, the attraction between the ions in the lattice is overcome and the ions become mobile. In this state they can carry an electric current in the same way as the mobile electrons in a metallic solid. Similarly, when an ionic substance is in aqueous solution the ions are free to move and the solution conducts electricity.

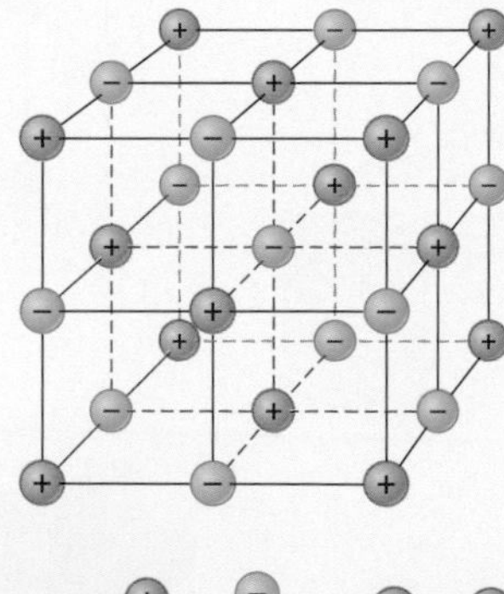

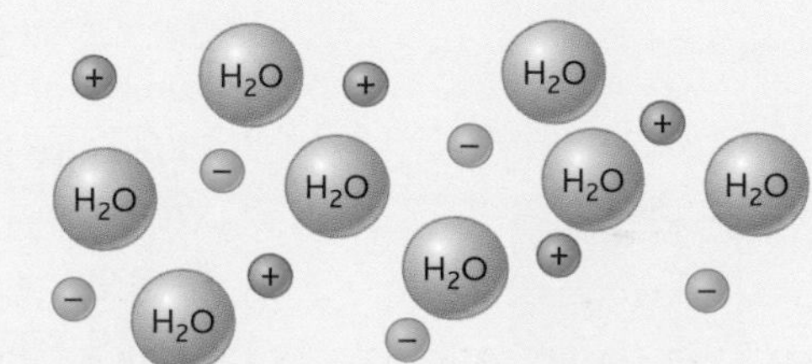

fig. 1.4.6 **The mobile ions in a molten ionic substance or an aqueous solution can conduct electricity, but in the solid form they cannot.**

Ionic substances differ from metals in one important way, namely that they are **brittle** – they break easily when hit. In contrast metals are generally both **malleable** (they can be hammered into sheets) and **ductile** (they can be drawn out into wires). This suggests that the particles in ionic substances are held in a regular structure and have no 'sea of electrons' as do metals. Again this is explained by a model of charged ions held in a regular lattice. A force could cause the ions in one layer in the lattice to shift relative to the other layers, putting ions of the same charge next to each other. The repulsive force that results would force the two layers apart so the ionic substance simply shatters (see **fig. 1.4.7**).

fig. 1.4.7 **Ionic substances shatter if they are hit or crushed – more evidence that they are made up of ions.**

Electrolysis

The behaviour of ionic substances during electrolysis is another clear piece of evidence for the existence of ions. For example, aqueous copper(II) chromate(VI), $CuCrO_4$, is an olive green solution. If a specially prepared solution undergoes electrolysis, a blue colour appears around the negative electrode (cathode) and a yellow colour appears around the positive electrode (anode). These changes can be explained by the existence of ions. Positive blue copper(II) ions, $Cu^{2+}(aq)$ are attracted to the cathode and yellow chromate(VI) ions, $CrO_4^{2-}(aq)$ are attracted to the anode. This and other similar demonstrations of electrolysis show the presence of ions in the solution.

fig. 1.4.8 **The colour changes seen at the electrodes during the electrolysis of copper(II) chromate(VI) demonstrate the presence of positive and negative ions.**

Electron density maps

Earlier in this chapter you looked at some ionic crystal structures revealed by electron density maps. These electron density maps provide more evidence for the existence of ions. When X-rays are passed through an ionic crystal onto photographic film, they are diffracted by the electrons and produce bright spots on the film. The bigger the ion (or atom) the more electrons it has and the brighter the spot it produces. By analysing the positions and intensities of the spots, experts can work out the **charge density** of the electrons in the crystal. This charge density can be defined as the amount of electric charge per unit volume and it is measured as electrons per cubic nanometre. Points of equal density are joined up to form contour lines which produce an electron density 'map'. This diffraction of X-rays in a clear and repeatable pattern is further evidence of the presence of ions in the crystal.

fig. 1.4.9 **Photos like these are not easy to interpret unless you are an expert – but they provide evidence for the existence of ions in compounds such as lysozyme, an anti-bacterial enzyme, shown here.**

Measurements taken from electron density maps allow different ions to be identified, and their distance apart in the lattice. All this adds to the evidence for the existence of ions.

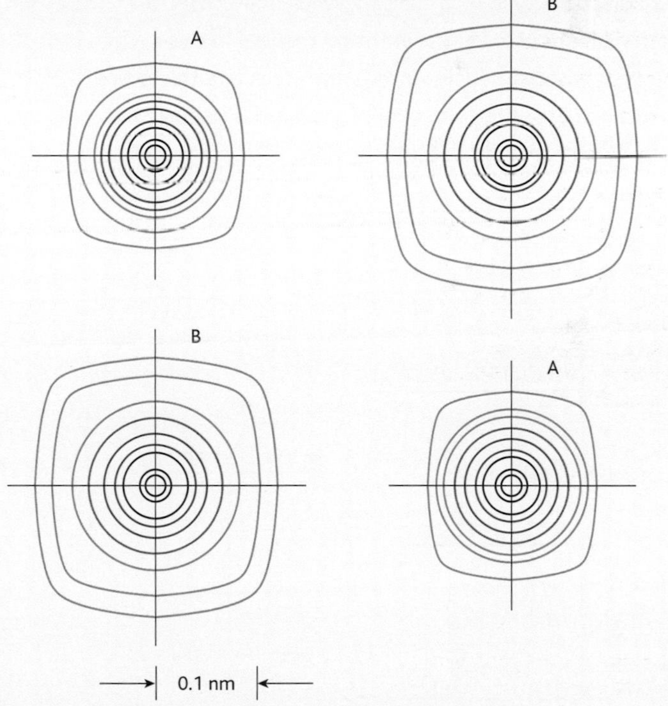

fig. 1.4.10 **Electron density maps for sodium chloride**

Questions

1 Ions are invisible. What evidence can you provide for their existence?

2 Look at the electron density map in fig 1.4.10. Which are sodium ions and which are chloride ions? What is the ionic radius of the sodium ion and the chloride ion? How did you make your decision?

Lattice energy

In chapter 1.2 you looked at bond enthalpies and the enthalpy changes involved in the formation of a compound. You saw how these can be used to predict, for example, whether a reaction will be exothermic or endothermic. You have also seen in this chapter how the formation of ionic bonds involves the transfer of electrons between atoms to form oppositely charged ions, and the coming together of these ions to form a lattice. The formation of ionic substances in this way is associated with energy changes. The formation of ions in the gaseous state from elements in the standard state is endothermic, while the formation of the lattice involves a release of energy - the **lattice energy** (sometimes known as lattice enthalpy). You can compare the lattice energy of an ionic compound with the bond enthalpy of a molecular compound. Lattice energies are negative because they relate to forming the lattice, while bond enthalpies are positive because they relate to breaking the bond.

The lattice energy of an ionic crystal can be defined as:

- The enthalpy of formation of one mole of an ionic compound from gaseous ions under standard conditions.

HSW Practical measurements of lattice energy

The lattice energies of substances can be calculated from a special type of enthalpy level diagram called a **Born–Haber cycle**. All the enthalpy changes in this cycle can be measured, which allows you to calculate the lattice energy, $\Delta H^{\ominus}_{lat}$ of the compound. **Figure 1.4.11** shows the Born–Haber cycle for sodium chloride.

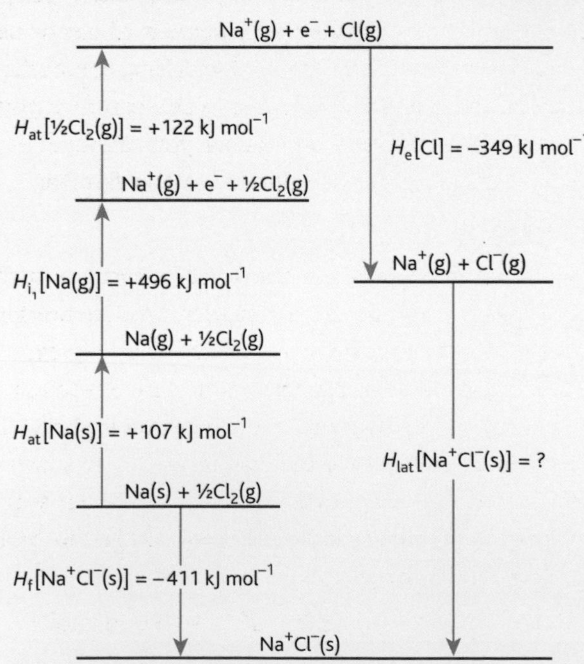

fig. 1.4.11 **The Born–Haber cycle for sodium chloride.**

The terms in the Born–Haber cycle are as follows:

- $\Delta H^{\ominus}_f[NaCl(s)]$ standard enthalpy change of formation of sodium chloride, measured using a bomb calorimeter.
- $\Delta H^{\ominus}_{at}[Na(s)]$ standard enthalpy change of atomisation of sodium, calculated from enthalpy changes of fusion and vaporisation and the specific heat capacity of sodium.
- $\Delta H^{\ominus}_{i_1}[Na(g)]$ first ionisation energy of sodium, found by spectroscopic measurements.
- $\Delta H^{\ominus}_{at}[Cl_2(g)]$ standard enthalpy change of atomisation of chlorine, found by spectroscopic measurements.
- $\Delta H^{\ominus}_e[Cl(g)]$ first electron affinity of chlorine, found by methods similar to those used to measure ionisation energies.

Given measured quantities for all these terms, the lattice energy of NaCl can be calculated:

$$\Delta H^{\ominus}_{lat}[NaCl(s)] = -411 - 107 - 496 - 122 - (-349) \text{ kJ mol}^{-1}$$
$$= -787 \text{ kJ mol}^{-1}$$

This value for the lattice energy of sodium chloride gives you some idea of the size of the attractive forces between the ions in sodium chloride – notice that it is of the same order of magnitude as the bond enthalpies given in chapter 1.2, comparing the strengths of ionic bonds and covalent bonds. However, in ionic lattices each ion is held by attractive forces to a number of other ions within the lattice. As a result the overall energy needed to separate them is much higher than that needed to break an individual covalent bond. This explains the high melting and boiling temperatures of ionic substances.

Table 1.4.2 shows some lattice energies for other compounds, also calculated from Born–Haber cycles.

Compound	Ionic radius of cation (nm)	Ionic radius of anion (nm)	Lattice energy ($kJ\ mol^{-1}$)
NaF	0.102	0.133	−918
NaCl	0.102	0.180	−780
NaBr	0.102	0.195	−742
NaI	0.102	0.215	−705
NaCl	0.102	0.180	−780
KCl	0.138	0.180	−711
MgF_2	0.072	0.133	−2957
CaF_2	0.100	0.133	−2630

table 1.4.2 Lattice energies for a range of compounds.

What affects lattice energies?

Looking at **table 1.4.2**, you can see that the lattice energy of a compound is affected by both the charge and the size of the ions. The effect of the ionic radius is shown by the trend in the sodium halides, in the chlorides of sodium and potassium, and in the fluorides of magnesium and calcium. The lattice energy becomes less negative as the size of the ions increases. In smaller ions the attractive force of the positive nucleus holds the outer electrons more tightly because they are closer to the nucleus.

You can also see that the lattice energy becomes more negative with ions of greater charge (Mg^{2+}, Ca^{2+}) because they attract other ions more strongly – compare the lattice energy of sodium fluoride with those of magnesium fluoride and calcium fluoride, which have a much greater magnitude.

Why is the lattice energy affected by the ion size and charge in this way? To answer this question, you need to know that the size of the force of attraction of one ion on another is explained by **electrostatic theory** and can be calculated using **Coulomb's law**:

$$\text{Force of attraction, } F = \frac{k\, q_1 q_2}{r^2}$$

where q_1 and q_2 are the charges on the ions, and r is the distance between them. This relationship shows you that increasing the charge q on either ion increases the attractive force F between two oppositely charged ions in the lattice. Decreasing the size of one or both ions decreases the distance r between them and so increases the attractive force F. This is why ionic radius is such an important factor in determining the strength of ionic bonds.

Predicting stability

You can use standard enthalpies of formation and Born–Haber cycles to help you predict the relative stabilities of different compounds, and indeed why particular ionic compounds exist. Why is sodium chloride always NaCl, never $NaCl_2$? Why don't you come across MgCl or $MgCl_3$? Theoretical lattice energies can be calculated for substances like these that do not exist or do not form ionic bonds. These can be used in a Born–Haber cycle to calculate the theoretical enthalpy of formation for these compounds – and this will usually show you exactly why the compound does not form! It usually reveals a very highly endothermic enthalpy of formation, because the ionisation energy for one of the electrons is so high. In the case of relatively simple compounds like $NaCl_2$ or $MgCl_3$, you know that the arrangement of electrons in the outer shell is not favourable. Using theoretical lattice energies and Born–Haber cycles can give you the same information – and allows you to make predictions about the compounds of less familiar elements.

The very high energies associated with substances such as $MgCl_3$ are because the third ionisation involves breaking into a noble gas configuration, and this takes a great deal of energy. So for example, in the case of sodium, the ionisation energy of the second electron is almost 10 times greater than that of the first. **Figure 1.4.12** shows a theoretical Born–Haber cycle for $NaCl_2$. You can see that NaCl is much more stable – the energy budget to create $NaCl_2$ would be huge!

$$Na(g) \rightarrow Na^+(g) + e^- \quad \Delta H^{\ominus}_{i1} = 496\ kJ\ mol^{-1}$$

$$Na^+(g) \rightarrow Na^{2+}(g) + e^- \quad \Delta H^{\ominus}_{i2} = 4563\ kJ\ mol^{-1}$$

$$\Delta H^{\ominus}_f[NaCl_2(s)] = 107 + (2 \times 122) + 496 + 4563 + (2 \times -349) + \Delta H^{\ominus}_{lat}[NaCl_2(s)]$$

If you estimate the lattice energy for $NaCl_2$ as being the same as that for $MgCl_2$, namely $-2526\ kJ\ mol^{-1}$:

$$\Delta H^{\ominus}_f[NaCl_2(s)] = +2186\ kJ\ mol^{-1}$$

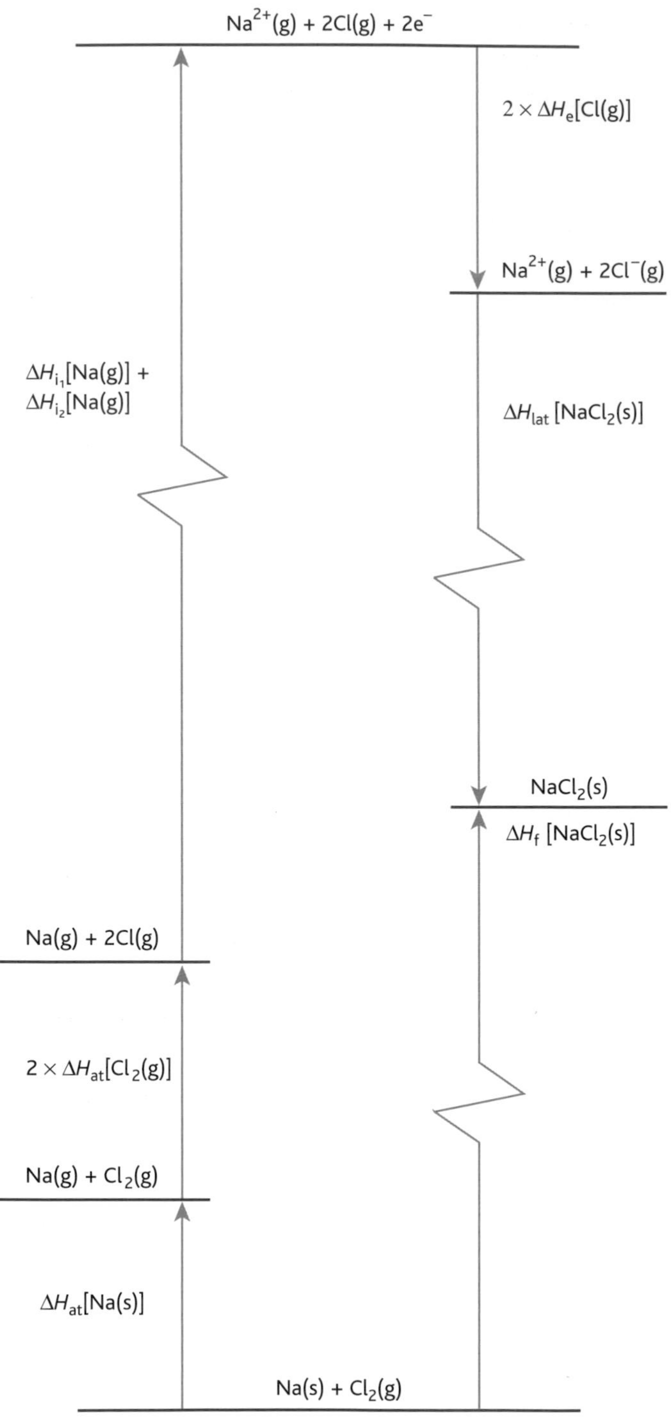

fig. 1.4.12 A Born–Haber cycle for the theoretical compound $NaCl_2$.

A closer look at lattice energies

By using Coulomb's law in conjunction with knowledge of the structure of ionic lattices, chemists can calculate theoretical lattice energies for ionic compounds. **Table 1.4.3** compares these theoretical values with the lattice energies obtained from Born–Haber cycles.

Compound	Lattice energy (kJ mol^{-1})	
	Born–Haber	Theoretical
NaF	–918	–912
NaCl	–780	–770
NaBr	–742	–735
NaI	–705	–687
AgF	–958	–920
AgCl	–905	–833
AgBr	–891	–816
AgI	–889	–778

table. 1.4.3 Lattice energies calculated from Born–Haber cycles and from theoretical models of attraction and repulsion between ions in crystal lattices.

In the case of the sodium halides, the lattice energies obtained in these different ways are very similar, agreeing within 3% or better. But in the case of the silver halides, agreement is much less good – why is this?

Polarisation in ionic bonds

Ionic bonds can be distorted by the attraction of the positive cation for the outer electrons of the negative anion. The electron density maps in **fig 1.4.13** can help you model this effect. **Figure 1.4.13(a)** shows a wholly ionic bond, with the electron clouds of the two ions quite separate. In **fig. 1.4.13(b)** you can see a situation where the electron cloud of the anion is attracted to the positive cation. If the distortion is great, it may lead to a charge cloud that begins to resemble that of a covalent bond (see **fig. 1.4.13(c)**). You will be looking at the electron density maps of covalent bonds on the next few pages.

The polarising power of a cation depends on its charge density, and this in turn depends on both its ionic radius and its charge. Cations with a small ionic radius are much more polarising than large cations because their charge density is higher – the effect of the positive nucleus is felt more strongly. Also cations with a large positive charge are more polarising than cations with a smaller charge, because again they have a stronger attraction for the outer electrons.

On the other hand, how readily an anion is polarised depends solely on its ionic radius which affects how tightly the electrons are held. The larger the anion, the more easily it is polarised.

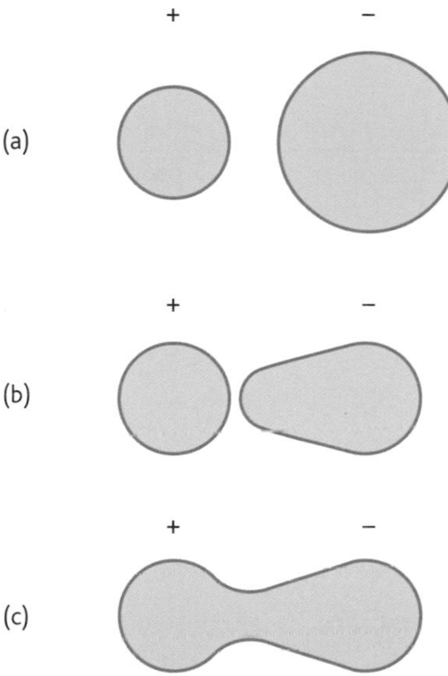

fig. 1.4.13 The distortion of an ionic bond. Distortion is favoured if the (positive) cation is small with a large charge, and the (negative) anion is large.

You can see this effect demonstrated clearly in the silver halides. **Table 1.4.3** shows that the theoretical lattice energy of the silver halides is considerably less negative than the lattice energy calculated from Born–Haber cycles. This suggests that the silver halides are *more stable* than a purely ionic model indicates. In other words, they show a considerable degree of covalent character. This is the result of the effect of the different relative electronegativities on the nature of the bonds. When the difference in electronegativity between the different ions in a crystal is high – one ion is much more electronegative than the other – the ionic model works well, and there is good agreement between the two values of lattice energies. When the difference in electronegativity is smaller, the bonding in the crystal has a considerable degree of covalent character. There is less attraction between the oppositely charged particles and more electron sharing. The melting temperatures of the silver halides are about 20% lower than those of the sodium halides, agreeing with the idea that they have a greater covalent character.

Why do these trends show up when you compare lattice energies calculated using a theoretical model with the values from a Born–Haber cycle using experimental values? The theoretical model assumes that all the ions are spherical and separate, and that the electron charge is always evenly distributed across the ion. Polarisation of the bond, a distortion in the shape of the ion and increasing covalent character in the bond all reduce the lattice energy and show up in the experimental values.

Questions

1 How does the ionisation energy of an atom affect the likelihood that it will undergo ionic bonding?

2 How does lattice energy differ from bond enthalpy?

3 How do the ionic radius and the ion charge affect the lattice energy of an ionic substance?

4 What trend can you see in the difference between the theoretical and Born–Haber values for the lattice energy of the silver halides? How do you explain it?

5 Using the method for $NaCl_2$ above as a model, and the data in **figs. 1.4.11** and **1.4.12** and **table 1.4.3**, calculate the enthalpy of formation associated with the following reactions:

a $Mg(s) + \frac{1}{2}Cl_2(g) \rightarrow MgCl(s)$ (ΔH_f negative)

b $Mg(s) + Cl_2(g) \rightarrow MgCl_2(s)$ (ΔH_f more negative)

c $Mg(s) + 1\frac{1}{2}Cl_2(g) \rightarrow MgCl_3(s)$ (ΔH_f positive)

Value	ΔH° (kJ mol^{-1})
Enthalpy of atomisation of magnesium	148
First ionisation energy of magnesium	738
Second ionisation energy of magnesium	1451
Third ionisation energy of magnesium	7733
Lattice energy of NaCl (use for MgCl)	–780
Lattice energy of $MgCl_2$	–2526
Lattice energy of $AlCl_3$ (use for $MgCl_3$)	–4500

table 1.4.4 Data for ΔH_f calculations.

Covalent compounds

Sharing electrons

Many elements do not form ionic compounds – the energy released in the formation of the lattice of ions would be insufficient to overcome the energy required to form the ions in the first place. These atoms use another method of achieving a full outer shell of electrons – electron sharing. This is especially true for elements in the middle of the periodic table, for which the loss or gain of three or four electrons would be required to form stable ions, which would require a great deal of energy.

Think about two hydrogen atoms approaching one another. Each atom consists of a single proton as its nucleus, with a single electron orbiting around it. As the atoms get closer together, each electron experiences an attraction towards the two nuclei, and the electron density shifts so that the most probable place to find the two electrons is between the two nuclei (see **fig. 1.4.14**). Effectively, each atom now has a share of both electrons. The electron density between the two nuclei exerts an attractive force on each nucleus, keeping them held tightly together in a **covalent bond**.

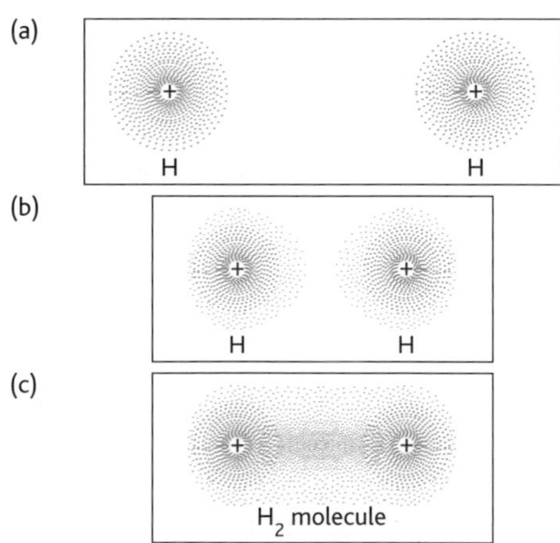

fig. 1.4.14 A covalent bond forming between two hydrogen atoms. (a) Two hydrogen atoms a large distance apart. (b) As the atoms approach, the electron density between them increases. (c) Eventually the electron density becomes greater between the two atoms than anywhere else in the neighbourhood.

The covalent bond represents a balance between the attractive force pulling the nuclei together (due to the electron density between the nuclei) and the repulsive force of the two positively charged nuclei pushing each other apart. For two hydrogen atoms, this balance of the attractive and repulsive forces occurs when the nuclei are separated by a distance of 0.074 nm. This distance is known as the **bond length** for the bond. As you have seen in chapter 1.3, the energy released as the two atoms come together to form the bond is the **bond enthalpy** (bond energy) – which is of course also the amount of energy required to break the bond. In the case of the H_2 molecule, the bond energy is 436 kJ mol^{-1}.

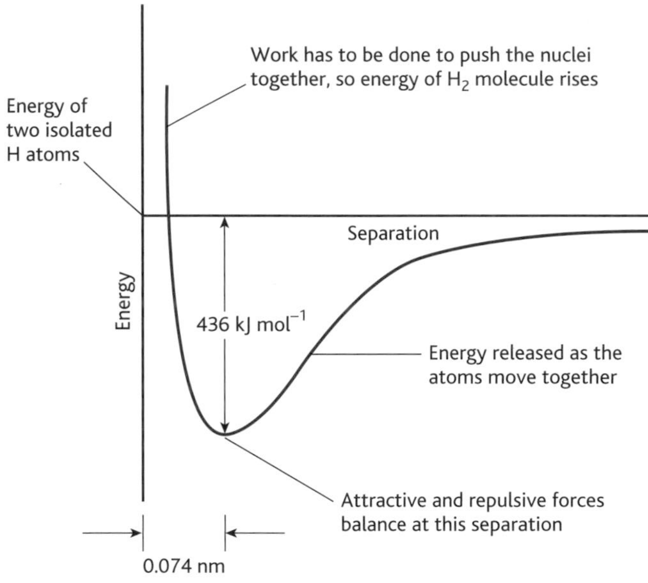

fig. 1.4.15 The energy of two hydrogen atoms at different separations.

A covalent bond arises from a shared pair of electrons. Covalent bonds form between atoms of non-metal elements such as Cl_2 and P_4, as well as in compounds such as H_2O and CH_4. A covalent bond is usually represented as a line, so that the hydrogen molecule is represented like this:

H—H

This is called the **structural formula** of the hydrogen molecule, which as you know has the molecular formula H_2.

As with ionic bonds, covalent bonds can be represented using dot and cross diagrams. Once again the octet rule can be applied in many cases, as **fig. 1.4.16** shows. Notice that it is possible for two atoms to share more than one pair of electrons – sharing two pairs results in a **double bond** (eg O=O), and sharing three pairs produces a **triple bond** (eg N≡N).

However, such diagrams are only used for 'electronic book-keeping' – they do not tell you anything about the shapes of molecules, and they do not represent the real positions of electrons.

Cl—Cl H—O—H

H—C—H (with H above and below C) O=O N≡N

fig. 1.4.16 Dot and cross diagrams can be used to show how the atoms in a covalently bonded molecule share electrons in order to obtain a full outer shell of electrons.

Dative covalent bonds

Sometimes both of the electrons that make up a covalent bond come from the same atom. This type of covalent bond is called a **dative covalent bond**. An example of such a bond occurs when ammonia is dissolved in a solution containing hydrogen ions, for which we can write:

$$NH_3(aq) + H^+(aq) \rightarrow NH_4^+(aq)$$

Figure 1.4.17 shows how a dot and cross diagram enables us to do the electronic book-keeping for this reaction. Notice that the NH_4^+ ion produced (called the ammonium ion) contains a nitrogen atom with a full outer shell of electrons, together with four hydrogen atoms, each of which also has a full outer shell.

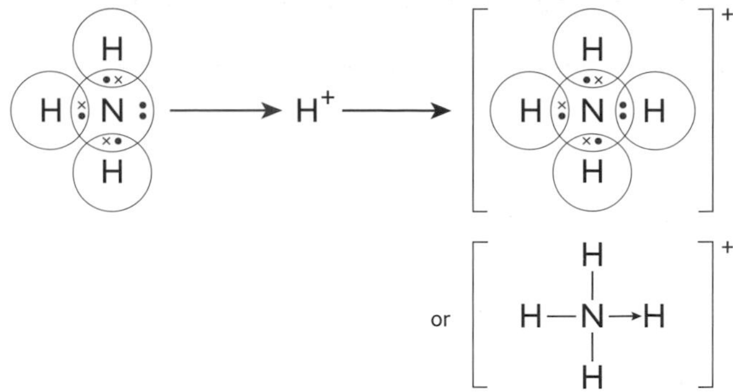

fig. 1.4.17 The formation of an ammonium ion. Notice that all the N—H bonds are equivalent so that in practice it is impossible to distinguish between them.

Another example of a dative covalent bond is in aluminium chloride, which consists of molecules of $AlCl_3$ in the vapour phase. As the vapour is cooled, however, pairs of molecules come together to form **dimers**, in which they are held together by dative covalent bonds (see **fig. 1.4.18**).

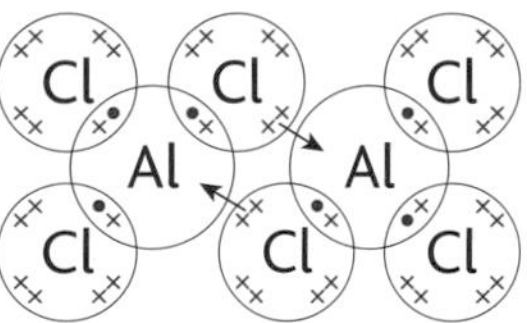

fig. 1.4.18 The formation of the Al_2Cl_6 dimer. Once again, this is only electronic book-keeping, and tells us nothing about the shape of the Al_2Cl_6 molecule.

Compounds that have unshared electron pairs readily form dative covalent bonds. It is important to realise that once they are formed, dative covalent bonds are exactly the same length and strength as any other covalent bond.

Dative covalent bonding can be used to explain the existence of different oxides. For example, carbon dioxide is the result of normal covalent bonding between carbon and oxygen, but carbon monoxide has two normal covalent bonds along with a dative covalent bond formed by electrons from the oxygen atom.

HSW Evidence for the nature of the covalent bond

Covalent bonds are strong bonds which arise from the electrostatic attraction between nuclei and the electrons between them. Many covalently bonded substances exist as discrete molecules. However, others form either **atomic crystals** or **molecular crystals**, and these are very useful for providing evidence of the nature of the covalent bond. In the same way that ionic bonds may show some covalent character, covalent bonds may be polarised and show some ionic character. You will be learning more about this in unit 2.

Giant structures

Giant atomic structures form distinctive atomic crystals. The lattice positions contain atoms held together by covalent bonds. The whole lattice can be thought of as a giant molecule. One of the best known of these is diamond, with carbon atoms held together by covalent bonds. Silicon(IV) carbide, SiC and silicon(IV) oxide, SiO_2, are examples of giant molecular structures. The crystals are very hard, with high melting temperatures demonstrating the great strength of the covalent bonds.

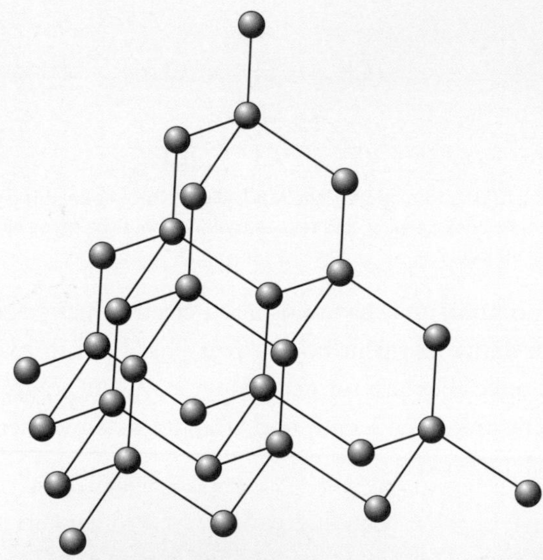

fig. 1.4.19 **The structure of diamond with all the covalent bonds holding the atoms in their lattice positions.**

Electron density maps and the shapes of covalent molecules

If you look at the electron density maps of simple covalent compounds, you will see that covalent bonds are highly directional. The shared electrons of a covalent bond, or a pair of unshared electrons (a **lone pair**), form a dense area of electronegativity. Each area of electronegativity repels the others, and this gives covalent molecules with more than two atoms a very definite shape in which the three-dimensional relationship between the atoms is constant (whether they are in the solid, liquid or gas phase). This spatial relationship provides more evidence for the nature of the covalent bond. It also governs the chemical and physical properties of molecules.

A very simple model for the shapes of covalent molecules is based on the idea that the outer shell electron pairs stay as far away from each other as possible to minimise the repulsive forces between them. This can be seen in the molecule $BeCl_2$, which has the electronic structure shown in **fig. 1.4.20(a)**. In this molecule, beryllium has only two pairs of electrons around it (it is one of the exceptions to the octet rule we met earlier). To minimise the repulsion between these pairs, they must be arranged so that they are on opposite sides of the beryllium atom. This gives the $BeCl_2$ molecule the shape shown in **fig. 1.4.20(a)** – it is a **linear** molecule.

The structure of BF_3, which has three pairs of electrons around the boron atom (another exception to the octet rule), gives the molecule the **trigonal planar** shape shown in **fig. 1.4.20(b)**.

Ammonia, water and methane all have four clouds of electronegativity and these affect the shape of the molecule. For example, you might expect water molecules to be linear but, due to the lone pairs of electrons on the oxygen atom, the covalent bonds are at an angle to each other (see **fig. 1.4.20(c)**).

The ability to work out the shapes of covalently bonded molecules will help you to understand their properties and the way they react with other chemicals. You will find out more about this in unit 2. The shapes of covalent molecules provide you with more evidence about the nature of the covalent bond and the way polarisation can take place within it.

(a)

Cl : Be : Cl

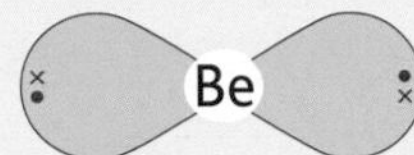

Cl—Be—Cl

(b)

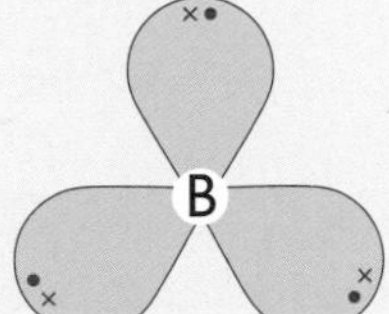

(c)

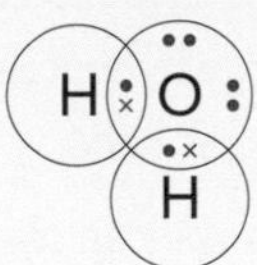

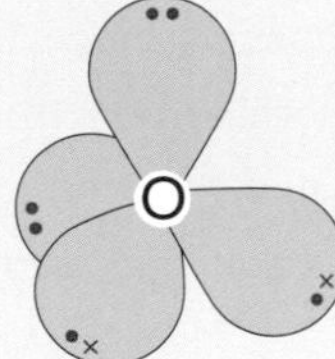

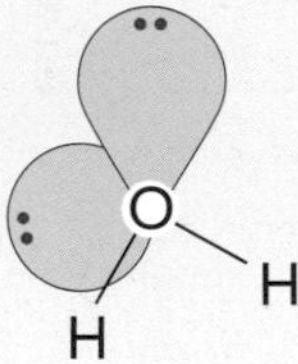

(d)

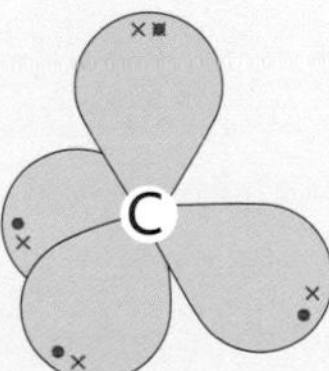

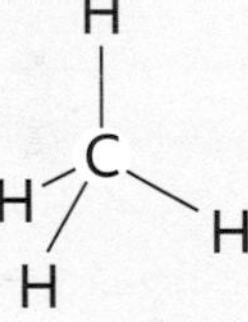

fig. 1.4.20 Electron density maps show the shapes of covalent molecules, which come about by minimising the repulsion between the outer electrons.

Questions

1. Which of the following compounds would you expect to be covalent and why?
   a HI **b** SO_2 **c** NaOH **d** NH_3
2. How does dative covalent bonding differ from ordinary covalent bonding?
3. Using dot and cross diagrams, show the covalent bonds in carbon monoxide and carbon dioxide.
4. Covalent bonds are very strong. How could you use giant atomic structures to compare the strength of covalent and ionic bonds?
5. How do electron density maps help develop an understanding of the nature of the covalent bond?

Metallic bonding

Metals are good conductors of heat and electricity. There is a special type of bonding between the atoms of a metal, known as **metallic bonding**. A simple model of a metal has a metal crystal containing positive metal ions surrounded by a 'sea' of mobile **delocalised electrons**. Metallic bonding is the strong attraction between the positive metal ions and the sea of delocalised electrons that surrounds them. This model of metallic bonding can be used to explain many of the properties of metals.

HSW Using the model of metallic bonding to explain the properties of metals

There are a number of typical metallic characteristics that are shown by most metals. The model of metallic bonding described above can be used to explain all of these characteristics:

- *Conducting electricity* – all metals conduct electricity well. The model of metallic bonding shown in **fig. 1.4.21** explains this because the delocalised electrons are free to move through the lattice under the influence of an electric field. Under normal conditions electron movement through the lattice is relatively random, but when a potential difference is applied the electrons tend to move away from the negative electrode which repels them and towards the attractive positive electrode.
- *High thermal conductivity* – the way metals conduct heat can also be explained by the model of metallic bonding with positive metal ions and a sea of delocalised electrons. The delocalised electrons move easily and so can transmit kinetic energy rapidly through the lattice. When a metal is heated, the electrons in the areas of high temperature have high kinetic energy and so move rapidly (and randomly) towards the cooler regions of the metal, transferring their energy to other electrons as they go. This explains the high thermal conductivity shown by metals.
- *High melting and boiling temperatures* – the melting and boiling temperatures of many metals are very high which suggests that the forces between metal atoms must be large because it takes a lot of energy to separate them. The simple model of metallic bonding seems to provide a reasonable explanation of this, with a lattice of positive ions held tightly together by the negatively charged electron 'glue'. The strong attraction between the positive ions and the negative electrons means it takes a great deal of energy to separate them.
- *Metals are malleable and ductile* – metals, unlike most solids, can be hammered into different shapes (they are malleable) and many of them can also be drawn out into wires (they are ductile). Our model suggests that the positive metal nuclei can move within the sea of electrons, and wherever they move they are still surrounded by a sea of negative electrons. This explains why metals can be hammered into shape and made to bend.

As you can see, the model of metallic bonding as positive metal ions in a sea of delocalised electrons provides what seems to be a good explanation for many of the typical metal characteristics. However, you cannot push the model too far – for example, the metals have a wide range of melting temperatures and our simple model does not provide an explanation for this.

However, not all metals have exactly the same properties and the metallic bonds obviously differ in strength in different metals. What might determine the strength of the metallic bonds in a metal lattice? The answer to this deepens our understanding of the metallic bond model.

As you saw earlier, the amount of electric charge in a given volume is called the charge density. It would seem logical that the higher the charge density of both the delocalised electron cloud and the ions in the metal lattice, the greater will be the electrostatic forces of attraction between the electron cloud and the ions in it.

On this basis, you would expect potassium (with large, 1+ ions) to have weaker metallic bonds than iron (with smaller, 2+ ions), since the ions and electron cloud in the potassium lattice will have a lower charge density than the ions and electron cloud in the iron lattice. The physical properties of potassium (soft, melting temperature 63°C) and iron (hard, melting temperature 1535°C) fit with this explanation.

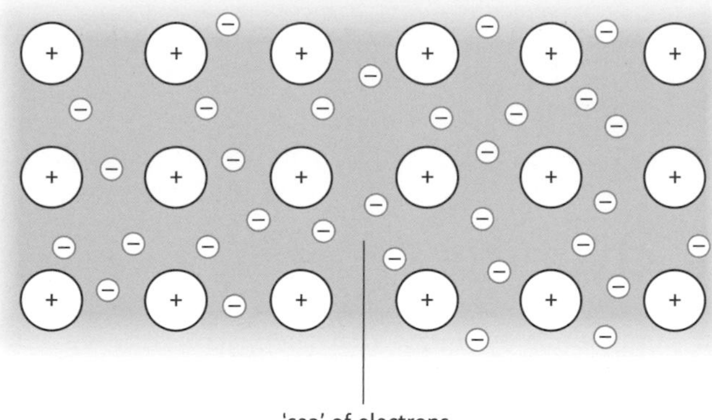

fig. 1.4.21 This simple model of a metal has a lattice of positive ions surrounded by delocalised electrons.

Using metals

The first known use of metals was around 10 000 years ago when copper was used for the first time. Steel was being produced and used in the Middle East 4000 years ago, but the use of iron did not become common in Europe until 2500 years ago.

Throughout the long history of the use of metals, people have taken advantage of their ductile and malleable properties to make jewellery and fine ornaments, wires and pots. The fact that metals conduct heat well means they have been and are still widely used in cooking pots. The electrical conductivity of metals was not widely used until much more recently, when people found ways of generating and using electricity in both domestic and industrial settings.

fig. 1.4.22 (a) Precious metals can be worked and shaped into jewellery and coins. (b) The delocalised electrons enable metals to conduct heat easily. (c) Much of modern life depends on the ability of metals to conduct electricity. (d) Strength, flexibility, lightness and resistance to corrosion make some metals ideal in the construction of aircraft.

Questions

1 How does the structure of the metallic bond within a metallic lattice help to explain why metals are malleable and ductile but ionic compounds are not?

2 Table 1.4.5 shows some physical properties of three metals. Use the metallic bonding model to explain these differences.

Metal	Melting temperature (°C)	Thermal conductivity (W cm^{-1} K^{-1})
Sodium	98	1.35
Magnesium	649	1.5
Copper	1083	3.85

table 1.4.5 Some physical properties of three metals.

Examzone

You are now ready to try the first Examzone test for unit 1 (Examzone Unit 1 Test 1) on page 248 which tests you on what you have learnt in the first four chapters of this book.

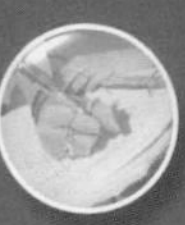

5 Introductory organic chemistry

About organic chemistry

'Organic' and 'inorganic' compounds are terms that chemists have used since the 1800s. Chemists now define **organic chemistry** as the study of carbon compounds (except very simple ones). Carbon forms a vast number of compounds (combined with hydrogen and other elements) due to the particular ability of carbon to form single, double and triple bonds, and to build up molecules with long chains and rings, all of which are the subject of organic chemistry. **Inorganic chemistry** is the study of all the 91 naturally occurring chemical elements and their compounds, including carbon and a few of its simple compounds such as carbon dioxide, carbon monoxide and the carbonates.

HSW Changing ideas about organic chemistry

In 1807 Jöns Jacob Berzelius observed that chemicals could be divided into two groups based on their behaviour on heating. Some substances melt or vaporise when they are heated, but when they cool down they return to their original state. Others – notably those from living organisms – burn or char on heating. Berzelius used these differences to classify chemicals as *inorganic* (those that melt) and *organic* (those from living organisms, that burn). 'Organic' comes from the word 'organism' which means living thing. Chemists believed that organic molecules could only be made in living bodies – they could not be **synthesised** (made in a laboratory).

In 1828 Friedrich Wöhler made ammonium cyanate by reacting silver cyanate with ammonium chloride. In a number of experiments he showed that the ammonium cyanate was exactly the same chemically as the urea (NH_2CONH_2) he had extracted from dog's urine. He had made an organic chemical from an inorganic source for the very first time. Wöhler wrote to Berzelius: 'I must tell you that I can make urea without the use of kidneys, either man or dog. Ammonium cyanate is urea.'

The first recorded synthesis of a new organic compound (as opposed to a naturally occurring one) is credited to Christian Schönbein in 1846. Like many advances in science, Schönbein's achievement resulted from a happy accident combined with a rigorous approach to thinking about his observations. Experimenting in his kitchen, as scientists often did in those days, Schönbein spilled a mixture of nitric and sulfuric acids. He grabbed his wife's apron to mop them up. He hung it on the door of the oven to dry – and when it dried, it exploded! The acids had reacted with the cellulose fibres in the cotton of the apron to produce nitrocellulose (known at the time as guncotton), an unstable explosive chemical which caused the deaths of several other chemists who attempted its synthesis.

Although the definition of organic and inorganic chemicals today is based on the composition and structure of their molecules, the early classification was sound in many ways. Organic molecules are the basis of living organisms and substances derived from them such as oil – and the names stuck.

fig. 1.5.1 Today we no longer classify chemicals as organic or inorganic depending on whether they burn or melt, but rather on whether they are carbon-based molecules.

The vast range of organic compounds

Carbon forms around 7 million compounds, far more than all of the other elements put together. Plastics and synthetic fibres, natural fibres such as wool and cotton, dyes, drugs, pesticides, flavourings and foodstuffs consist largely of organic compounds. The complex structural molecules that make up living cells, and the enzymes that control the reactions within them, are also organic chemicals, as is oil and all the oil-based products that we use. How do chemists get to grips with all these organic chemicals? Fortunately the very properties of carbon that make such diversity possible also allow us to divide its compounds into distinct types, or families, making them much easier to study.

fig. 1.5.2 The food, clothing and even the cells of the people's bodies all consist of organic chemicals, chains of carbon atoms combined in various ways with hydrogen and other elements.

One thing to be clear about is that organic chemistry is not the study of organic food production! Organic food is produced without the use of many inorganic pesticides and fertilisers, but all food is made up of organic chemicals, whether produced 'organically' or not.

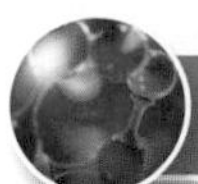

HSW Medicines old and new

Many drugs are organic compounds. A surprising number have true 'organic' origins as they were originally extracted from plants or animals. For example, in the past, the anal glands of beavers were used to relieve headaches and other pains. The raw pain-relieving chemical in aspirin, salicylic acid, comes from willow bark. Beavers eat willow bark and the salicylic acid becomes concentrated in their anal glands, which is why they made effective painkillers.

In time chemists analysed willow bark and leaves and found the active organic ingredient. Then the compound was synthesised and improved on to give us acetylsalicylic acid, or aspirin that is widely used today. In the twenty-first century most drugs are discovered in rather more scientific ways and more directly by the work of organic chemists. Synthetic versions of naturally occurring compounds, along with totally new synthetic organic molecules, are designed and made using a variety of research techniques.

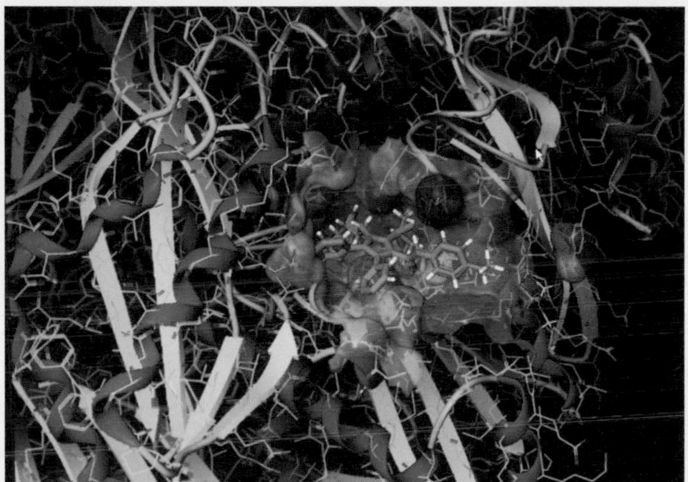

fig. 1.5.3 New drugs are designed using computer-generated images.

Questions

1 How and why has the definition of organic chemistry changed over time?

2 Who do you think is most likely to be involved in the following, organic or inorganic chemists? Explain your answers.

- a increasing yields of the fertiliser ammonium nitrate
- b developing new biodegradable plastic bags
- c developing new alloys for Formula 1 cars.

Hazard and risk in organic chemistry

The difference between hazard and risk

Many organic substances, including those you might come across in the school lab, require special handling because they are hazardous and carry a risk of harm to people using them. In everyday conversation the words hazard and risk may be taken to mean much the same thing – that something is potentially dangerous. However, in chemistry these words have very specific meanings:

- *Hazard* – the hazard presented by a substance or an activity is its potential to do harm. This potential is absolute. For example, some chemicals are flammable and some are toxic. Their tendency to burn, or to poison, are tested and calculated and will always be the same.
- *Risk* – the risk associated with a particular hazard is the chance that it will actually cause harm. Risk is affected by a number of things, in particular the nature of the hazard involved and the level of exposure to it. The level of exposure in turn is dependent on factors such as the expertise of the person working with the chemical, the volumes being used, the conditions in which it is used and the protective clothing and equipment available. A hazardous substance can be safe to use if the risks are minimised. For example, you know that petrol is hazardous – it is flammable and poisonous. However, the risk of coming to harm when you fill a car with petrol is very small. On the other hand, if you spill petrol on your clothing and then light a match, the risk of harm is high. The hazard presented by petrol is the same however it is used – but the risk of harm changes considerably depending on how the petrol is handled.

fig. 1.5.4 The hazard associated with using chemicals in the laboratory is carefully and rigorously assessed. These internationally recognised Hazchem symbols show the hazard clearly.

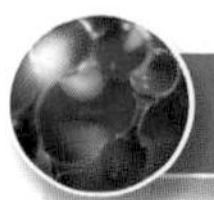

HSW Managing hazard, calculating risk

It isn't possible to eliminate risk from life completely – everything you do carries some element of risk, however small. However, in a chemistry lab you are more likely to be exposed to hazardous materials than in many other places. As a result, people in labs take precautions, use safety features and behave with care to reduce the risks. Because of this laboratories are generally very safe environments. In contrast, people jump into cars and drive every day with little or no thought for the hazardous nature of what they are doing – and often make little effort to reduce the risks.

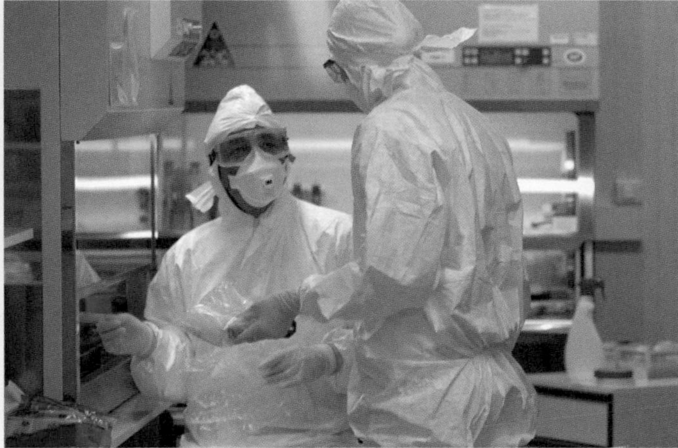

fig. 1.5.5 Safety features in a laboratory reduce the risks in an environment which contains a number of hazards.

There are a number of hazards particularly associated with organic chemicals. Many of these chemicals are poisonous and/or carcinogenic (they cause cancer). Many are flammable. As a result, when working with organic chemicals it is very important to carry out a full **risk assessment**. The purpose of this is to identify the risks of using hazardous chemicals and reduce them as far as possible.

To carry out a risk assessment you follow the steps below:

1 Identify any hazards of the chemicals you want to use and the procedure you plan to carry out.

2 Quantify the risk associated with the way you want to use the hazardous substance, eg how much of the hazardous substance do you need to use, will it be used by trained personnel, are fume cupboards available?

3 Identify who is at risk.

4 Identify any control measures which you can put in place to reduce the risk.

5 Quantify the risk that remains and decide whether it is now at an acceptable level. If so, record your risk assessment and carry out your procedure.

Ways of reducing risk

How can you reduce the risk associated with using hazardous chemicals? Some strategies are listed below:

- *Working on a smaller scale* – for example, there is less risk of inhaling fumes with smaller amounts of chemicals because there are fewer fumes. When the quantities are small it is easier to contain the reaction in closed apparatus, and to carry out the reaction in a fume cupboard. If heat is given out (exothermic reaction) this will be less for smaller quantities of reactants, and smaller quantities are easier to transfer from one container to another without spillage.

- *Taking specific precautions or using alternative techniques depending on the properties of the hazardous substances you are using* – for example, one way of reducing risk is to use the lowest possible concentration of a solution to achieve a particular reaction. Using a low concentration considerably reduces the risks for a hazardous solute. The lowest suitable concentration of any particular reagent will vary from reaction to reaction, eg in many laboratory reactions 0.4 mol dm^{-3} or even 0.1 mol dm^{-3} acids and alkalis are quite adequate. At these concentrations the chemicals are **irritant** rather than **corrosive**, so the hazard level is reduced. This in turn reduces the level of risk. However, there are other reactions, particularly at AS and A2 level, that require 1 mol dm^{-3} and even 5 mol dm^{-3} solutions to be used. Although the hazard level is higher, the risk can be kept low by careful planning and risk assessment.

- *Careful use of safety measures* – such as fume cupboards to remove toxic or flammable fumes, and personal protection such as safety goggles, considerably reduce the risk of carrying out reactions using hazardous chemicals.

- *Changing the conditions under which a reaction takes place* – for instance, lowering the temperature of a reacting mixture will slow the reaction down. This can substantially reduce the risks of the reaction mixture overheating and/or excessive fumes being given off, though cooling may change the equilibrium position and so affect the proportion of reactants and products in the final mixture. For example, the reaction between sodium thiosulfate and dilute hydrochloric acid is often used to show the effect of increasing temperature on the rate of a reaction. However, the sulfur(IV) oxide given off can trigger asthma attacks in vulnerable individuals. By carrying out the experiment in closed vials and reducing the temperature immediately after readings have been taken, the amount of gas given off can be greatly reduced and almost all students can carry out the investigation without suffering any ill effects.
- *Using alternative methods with less hazardous substances* – sometimes it is possible to substitute chemicals and still study the same basic reaction. In many cases the alternative chemicals are not as effective as the original, more hazardous ones. The reaction may be slower or the yield may be lower but if the risk is also substantially lower these disadvantages are worthwhile. CLEAPSS (the Consortium of Local Education Authorities for the Provision of Science Services) is an advisory service providing guidance on practical science teaching in schools and colleges. They have produced a list of alternative compounds which can be used in organic chemistry. For example, tetrachloromethane was the best solvent to make a non-aqueous bromine solution because the solution would keep for a long time. However, tetrachloromethane has a high hazard rating (it is toxic). Cyclohexane can be used instead if it can be stored out of the light (eg in a dark bottle or in a clear bottle in a labelled cardboard box). However, plastic screw tops or bungs will be affected if it is stored for over a month. In a clear bottle, noticeable deterioration will occur within two days. So there is often a safer alternative, even though it may not perform quite as well.

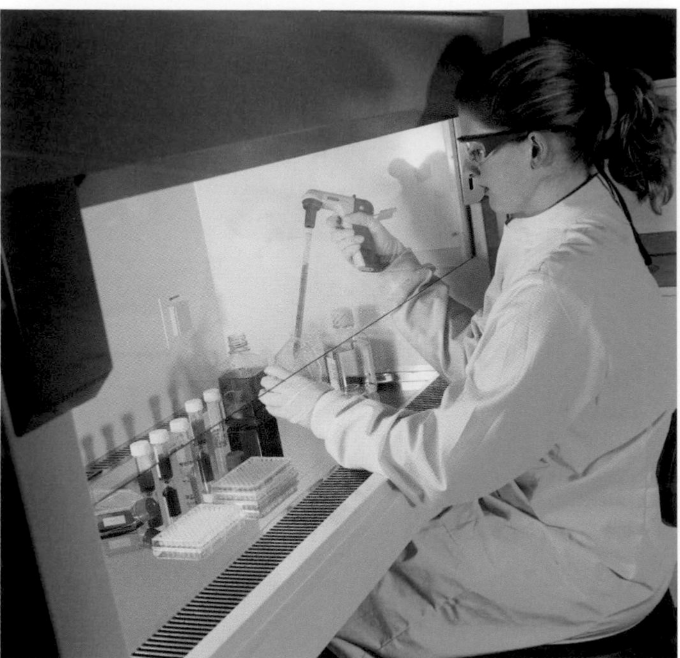

fig. 1.5.6 Whether organic or inorganic, the risk of using hazardous chemicals is greatly reduced if they are handled carefully.

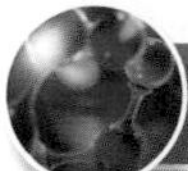

HSW Pesticides and comparative risks

Providing food to feed the population of the world is not an easy task, made more difficult by the fact that insects and fungi destroy up to 30% of the food that is grown. To prevent insect pests from destroying crops as they grow, farmers around the world use **pesticides** – chemicals that kill the pests.

There are two main classes of pesticide used. **Natural pesticides** are derived from plants. Many plants have evolved chemicals in their leaves and flowers that are toxic to insects and other pests. People have extracted some of these chemicals and used them as pesticides. The best-known natural pesticide is pyrethrin, an organic compound originally extracted from pyrethrum flowers. Chemists have made a number of closely related chemicals that are even more effective. Natural pesticides like pyrethrum are usually broken down relatively quickly in the environment both by sunlight and within the bodies of many animals. Pyrethrum is also effective against malaria mosquitoes, which cause disease and death to millions of people every year. However, there may be health risks to the people who use it – in the US it has been classified as a likely human carcinogen by the Environmental Protection Agency. Because low levels of pyrethrin are broken down in the human body, the best way to reduce risk from the pesticide is to spray the lowest effective amount.

Wearing a mask while spraying helps to avoid any risk of inhaling the chemical. However, the risks of starvation and malaria are far higher than any potential risks from pyrethrin sprays. On the other hand, there are other much more effective pesticides that can kill many more pests and so save many more people from starvation.

Synthetic pesticides can be very effective at destroying pests – often organic compounds, they help prevent the destruction of millions of tonnes of food every year. However, they can be expensive, and there are some significant risks associated with their use. The chemicals are often very persistent in the environment. They are not broken down readily in biological systems and so they build up in food chains until they reach a level where they are harmful to humans or other animals. They can also be toxic to the farmers who work with them. This is a particular problem in the poorer countries of the world, where pesticides are often still applied by hand. In China, for example, up to 500 farmers die each year from acute pesticide poisoning. Careful use of the chemicals, wearing protective clothing and breathing equipment – even a simple mask – can reduce these risks. If several farmers get together they may be able to afford a vehicle to spray the pesticide.

The comparative risks of natural and synthetic pesticides can be hard to balance. Applying synthetic pesticides gives good pest control – but at a high price in terms both of cost and of toxicity to the farmer and the environment. Natural pesticides are less toxic to the environment and may be cheaper – but they are still hazardous to those who use them, and less effective as pesticides, so make a more limited contribution to solving the world's food shortage. It is a dilemma which has not yet been solved.

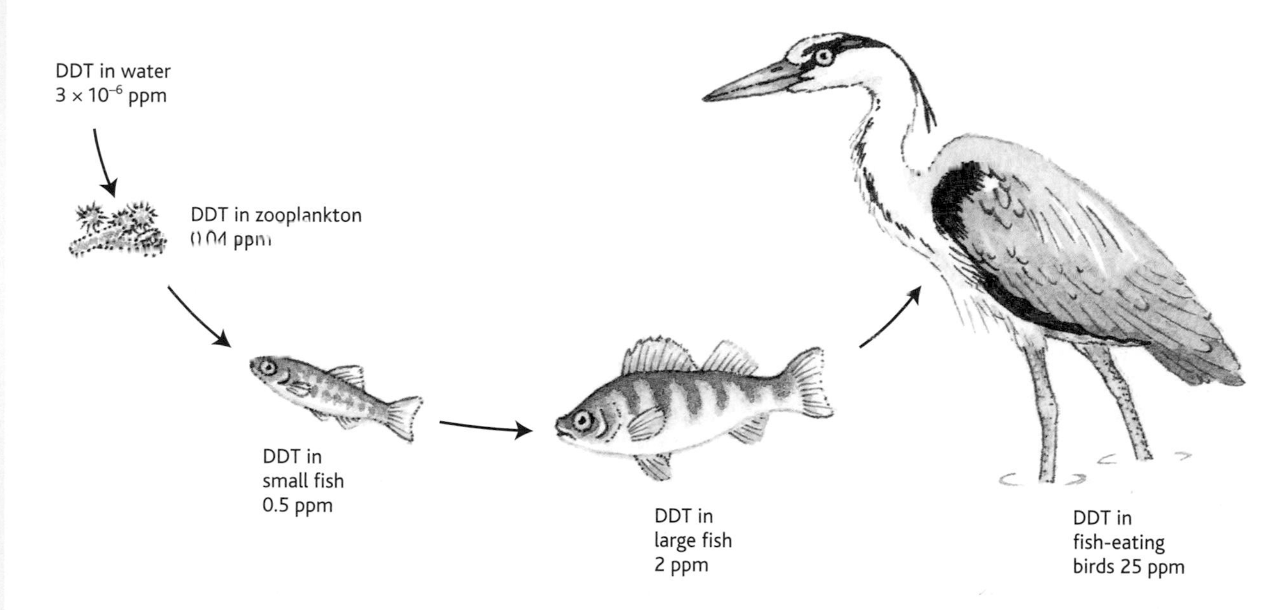

fig. 1.5.7 DDT is a very effective insecticide, and its use brings many benefits to people, especially in the control of the mosquitoes that carry malaria. However, it builds up in food chains so the risk to the environment is very high and the chemical is now only used when there is no alternative.

Questions

1. What is the difference between hazard and risk?
2. Why are chemical laboratories relatively safe places to work?
3. What ethical issues would you need to consider when deciding whether to use natural or synthetic pesticides on a large scale?

The properties of the carbon atom

Why does carbon form four covalent bonds?

Carbon forms four covalent bonds by sharing electrons with other atoms, and the arrangement of the four bonds around a carbon atom is tetrahedral or very nearly so. The electronic structure of the carbon atom in the ground state is $1s^22s^22p^2$, which means that the carbon atom needs to gain four electrons to achieve a noble gas configuration (see **fig. 1.5.8**). Once the four bonds are made, the outer shell electrons form a fully shared octet, the inner shell is complete and there are no lone pairs or empty orbitals, so carbon atoms do not bond to more than four other atoms.

carbon atom (2,4)

methane C(2,8) same electronic arrangement as neon

bond angle 109.5° tetrahedral

fig. 1.5.8 In methane, carbon is combined with hydrogen to achieve a stable octet of electrons in the outer shell.

Carbon atoms are unique in their ability to form covalent bonds with other carbon atoms and with other non-metals at the same time. Carbon forms very strong bonds with itself, which can be single, double or triple bonds (see **table 1.5.1**). Carbon also forms very strong bonds with hydrogen and almost all organic molecules contain at least one carbon–hydrogen bond. **Table 1.5.1** shows the average bond energies of carbon atoms with other carbon atoms, with hydrogen and with oxygen. The Si—Si and Si—H bond energies are given for contrast – silicon is the most similar element to carbon and yet the bond energies are very different.

Bond	Average bond enthalpy (kJ mol^{-1})
C—C	+347
C=C	+612
C≡C	+838
C—H	+413
C—O	+358
Si—Si	+226
Si—H	+318

table 1.5.1 Carbon forms very strong bonds.

The bonding of carbon and the strength of its bonds produce a great diversity of organic compounds. They can lead to the formation of chains of carbon atoms which may be thousands of carbon atoms long, with many different groups of other elements attached. Carbon can also form complex ring structures.

butane

cholesterol (skeletal formula)

linoleic acid, a long-chain fatty acid

fig. 1.5.9 As these few examples show, organic molecules come in a variety of shapes and sizes.

Organic families

Classifying organic compounds

The millions of different organic molecules can be classified in various ways. The first major division is based on the arrangement of the carbon chain itself:

- **Aliphatic** molecules contain straight- or branched-chain carbon skeletons, eg

```
                                  H
                                  |
                               H—C—H
                                  |
    H  H  H                 H     |  H  H
    |  |  |                 |     |  |  |
 H—C—C—C—OH              H—C—C—C—C—H
    |  |  |                 |  |  |  |
    H  H  H                 H  H  H  H
```

propan-1-ol

2-methylbutane

- **Alicyclic** molecules consist of closed rings of carbon atoms which may contain single or multiple carbon–carbon bonds.
- **Arenes**, which are covered in the A2 course, are all derived from the **benzene** molecule and contain a benzene ring with six carbon atoms in their structure.

These different molecular arrangements have a marked effect on the reactions of organic compounds. Another important factor is the presence of double and triple bonds in the structure. You will remember from your work on fats at GCSE level that an **unsaturated** fatty acid containing a carbon–carbon double bond is much more reactive than a **saturated** fatty acid containing only carbon–carbon single bonds.

As well as these two main divisions of aliphatic compounds and arenes, organic compounds fall into a range of families or **homologous series**.

Families of organic chemicals can be identified by the possession of a particular **functional group**. A functional group is an atom or group of atoms that is typical of a particular organic family and which determines the chemical properties of the molecule. The functional group has the same distinctive effect on the properties of all the molecules in the family. However, the functional group is also influenced by its environment, eg as the hydrocarbon chain gets bigger it has an increasing effect on the chemistry of the molecule and the influence of the functional group gets less as a result.

The different functional groups

The simplest organic molecule is a chain of carbon atoms joined by single bonds, with hydrogen being the only other atoms in the molecule. The homologous series with this structure is the **alkanes** (see chapter 1.6) and an example of an alkane is propane.

Each homologous series has a **general formula** which describes the number of carbon atoms and their relationships to the other atoms – eg the general formula of the alkanes is C_nH_{2n+2}.

```
    H           H  H           H  H  H
    |           |  |           |  |  |
 H—C—H       H—C—C—H       H—C—C—C—H
    |           |  |           |  |  |
    H           H  H           H  H  H
```

methane

ethane

propane

Another homologous series of organic compounds is similar to the alkanes, but all the members have an -OH functional group. This series is known as the **alcohols** – an example is propan-1-ol. (The way chemists name organic molecules is discussed later in this chapter.)

```
    H            H  H              H  H  H
    |            |  |              |  |  |
 H—C—OH       H—C—C—OH         H—C—C—C—OH
    |            |  |              |  |  |
    H            H  H              H  H  H
```

methanol

ethanol

propan-1-ol

If the -OH functional group of the alcohols is replaced by -COOH we get another organic family – the **carboxylic acids**. An example of this group is propanoic acid.

```
                   H                 H  H
      O            |     O           |  |     O
     //            |    //           |  |    //
 H—C          H—C—C            H—C—C—C
     \             |    \            |  |    \
      OH           H     OH          H  H     OH
```

methanoic acid

ethanoic acid

propanoic acid

As you have seen, the properties of organic families are determined first and foremost by their functional group, with the shape and size of the carbon chain also affecting how the compound reacts. The number of double and triple bonds between the carbon atoms in the carbon chain also has its effect. Alkanes and alcohols have a single bond between each of their carbon atoms. Other homologous series are distinguished by the double or triple bonds between their carbon atoms. The **alkenes** (see chapter 1.7) have at least one carbon–carbon double bond and another family, the **alkynes**, have a carbon–carbon triple bond. As you would expect, the alkenes and alkynes show greater reactivity than the alkanes.

Homologous series	General formula	Functional group	Example
Alkanes	C_nH_{2n+2}	—C—H (with single bonds above and below C)	CH_3CH_3 (ethane)
Alkenes	C_nH_{2n}	>C=C<	CH_2CH_2 (ethene)
Alkynes	C_nH_{2n-2}	—C≡C—	CHCH (ethyne)
Alcohols	$C_nH_{2n+1}OH$	—O—H	C_2H_5OH (ethanol)
Halogenoalkanes	$C_nH_{2n+1}X$ X is a halogen	—X	CH_3Cl (chloromethane)
Aldehydes	RCHO	—C(=O)H	CH_3CHO (ethanal)
Ketones	RCOR	C—C(=O)—C	CH_3COCH_3 (propanone)
Carboxylic acids	C_nH_{2n+1} COOH	—C(=O)OH	CH_3CH_2COOH (propanoic acid)

table 1.5.2 Some of the main functional groups and the organic homologous series to which they belong.

fig. 1.5.10 Different numbers of carbon atoms and different functional groups make a range of organic compounds with different physical as well as chemical properties. Rhubarb contains ethanedioic acid, which is solid at room temperature. Propane-1,2,3-triol is a viscous (thick) liquid, and trichloromethane is a non-viscous (runny) liquid. The methane you use in a Bunsen burner is a gas at room temperature.

Describing organic compounds

Representing organic compounds

To understand the chemistry of a compound you need to know its chemical make-up – the numbers of atoms of different elements which are involved and the way in which they are arranged. The structural formula of an organic compound is the first step to determining its functional groups and unravelling its chemistry.

In chapter 1.1 you met the terms empirical formula and molecular formula. For inorganic compounds the molecular formula is usually the most useful. However, organic molecules are often relatively large and the types of bonds between carbon atoms have a big effect on the chemical nature of the compounds.

For aldehydes, ketones and carboxylic acids, R represents either a hydrogen atom or an alkyl group C_nH_{2n+1} (see page 105).

For organic molecules the **structural formula** is often used, as it shows not only the numbers of atoms present but also the way in which they are arranged relative to each other. However it doesn't tell you anything about the bonds between the different atoms. The structural formula of propan-1-ol can also be written with some of the bonds shown to give you more information about the molecule. This is the **displayed formula**. Here you can see that it shows both the relative placing of the atoms and the number of bonds between them:

propan-1-ol

empirical formula	C_3H_8O
molecular formula	C_3H_8O
structural formula	$CH_3CH_2CH_2OH$ or $C_2H_5CH_2OH$
displayed formula	(see below)

```
      H   H   H
      |   |   |
  H — C — C — C — O — H
      |   |   |
      H   H   H
```

Sometimes even the displayed formula is not detailed enough because you need to know not only what type of bonds are present but also how they are arranged in space. A more detailed version of the displayed formula can show the shape of a molecule more accurately by showing the orientation of the bonds:

propan-1-ol

(wedge) indicates a bond sticking out of the plane of the paper

(hashed) indicates a bond sticking into the plane of the paper

The type of formula that you use depends on the information you need. Empirical, molecular and structural formulae are commonly used. **Space-filling models** can also be used to model a molecule in three dimensions. These may be produced using computers (especially in the case of complex molecules) or using modelling kits which connect together wooden or plastic spheres.

Computer-generated space-filling models are particularly useful to scientists for research tasks such as drug design. Often the shape of a molecule is crucial to the way it works in a biological system, so the ability to create models like the one in **fig. 1.5.11** can be vital. In school laboratories space-filling models are usually much simpler!

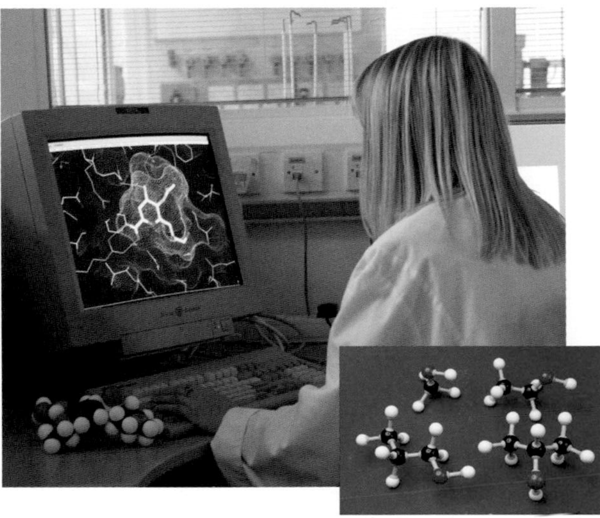

fig. 1.5.11 Space-filling models like this, whether produced by a computer or simply by fitting coloured spheres together, are an attempt to represent accurately the extent of the electron cloud in a molecule.

Finally, there are times when you may use **skeletal formulae**. These are more difficult to interpret – in a skeletal formula you do not show any hydrogen or carbon atoms. All that is left is a carbon skeleton, with the carbon atoms represented by the junctions between the bonds, and functional groups are shown. Skeletal formulae are most frequently used for ring compounds.

Questions

1 Explain the wide range of organic compounds, with reference to the electronic structure of the carbon atom.

2 For each of the following compounds state which homologous series of organic compounds it belongs to and give the empirical, molecular, structural, displayed and skeletal formulae:
a ethanol **b** propanoic acid.

3 What are functional groups and why are they so important in organic chemistry?

Naming organic molecules

Naming inorganic compounds is usually quite straightforward because small numbers of atoms are involved. However, the naming of organic molecules is rather different. The name of an organic compound needs to show the homologous series to which it belongs, the number of carbon atoms in the molecule and whether the chain is straight, branched or in a ring, and to also include any additional atoms that may be in the molecule.

The IUPAC system

A rigorous system of naming is used for both organic and inorganic compounds, according to rules drawn up by **IUPAC** (the International Union of Pure and Applied Chemistry). The system is very useful, although complex organic compounds can end up with names that are longer and more involved than the formula of the compound! However, this is rarely a problem because often these very large molecules – such as carbohydrates, amino acids and fats – have standardised common names that are easier to remember. It is well worth taking the time to get to grips with the way systematic names are built up, because then the name of an organic compound will give you a considerable amount of information about its structure and therefore the way in which it is likely to react.

Naming aliphatic compounds

You have seen that aliphatic organic compounds are made up of straight and branched chains. For an aliphatic compound, the first part of the name refers to the number of carbon atoms in the carbon chain or, if the molecule is branched, the number of carbon atoms in the longest carbon chain (see **table 1.5.3**).

Prefix	Number of carbon atoms in the main carbon chain
meth-	1
eth-	2
prop-	3
but-	4
pent-	5
hex-	6
hept-	7
oct-	8
dec-	10
dodec-	12
eicos-	20

table 1.5.3 The principles of naming carbon skeletons.

The second part of the name, the suffix, refers to the functional group of the homologous series to which the molecule belongs (see **table 1.5.4**). For example, propane is an alkane with three carbon atoms in the chain, while propyne is an alkyne which also has three carbon atoms. Propanol is a three-carbon alcohol, and propanal is a three-carbon aldehyde. Equally, ethene has two carbon atoms, while pentene has five carbons and dodecene has 12, all being members of the alkene homologous series.

Suffix	Homologous series
-ane	alk**ane**
-ene	alk**ene**
-ol	alcoh**ol**
-oic	carboxy**lic** acid
-al	**al**dehyde
-one	ket**one**

table 1.5.4 The suffix tells you which functional group the compound belongs to.

The organic families other than the alkanes contain double or triple bonds or other functional groups. Numbers are used to show where in the carbon chain the functional group is. For example, in an alkene the parent or main carbon chain is taken to be the longest one in which the double bond occurs. It is numbered from the end that gives the first carbon of the double bond the lowest possible number, eg:

```
H           H
 \         /
  C¹ = C²    H   H
 /        \  |   |
H          C³ - C⁴ - H
           |    |
           H    H
```

but-1-ene

```
H          H
 \        /
  C¹ = C²           H
 /       \         /
H         C³ = C⁴          H
         /       \        /
        H         C⁵ = C⁶       H
                 /       \     /
                H          C⁷
                          /  \
                         H    H
```

hepta-1,3,5-triene

Many organic molecules are not simple straight chains but have branches that are often **alkyl groups**. An alkyl group is an alkane molecule that has lost a hydrogen and joined to another carbon chain, so it has the general formula C_nH_{2n+1}. Typical alkyl groups include:

methyl	CH_3–	
ethyl	C_2H_5–	or CH_3CH_2–
propyl	C_3H_7–	or $CH_3CH_2CH_2$–

The position of a side chain is again indicated by the lowest possible number, to show to which carbon atom in the parent chain it is attached, eg:

```
        H
        |
    H — C — H
        |
    H   |   H
    |   |   |
H — C — C — C — H
    |   |   |
    H   H   H
```

2-methylpropane

```
                  H
                  |
              H — C — H
                  |
              H — C — H
                  |
    H   H   H     |     H   H
    |   |   |     |     |   |
H — C — C — C  —  C  —  C — C — H
    |   |   |     |     |   |
    H   H   H     H     H   H
```

3-ethylhexane

If more than one side chain is attached then the name of the compound includes the side groups in *alphabetical order*, regardless of the number of the carbon atom on which they are found. So the order for referring to the smaller alkyl groups is:

- butyl, ethyl, hexyl, methyl, pentyl and propyl.

If two side chains are attached to the same carbon atom then the number of the carbon atom is repeated in the name (see below). If there are two methyl groups the compound is described as dimethyl, and if there are three methyl side chains it becomes trimethyl, etc.

```
       CH3   CH2—CH3
       |     |
CH3 — CH — CH — CH2 — CH3
```

3-ethyl-2-methylpentane

```
CH3                   CH3
   \                  |
    C == CH — CH2 — CH — CH2 — CH3
   /
CH3
```

2,5-dimethylhept-2-ene

```
       CH3
       |
CH3 — C — CH2 — CH2 — CH3
       |
       CH3
```

2,2-dimethylpentane

You can see that the systematic naming of organic compounds allows you to identify the functional group, as well as any side chains present and where they are found within the molecule. In most cases throughout this section and the work you will do in organic chemistry later in this course, organic chemicals will be given their IUPAC names, although there will be times, particularly for very large molecules, when common names will be used instead.

The shapes of molecules

Isomerism in organic molecules

Another complication that contributes to the large number of organic compounds is that the vast majority of organic compounds have two or more **isomers**. Isomerism occurs when two or more compounds have the same molecular formula but the atoms are connected together differently. All organic molecules containing four or more carbon atoms show isomerism, and many smaller molecules do too. There are several types of isomerism and organic chemicals exhibit all of the different types. In this chapter you will be looking at structural isomerism. In chapter 1.7 you will meet geometric isomerism.

Formula	Number of isomers
C_8H_{18}	18 (one of these is a major constituent of petrol)
$C_{10}H_{22}$	75
$C_{20}H_{42}$	366 319
$C_{40}H_{82}$	6.25×10^{13} (estimated!)

table 1.5.5 Some of the larger organic compounds have vast numbers of isomers, contributing to the enormous number and variety of organic chemicals.

Structural isomerism

Structural isomerism is probably the simplest form of isomerism to understand. For example, the molecular formula C_4H_{10} gives two isomers, butane and 2-methylpropane (see **fig. 1.5.12**). These **structural isomers** remain part of the same homologous series although their boiling temperatures differ considerably. Butane is a straight-chain isomer, while 2-methylpropane is a branched-chain isomer. The differences in boiling temperatures and other physical properties of structural isomers can be striking. They are brought about by the different shapes of the molecules which affect the intermolecular forces between the molecules due to differences in the way the molecules can pack together.

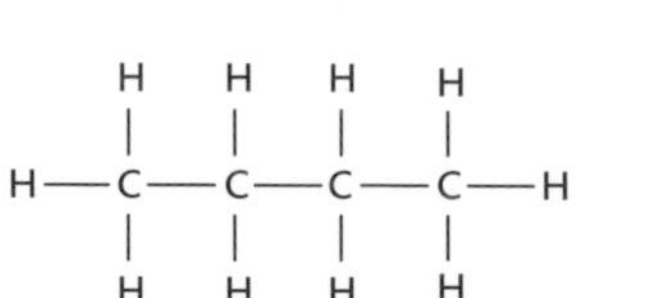

butane:
boiling temperature –0.4°C

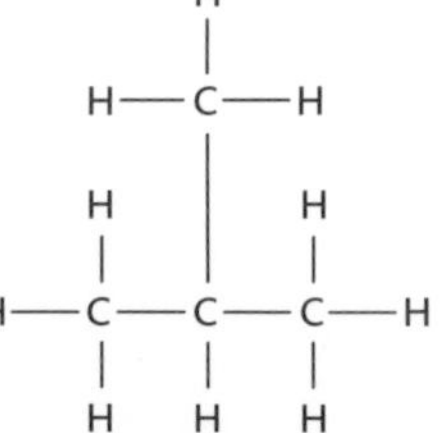

2-methylpropane:
boiling temperature –11.6°C

fig. 1.5.12 The two isomers of C_4H_{10} are members of the same homologous series but have different physical properties.

Another example of structural isomerism is shown by compounds with the molecular formula C_2H_6O. In this case the two structural isomers are members of different organic families and have very different properties – ethanol is an alcohol and methoxymethane belongs to a homologous series called the **ethers** (see **fig. 1.5.13**).

ethanol:
boiling temperature 78.6 °C

methoxymethane:
boiling temperature –24.9 °C

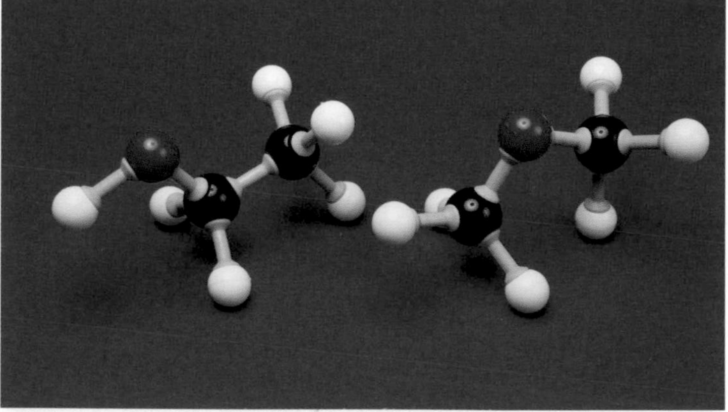

fig. 1.5.13 The two structural isomers of C_2H_6O are members of different organic families and have very different physical and chemical properties as a result of their different structures.

Some pitfalls in drawing isomers

If you know the molecular formula of an organic compound, you can work out the different structural isomers by drawing them out – and then you can name them! However, there are some common mistakes when drawing structural isomers.

Isomers often involve branched chains. Many people make the mistake of drawing 90° bends in a carbon chain rather than branches, thinking that this forms a different isomer. Another easy mistake to make is to draw the same molecule in two different ways and think they are isomers when in fact, if you look at them carefully, they are the same representation drawn in a different orientation (see **fig. 1.5.14**). Making models is the easiest way to get isomers right!

hex-2-ene

hex-2-ene

2-methylpent-2-ene

4-methylpent-1-ene

2,3-dimethylbut-1-ene

2-methylpent-2-ene

fig. 1.5.14 Structural isomers of hexene – some common mistakes when drawing isomers.

Stereoisomerism

Another form of isomerism is **stereoisomerism** (isomerism in space). It arises when the three-dimensional arrangement of the bonds in a molecule allow different possible orientations in space. This results in two different forms – they look similar but however hard you try you cannot superimpose the mirror image of one onto the other. The two isomers are called **stereoisomers** and you will be looking at these in more detail in chapter 1.7.

Questions

1 Name and write out the full structural formulae for the following compounds:
   a $CH_3(CH_2)_2Cl$ b $CH_3C(CH_3)_2CH(CH_3)CH_3$

2 What is an isomer? Why are isomeric forms so common in organic chemistry?

3 Draw out the three structural isomers of pentane.

4 Decide whether the members of the pairs of molecules in **fig 1.5.15** are identical compounds, are isomers or are chemically unrelated.

5 Give all the possible structural formulae for C_7H_{16}.

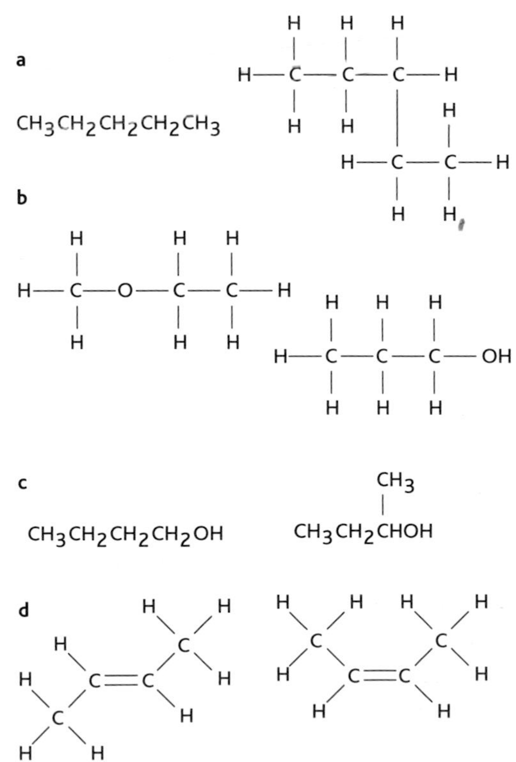

fig. 1.5.15 Isomers or not?

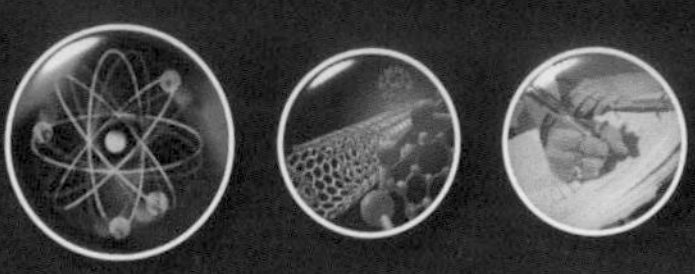

6 The alkanes – a family of saturated hydrocarbons

What are hydrocarbons?

The **hydrocarbons** are a group of organic compounds containing several homologous series. The common feature of hydrocarbons is that their molecules contain only the elements carbon and hydrogen. They can be divided into the **aliphatic hydrocarbons** (straight- and branched-chain molecules), the **alicyclics** (hydrocarbons with closed rings) and the **arenes** (based on benzene, with rings of carbon atoms stabilised by delocalisation of electrons). In this chapter you are going to concentrate on the aliphatic hydrocarbons. However, all these hydrocarbons have certain properties in common:

- They are insoluble in water.
- They all burn, and in sufficient oxygen they give carbon dioxide and water as the only combustion products.

Virtually all of our usable supplies of hydrocarbons come from fossil fuels – coal, petroleum and natural gas, which yield pure hydrocarbons after varying degrees of processing. The aliphatic hydrocarbons can be classified as belonging to one of three families – the **alkanes**, the **alkenes** and the **alkynes**. They have a very wide variety of uses from fuels to the raw materials for an enormous number of industrial processes. In this chapter you will be concentrating on the alkanes.

General properties of the alkanes

The alkanes are a family of **saturated** hydrocarbons – they have no double or triple bonds between the carbon atoms, so they contain the maximum amount of hydrogen possible. The general formula for the alkanes is C_nH_{2n+2} and they can occur as both straight- and branched-chain molecules.

Straight-chain alkanes form a classic homologous series with predictable physical properties. However, there is considerable structural isomerism in the alkanes, as you saw in chapter 1.5, and this makes predicting the properties of a particular chemical from its molecular formula harder than it might at first appear.

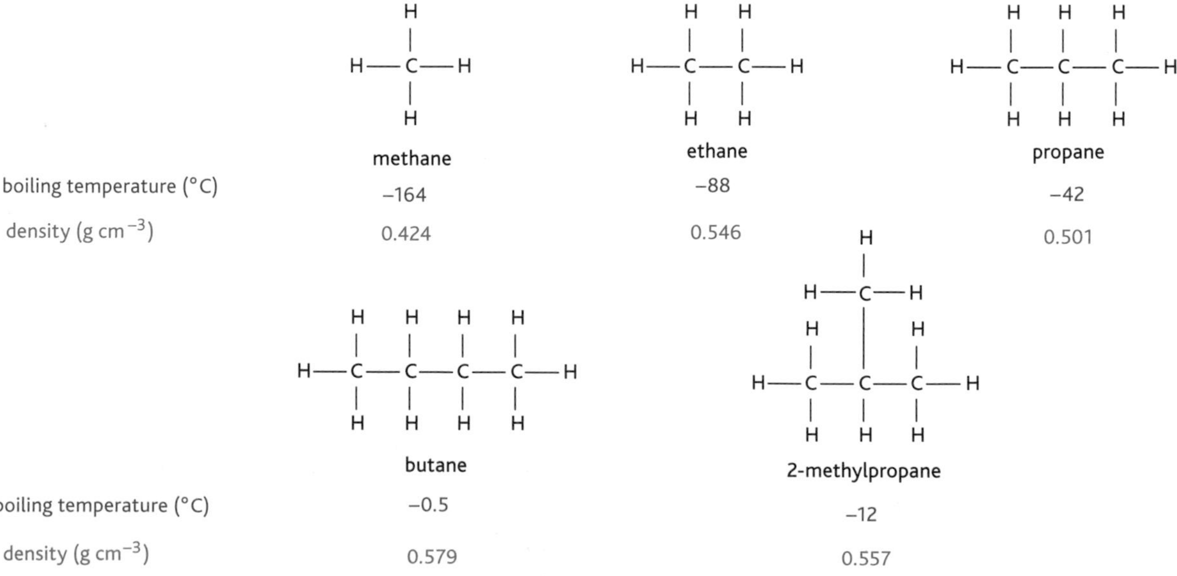

fig. 1.6.1 Structural isomers of the alkanes. The molecule shapes affect their physical properties. However, clear trends can be seen in the physical and chemical properties of the straight-chain alkanes.

Where are alkanes obtained from?

The single most important source of the alkanes is the fossil fuels. In prehistoric times a far greater proportion of the Earth's surface was covered with water than is the case today. Minute animals and plants lived in these seas. As they died and sank to the bottom, deep layers of rich, decomposing material formed which became encapsulated in rock. Over great lengths of time, exposed to immense heat and pressure deep in the Earth, this trapped decaying matter has formed what we know today as **crude oil** or **petroleum** ('rock oil'). It is frequently found with another product of the same decomposition process, **natural gas**, which is largely methane.

The fossilised remains of land plants form coal by a similar process. In a mysterious extinction not unlike the one that brought an end to the age of the dinosaurs, giant prehistoric 'fern forests' were largely destroyed. The remains of these plants, fossilised after millions of years buried underground, form coal. Occasionally a clear imprint of an ancient fern leaf is seen on a piece of coal.

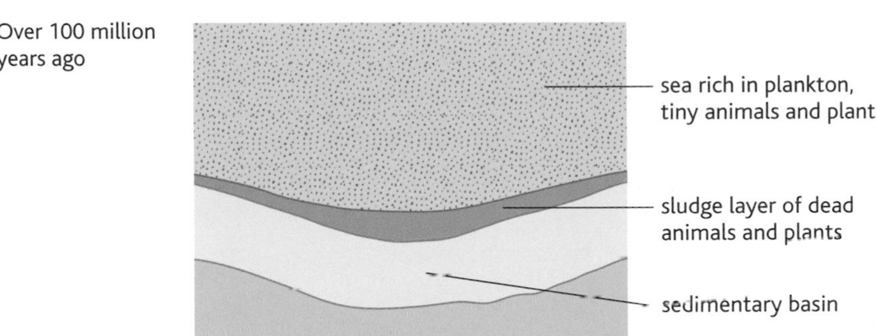

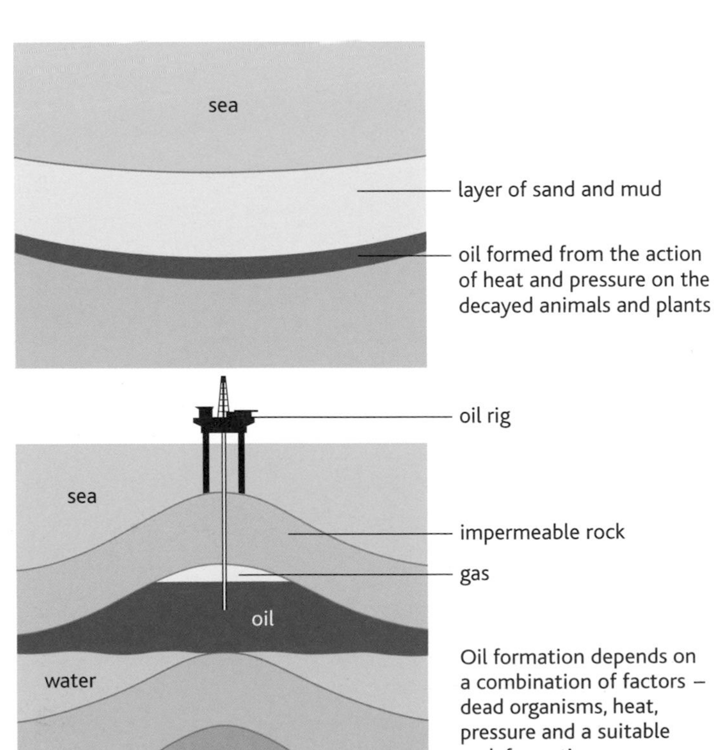

fig. 1.6.2 Fossil fuels are used for heating our homes, to generate electricity and as raw materials for plastics, pharmaceuticals and countless other chemicals.

HSW The economic importance of oil

Modern life depends on crude oil, which provides fuel for transport and for generating electricity. It is the source of raw materials for many of the major chemical industries around the world. However, crude oil and natural gas are found in only a limited number of places on the planet, and very large deposits are relatively rare. The majority of the world's oil supplies are found in politically sensitive areas of the world. The supply of oil and its price affect the whole global economy.

Until the 1970s oil was regarded as a cheap and plentiful resource. Then the Middle Eastern countries that supply much of the crude oil to the UK decided to produce less and also to increase their prices. Ever since then oil production and prices have fluctuated (see **fig. 1.6.3(a)**). As you can see in **fig. 1.6.3(b)**, oil prices also vary from month to month and even from day to day, and predicting these changes can make fortunes – and lose them – in the financial markets. The overall trend in oil prices has been an increase, having a knock-on effect on the cost of transport and of the raw materials for many industries. Add in the fact that fossil fuels are a finite resource and you can see to what degree the global economy is affected by the price and availability of oil – and the power held by countries that hold rich oil reserves.

In 2008, oil prices rose rapidly due to instability in world financial markets and problems in some of the major oil-supplying countries. The long-term impact of these changes on the world economy will become clear in time.

Rank	Country	Proved reserves (billion barrels)
1	Saudi Arabia	264.3
2	Canada	178.8
3	Iran	132.5
4	Iraq	115.0
5	Kuwait	101.5
6	United Arab Emirates	97.8
7	Venezuela	79.7
8	Russia	60.0
9	Libya	39.1
10	Nigeria	35.9
11	United States	21.4
12	China	18.3
13	Qatar	15.2
14	Mexico	12.9
15	Algeria	11.4
16	Brazil	11.2
17	Kazakhstan	9.0
18	Norway	7.7
19	Azerbaijan	7.0
20	India	5.8
Top 20 countries		1224.5 (95%)
Rest of world		68.1 (5%)
World total		**1292.6**

table 1.6.1 These figures show the countries with the biggest reserves of crude oil, (from the US Energy Information Administration, 2005). 'Proved reserves' are estimated to be worth extracting with present technology and prices.

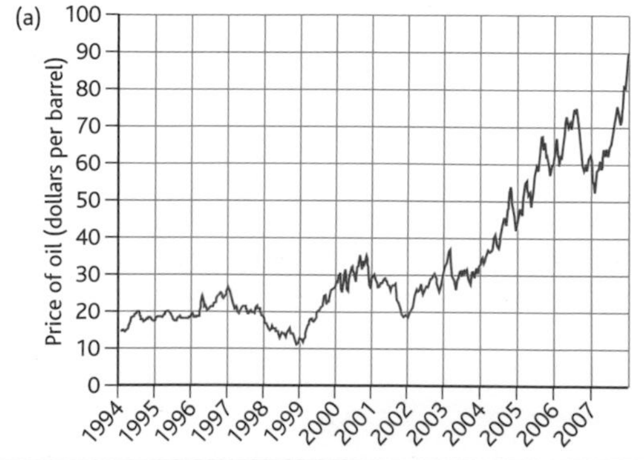

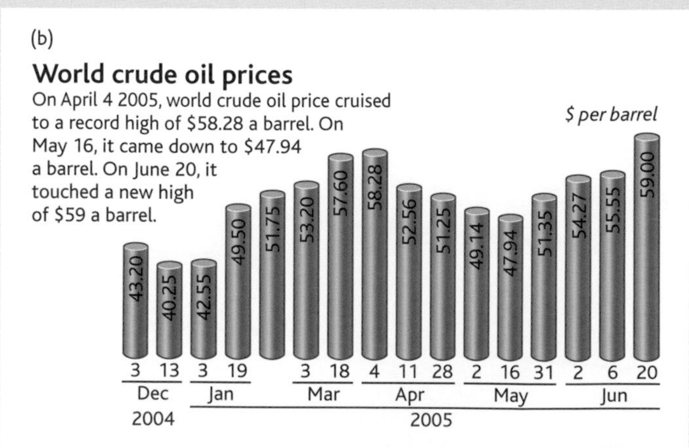

fig. 1.6.3 Oil reserves and the prices charged for this vital resource have a major effect on world economies. These graphs show how the price fluctuates in both the long and short term.

Using crude oil

The crude oil extracted from the ground is of little use without processing. It is a mixture of many different hydrocarbons, many of them alkanes along with some alkenes and alkynes. The basic process by which petroleum (crude oil) is turned from a dark, thick, smelly but very valuable liquid into useful chemicals with a whole variety of properties is known as **primary distillation**, an industrial version of the fractional distillation you can carry out in the laboratory. Petroleum is boiled and the vapours are cooled and liquefied at particular temperatures. The liquid collected over each range of temperatures is known as a **fraction**. This process provides five major fractions (see **fig. 1.6.4**). It can also be used in further, more precise, fractional distillation processes to give a pure yield of a particular alkane.

fig. 1.6.4 The oil industry locates deposits of crude oil in the ground, extracts it and processes it to supply fuels and many other organic source materials.

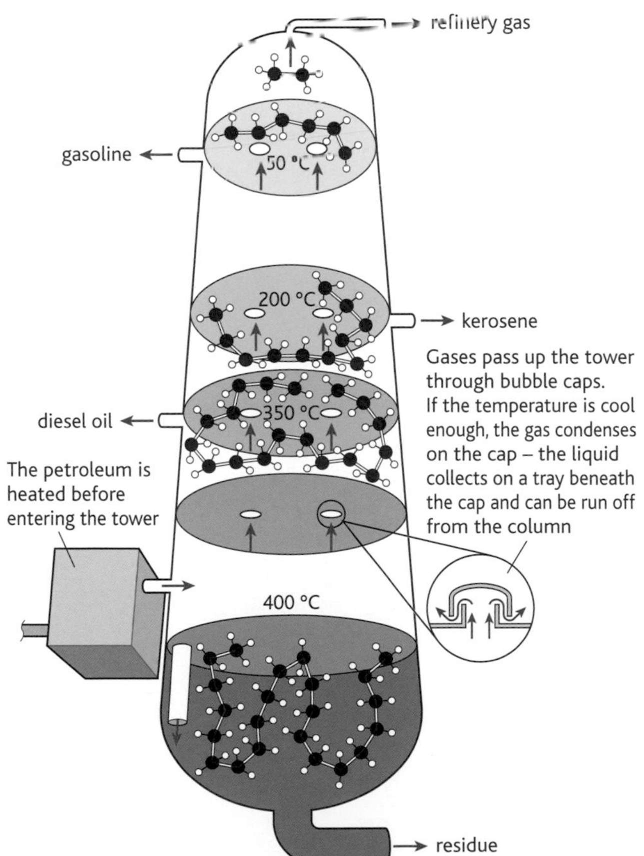

The main fractions that result when crude oil is distilled are:

- 1–2% **refinery gas** – the lightest fraction of crude oil with the lowest boiling point. Contains the gaseous alkanes (1–4 carbon atoms in the chain, C_1–C_4) with a large proportion of methane. Used mainly as a fuel or as a starting point for other organic syntheses.
- 15–30% **gasoline** – a mixture of liquid hydrocarbons (C_5–C_{10}) which is widely used as fuel for the internal combustion engines that power cars. This is a mixture of straight- and branched-chain alkanes, and the more branched molecules there are the more useful the fuel. Some gasoline is also used in the production of other organic chemicals, although this usually involves further processing such as catalytic cracking.
- 10–15% **kerosene** – mainly C_{11} and C_{12} fractions, this is largely used as fuel for aircraft engines. It can also be 'cracked' to provide other useful chemicals such as gasoline.
- 15–20% **diesel oil/gas oil** – used both in the diesel engines suggested by the name and as a fuel for industrial boilers. Diesel oil can also be further split up in a catalytic cracker to yield other useful fractions.
- 40–50% **residue** – this complex mixture of hydrocarbons is very viscous and has a high boiling temperature. It can be used as fuel for the furnaces of power stations or large ships, or it can be further fractionated to yield lubricating oils and waxes and a solid material which we know as the bitumen used to surface roads. This final distillation has to be carried out in a vacuum to avoid the need for high temperatures which would 'crack' or break open rather than separate the components.

Questions

1. What is an alkane?
2. Which of the following belongs to the homologous series of alkanes? C_8H_{16}, C_9H_{20}, $C_{15}H_{30}$, $C_{30}H_{58}$
3. Why are fossil fuels regarded as a finite resource?
4. Explain why the different fractions of crude oil are collected at different temperatures.

Making the most of crude oil

Different fractions in demand

Crude oil from different sources contains slightly varying proportions of the different fractions. While the lighter fractions produced during the fractional distillation of crude oil are in great demand both as fuels and as the raw materials for the chemical industry, some of the heavier fractions are of less use – and they can make up to 50% of the products, depending on the type of crude oil (see **fig 1.6.5**). The demand for the lighter fractions is enormous, and there is a constant pressure to find ways to make the heavier fractions more useful.

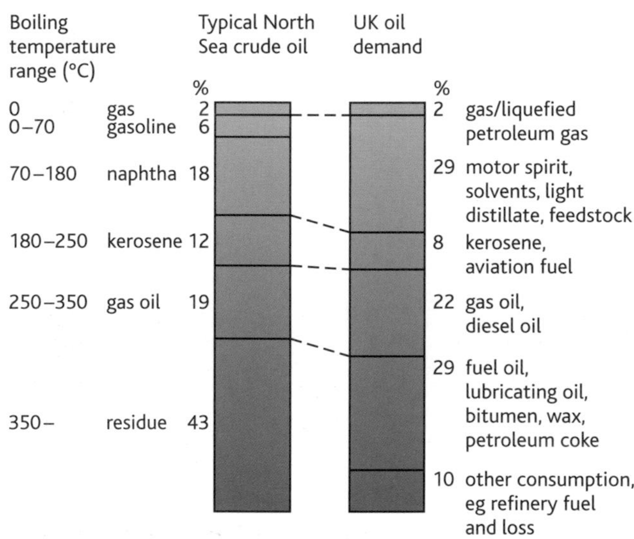

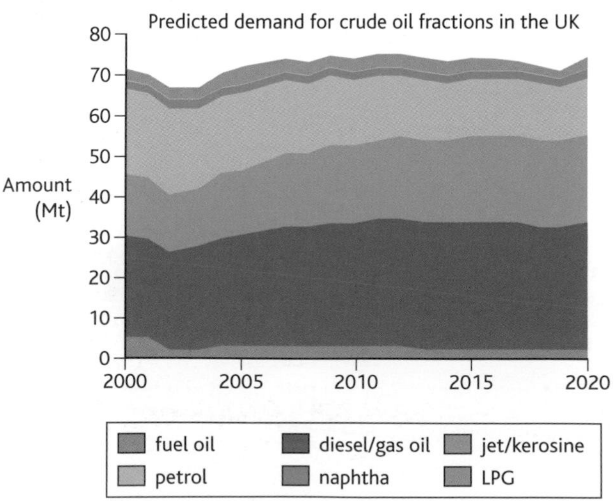

fig. 1.6.5 There is a mismatch between the demands of the market and the fractions obtained from crude oil.

To address this imbalance, scientists have developed a way of treating the heavier, long-chain fractions of crude oil and breaking them down (**cracking** them) to produce more of the lighter compounds. Heating long-chain alkanes to high temperatures causes the molecules to split and form shorter-chain molecules. Some of these are alkanes which can then be used, eg in gasoline. Others will be alkenes, particularly ethene, which is extremely useful. Apart from combustion, the alkanes are a very unreactive family of compounds and so they are of limited use as a starting point for the petrochemical industry. Ethene, on the other hand, contains a double bond which makes it much more reactive and so very suitable as a starting point for synthetic reactions (see chapter 1.7).

Producing the high temperatures needed to crack the heavy petroleum fractions would be very expensive. By using catalysts the process can be carried out at a much lower temperature. It is known as **catalytic** or **cat cracking** and is carried out in a **cat cracker**. Crystalline aluminosilicates (zeolites) are commonly used as catalysts in the process.

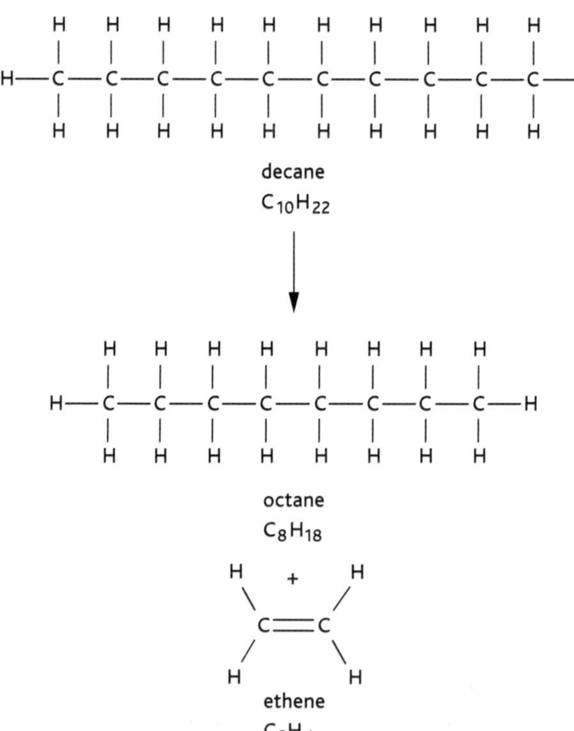

fig. 1.6.6 The effect of catalytic cracking on large alkane molecules.

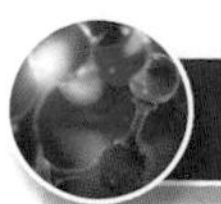

HSW 'Knocking' and the need for catalytic reforming

The alkanes in gasoline, the crude oil fraction used as petrol, have carbon chains which are usually 5–10 carbon atoms long. The power is produced in engine cylinders by the explosive combustion of gasoline in the cylinders.

The smooth running of a car engine depends on the explosion in the cylinder occurring at just the right moment so that energy wastage as heat is minimal and the maximum amount of energy is transferred to the pistons. If the explosion occurs too early the pistons are jarred and the engine not only makes disconcerting knocking noises but also loses power. This 'knocking' is particularly common when fuels with a large proportion of straight-chain alkanes are used. Heptane, octane and nonane in particular ignite very easily. Branched-chain alkanes such as 2,2,4-trimethylpentane are ignited less readily so their combustion is much more controllable. Gasoline mixtures high in these branched-chain alkanes make much more efficient fuels than mixtures high in straight-chain molecules.

To indicate the proportion of branched- to straight-chain molecules, different mixtures are given an **octane rating**. The octane rating of 2,2,4-trimethylpentane (old name isooctane) is set at 100, whilst that of straight-chain heptane is set at 0. A fuel can be tested by comparing it with known mixtures of heptane and 2,2,4-trimethylpentane in a test engine. The ratio of heptane to 2,2,4-trimethylpentane is varied in the comparator mixture until it has the same ignition properties as the sample of fuel being tested. The percentage of 2,2,4-trimethylpentane in the mixture is then taken as the octane rating of the fuel. Other methods for measuring an octane number involve detailed chemical analysis of fuels. Octane ratings are shown on petrol pumps by a Research Octane Number, or RON set by measuring the performance of the fuel under set conditions in a research engine. Ordinary unleaded petrol is 95 RON, while super unleaded is 98 RON.

Most modern cars have relatively high-performance engines and require high-octane fuels. To meet the demand there are two methods of preventing 'knocking'. Historically tetraethyllead(IV), $Pb(C_2H_5)_4$ was added to gasoline to retard its ignition. This caused lead pollution from car exhausts, which can potentially cause health problems so most countries have eliminated the use of 'leaded' petrol. By 2007 only 17 countries in the world still used leaded petrol – and the number is falling all the time. The other alternative is to produce gasoline mixtures that are artificially high in branched-chain alkanes. This can be done as a result of **catalytic reforming**, which is very similar to catalytic cracking. It involves breaking up straight-chain molecules in the heavier oil fractions and reforming them into new isomers with branched chains. A platinum catalyst is used, so the process is sometimes referred to as 'platforming'. Since the withdrawal of leaded petrol, catalytic reforming has become central to the production of motor fuels. Reforming also provides a supply of ring hydrocarbons (arenes) for the chemical industry.

fig. 1.6.7 High-performance cars like this have always needed high-octane fuels to prevent premature ignition in their high-compression cylinders. But as all cars become more sophisticated, their engines need more branched-chain alkanes in their fuel to prevent unwanted loss of power.

Questions

1 Compare the processes of catalytic cracking and catalytic reforming.

2 What is meant by the term 'knocking'?

3 There are two main ways of overcoming the problems of knocking. Outline each method, explaining the advantages and disadvantages. Which method is used almost exclusively in the UK and why?

The chemical properties of the alkanes

The old, non-systematic name for the alkanes was the **paraffins**. This came from the Latin *parum affinis* which means 'little affinity' and it described the alkanes as a family very well. Apart from combustion, they are very unreactive, having few reactions with other elements or compounds. This makes the alkanes very useful. They are non-corrosive with metals, which makes them good lubricating oils. They are also harmless to your skin, yet protect it – which is why petroleum jelly is effective at preventing your skin from chafing.

At room temperature the alkanes are unaffected by concentrated mineral acids such as sulfuric acid or concentrated alkalis such as sodium hydroxide solution. They are not affected by oxidising agents such as potassium manganate(VII) and they do not react with even the most reactive metals. The reason for this lack of reactivity is that both C—C and C—H bonds involve a very even sharing of electrons, since the electronegativities of carbon and hydrogen are very close. This means that the bonds in the molecules of the alkanes are not polar to any extent, and so there are no charges to attract other polar or ionic species. Almost all of the reactions of the alkanes occur due to the formation of **free radicals**, which contain an unpaired electron. As a result they have high activation energies, but once this barrier to reaction is overcome they proceed to react very rapidly in the gas phase. Before you move on to look in further detail at the reactions of the alkanes you can find out more about these free radicals.

Breaking bonds

As you have seen, chemical reactions involve the breaking and making of bonds between atoms. Breaking bonds is also known as **bond fission**. In organic chemistry we almost always talk about the breaking and making of covalent bonds with varying degrees of polarity or ionic character. Within a covalent bond, two electrons are shared between two atoms. When that bond is broken during a chemical reaction there are two ways in which the electrons may be shared out.

Homolytic fission

Homolytic fission involves the equal sharing out of the electrons in the bond, so that each of the participants in the bond receives one electron when the bond splits.

To show exactly what's going on in chemical reactions, chemists use curly arrows. The most common use of curly arrows is to show the movement of pairs of electrons (see Chapter 1.7). However, in homolyic fission, because just one electron is moving, a curly half-arrow is used (as seen in **fig. 1.6.8**) – note how the head of this arrow only has a single line head rather than two lines. The unpaired electron gained by each atom is indicated by a dot.

$$H^{\bullet} + Cl{-}Cl \rightarrow H{-}Cl + Cl^{\bullet}$$

The important point is that although neither atom has an overall charge, these free radicals with their unpaired electrons are *extremely* reactive. This is because the unpaired electron has a very strong tendency to pair up with an electron from another substance. The reaction of one free radical with another substance usually results in the formation of a further free radical. The equal sharing of the electrons of homolytic fission usually occurs in situations where there is little or no ionic character in the covalent bond.

$$Cl^{\bullet} + CH_4 \rightarrow CH_3^{\bullet} + HCl$$

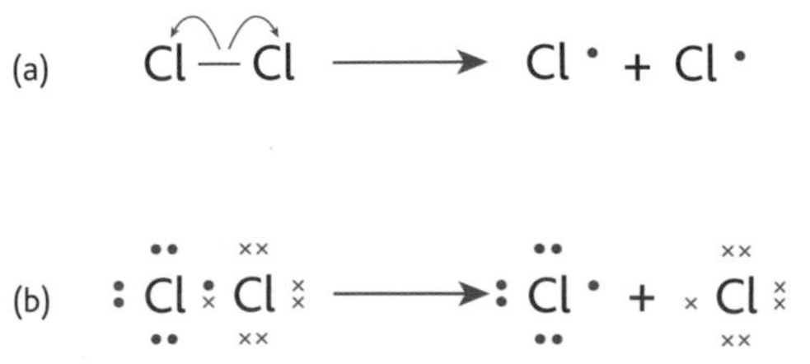

(a) $Cl{-}Cl \longrightarrow Cl^{\bullet} + Cl^{\bullet}$

(b) $:\ddot{Cl}:Cl \longrightarrow :\ddot{Cl}^{\bullet} + {}_{\times}Cl$ (dot and cross diagram)

fig. 1.6.8 Homolytic fission gives each new species one electron from the bond. These free radicals can be shown (a) using dots to represent the unpaired electrons and a curly half-arrow to represent the movement of the single electron, or (b) by the dot and cross diagrams you are already familiar with.

Heterolytic fission

Heterolytic fission involves an unequal sharing of the electrons of the covalent bond, so that both electrons go to one atom. This results in two charged particles – the atom receiving the electrons gaining a negative charge and the other atom gaining a positive charge. Heterolytic fission is usually seen when the covalent bond already has a degree of polarity.

$$H—Cl \rightarrow H^+ + Cl^-$$

$$\overset{\delta+}{H}—\overset{\delta-}{Cl} \longrightarrow H^+ + Cl^-$$

$$H:\ddot{\underset{..}{Cl}}: \longrightarrow [H]^+ + [:\ddot{\underset{..}{Cl}}:]^-$$

1p 0e 17p 18e

$$H \times\!\!\bullet\, Cl \longrightarrow H^+ + Cl^-$$

$$H—Cl \longrightarrow H^+ + Cl^-$$

fig. 1.6.9 Different representations of heterolytic fission

HSW Burnt toast, free radicals and the cancer connection

Free radicals are an important element of our model for many organic reactions, and they are obviously of great importance for that reason. But free radicals also make guest appearances in the media on a regular basis. This is as a result of their role in several common human problems.

Cancer is one of the major causes of death in the developed world. Cancer occurs when the normal growth-controlling mechanisms of a cell break down, so that rapid reproduction of small unspecialised cells takes place. This results in the formation of a tumour or growth which may itself cause serious illness or death by filling up and destroying a vital organ. More commonly, small pieces will break off the tumour, travel around the body in the bloodstream and lodge in other places to grow again.

Many different substances are thought to be responsible for the cellular changes that bring about the loss of control over cell growth – they are known as **carcinogenic** substances. Free radicals in the body are now considered to be one of the culprits. The number of free radicals in the body can be reduced by cutting down on the intake of food high in free radicals – burnt toast and charred food from the barbecue amongst others. Equally, certain vitamins are very useful in enabling the body to 'mop up' free radicals which may form.

A plentiful supply of fresh fruit and vegetables containing vitamins A, C and E in the diet helps the enzyme superoxide dismutase to deactivate free radicals and so prevent any damage they may cause.

fig. 1.6.10 The extreme reactivity of free radicals which enables so many organic reactions to occur is also the reason that they are so potentially damaging to the cells of your body when they occur in excess.

Questions

1 Why are alkanes so unreactive?

2 What is the difference between homolytic and heterolytic fission?

3 Explain a benefit of eating fruit and vegetables with reference to reactions demonstrated in the alkanes.

The reactions of the alkanes

The alkanes have just two common types of reaction, occurring when they are heated or in the presence of halogens. These are both considered in more detail below.

Heating alkanes

Cracking

When alkanes are heated to high temperatures in the absence of air they split into smaller molecules. The thermal decomposition of methane yields finely powdered carbon, which is used in car tyres and to make artificial diamond coatings, and hydrogen, which is used as a raw material for the chemical industry:

$$CH_4(g) \rightarrow C(s) + 2H_2(g)$$

Cracking is thermal decomposition that involves the breaking of carbon–carbon bonds to form smaller molecules (see page 112). The cracking of ethane gives ethene, one of the most important raw materials of the chemical industry, as well as hydrogen. This process involves the breaking of the C—C bond followed by the formation of the C=C bond:

$$C_2H_6(g) \rightarrow CH_2{=}CH_2(g) + H_2(g)$$

Combustion

When alkanes are heated in a plentiful supply of air, combustion occurs. Alkanes are energetically unstable with respect to their oxidation products, water and carbon dioxide, so once lit they will burn completely. Alkanes only burn when they are in the gaseous state so solid and liquid alkanes tend to be less flammable than gaseous ones – they must be vaporised before they will burn. You can see this in the burning of a candle flame – the solid wax melts as the wick burns. The liquid wax soaks up through tiny channels in the wick, vaporises, mixes with the surrounding air and then burns. This releases more heat to keep these processes going. Convection currents are produced in the air around the hot flame and these carry the products of combustion away from the candle. This allows more air to mix with the newly vaporised wax.

The combustion of the alkanes is of great importance in our way of life – it is used to generate electricity, to fuel fires in the home, to provide central heating,

fig. 1.6.11 The heat from the flame melts the wax, which then vaporises from the wick where it burns.

for cooking and for transport. In all these examples the central process is the transfer of energy from the exothermic combustion reaction into heat and light energy as well as into the bonds in the combustion products.

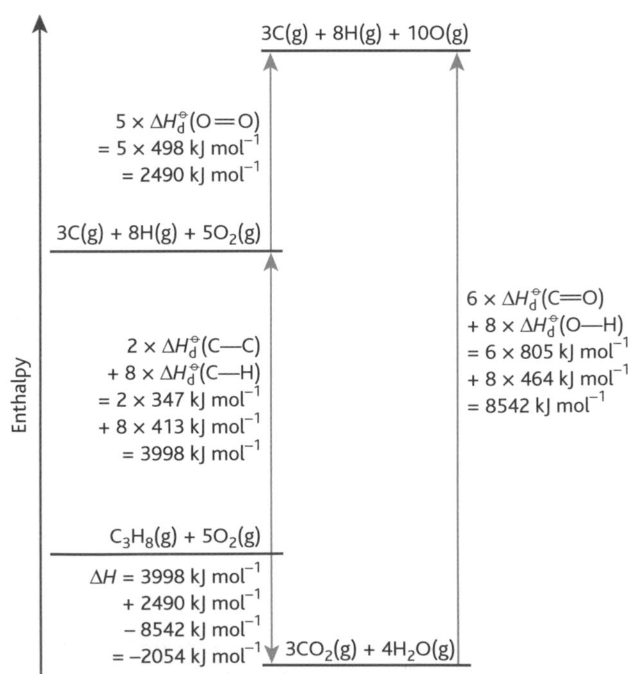

(Note that this is not the *standard* enthalpy of combustion, since the water formed is in the gaseous state, not a liquid.)

$$C_3H_8(g) + 5O_2(g)\ 3CO_2(g) + 4H_2O(l)$$

fig. 1.6.12 When propane burns in air, the energy required to break apart the propane and oxygen molecules is more than compensated for by the energy released when the bonds in carbon dioxide and water are formed.

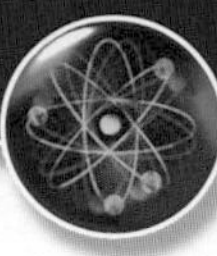

As with all exothermic changes, the release of energy results from the fact that the energy needed to break the bonds in the chemical reaction of combustion is less than the energy returned when the new bonds are made as the products of combustion form – see **fig. 1.6.12**.

Methane (natural gas), propane and butane are all commonly used as fuels. In many areas natural gas is piped into homes. Where this is not the case, canisters of propane can be supplied. The great advantage of propane over methane is that it can be liquefied readily under pressure and at low temperatures, so that a large quantity of gas can be stored in a very small space. The liquid alkane will not burn until it is returned to the gaseous state and so it can be stored and transported relatively easily and safely. This **liquid petroleum gas (LPG)** is transported to countries such as Korea in enormous refrigerated ships and is then stored under pressure in vast tanks underground. It is transferred to pressurised cylinders to be used for cooking and heating in the home. In contrast, in the UK the main use of propane is for leisure purposes – camping and BBQs!

Although the combustion of the alkanes brings huge benefits to people around the world, it is also central to some major environmental problems which you will consider at the end of this chapter.

fig. 1.6.13 Cylinders of liquid butane are a common sight on caravan holidays as well as on boats. In cold climates propane (boiling temperature –42 °C) is used in preference to butane (boiling temperature –0.4 °C) because on a cold morning the liquid butane would not vaporise.

HSW Alkanes for heating – risk awareness

If an alkane is burnt without plenty of oxygen, **incomplete combustion** occurs. The products of an incomplete combustion reaction may include carbon or the potentially fatal gas carbon monoxide. Many people who use gas heaters and boilers may be unaware of the potential hazard of an inadequate supply of oxygen to a heater, or inadequate venting of the waste gases. Every year there are tragic examples of deaths resulting from carbon monoxide poisoning. This risk can be reduced greatly by the proper maintenance of gas appliances. You can also buy carbon monoxide monitors which give a warning when the gas is produced in the room. Landlords need a safety certificate which guarantees that their gas appliances have been checked, something all tenants should be aware of.

Questions

1 What happens to alkanes when they are heated to high temperatures:
  - **a** in the absence of air
  - **b** in the presence of air?

  Give chemical equations for examples of these reactions.

2 Use the example of a gas fire in a student's room to produce a short report or poster explaining the difference between hazard and risk to year 9 pupils.

Reactions of the alkanes with chlorine

The alkanes react with chlorine, but only with an input of energy in the form of sunlight or ultraviolet light. For example, methane and chlorine do not react in the dark, but in sunlight they react explosively to produce chloromethane and hydrogen chloride. The light provides the input of energy needed to break the hydrocarbon bonds. This, combined with the very rapid reaction that follows, is typical of a reaction that takes place by a free-radical mechanism. This is an example of a **substitution reaction**, with chlorine atoms substituting for the hydrogen atoms in the methane molecule.

The Cl—Cl bond is easier to break than the C—H bond. Light provides the energy to split the chlorine molecules into atoms – in other words to **initiate** the reaction. The splitting of the chlorine molecules is an example of homolytic fission – the chlorine atoms formed are free radicals, and so are extremely reactive.

$Cl_2 \rightarrow Cl^{\bullet} + Cl^{\bullet}$
$\Delta H = +243\ kJ\ mol^{-1}$

Chlorine free radicals react with methane molecules, combining with one of the hydrogen atoms to form hydrogen chloride and another free radical:

$CH_4 + Cl^{\bullet} \rightarrow CH_3^{\bullet} + HCl$
$\Delta H = -19\ kJ\ mol^{-1}$

The methyl free radical then reacts with another chlorine molecule to form chloromethane and a chlorine free radical:

$CH_3^{\bullet} + Cl_2 \rightarrow CH_3Cl + Cl^{\bullet}$
chloromethane

$\Delta H = -103\ kJ\ mol^{-1}$

The process is then repeated hundreds of times in these **propagation** reactions that produce another free radical. This rapid and repeated propagation results in a **chain reaction**, an explosive process.

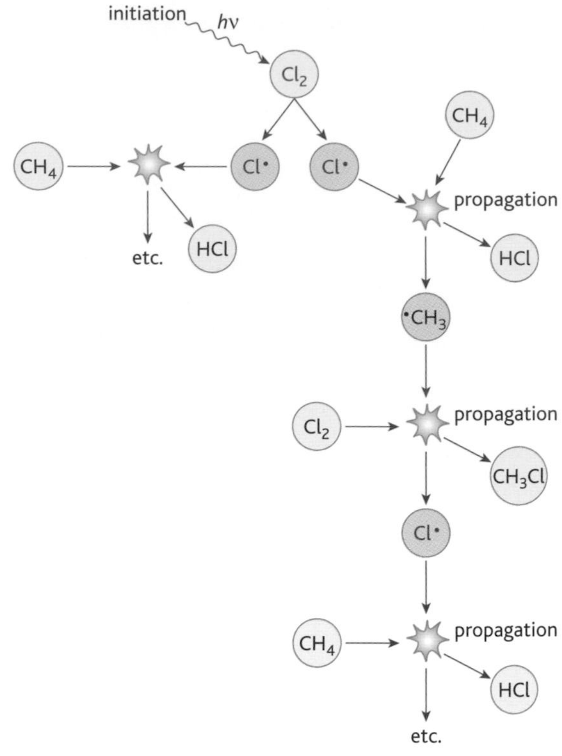

fig. 1.6.14 The reaction of methane with chlorine is a free-radical chain reaction.

The propagating steps of this reaction continue until there is a **termination** step. This is a reaction between two free radicals – a highly exothermic process. A termination step happens every few thousand reactions!

$Cl^{\bullet} + Cl^{\bullet} \rightarrow Cl_2$	$\Delta H = -243\ kJ\ mol^{-1}$
$CH_3^{\bullet} + Cl^{\bullet} \rightarrow CH_3Cl$	$\Delta H = -346\ kJ\ mol^{-1}$
$CH_3^{\bullet} + CH_3^{\bullet} \rightarrow C_2H_6$ ethane	$\Delta H = -347\ kJ\ mol^{-1}$

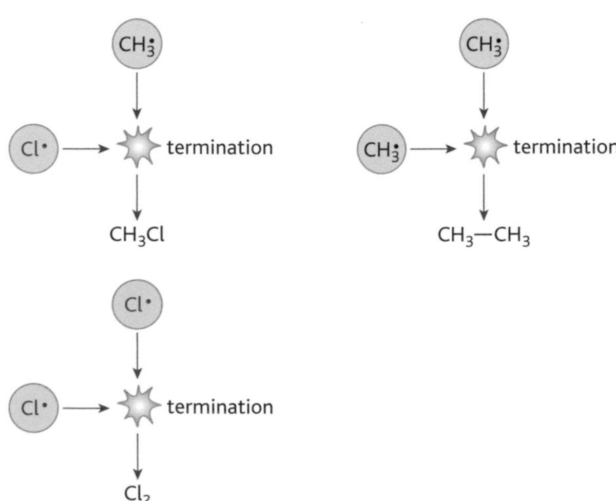

fig. 1.6.15 The termination of a chain reaction.

If the supply of chlorine is limited, the net result of this reaction is lots of chloromethane, CH_3Cl and hydrogen chloride, and relatively little ethane. However, if there is a plentiful supply of chlorine then further substitutions of the methane will take place to give di-, tri- and tetrachloromethane as follows:

$CH_4 + Cl_2 \rightarrow CH_3Cl + HCl$; $CH_2Cl_2 + Cl_2 \rightarrow CHCl_3 + HCl$
$CH_3Cl + Cl_2 \rightarrow CH_2Cl_2 + HCl$; $CHCl_3 + Cl_2 \rightarrow CCl_4 + HCl$

Some of these halogenated alkanes (halogenoalkanes) are well-known chemicals. Trichloromethane or chloroform, $CHCl_3$ was one of the first anaesthetics to be used for surgical operations and to ease the pain of childbirth, whilst tetrachloromethane, CCl_4 was widely used as a solvent until its carcinogenic properties were recognised.

Questions

1 Under what conditions do the alkanes react with chlorine?

2 Explain, using chemical equations, the mechanism of the reaction between hexane and chlorine, showing why the reaction conditions are so important.

3 What type of experiments would you like to see to help to confirm the effects of antioxidants such as vitamin E in preventing ageing?

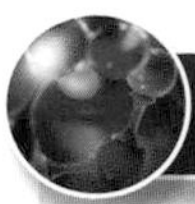

HSW Antioxidants and ageing

A growing body of experimental evidence shows that free radicals exist in the human body where they are responsible for many negative effects. As you saw earlier they are linked to the development of cancer. They have also been linked to heart disease and very specifically to the ageing process. For example, your skin changes very noticeably as you age. The smooth, elastic appearance of young skin is due in part to collagen, a flexible protein. One theory of ageing is that free radicals in the body attack the cross-links between collagen fibres, making them stiffer and less flexible, which in turn affects the appearance of the skin.

Free radicals also seem to affect the biochemistry of cells and may even attack the DNA in the nuclei of cells. This could be how they cause cancers.

To try to counteract these effects, people are increasingly turning to **antioxidants** – compounds that 'mop up' free radicals. Vitamin E is a well-known example of a natural free-radical inhibitor. Many people take antioxidants as dietary supplements and use them in cosmetics. Do they work? There is evidence that people deficient in vitamin E show skin changes similar to those of ageing, so maintaining healthy levels of such vitamins in the diet should have a protective effect. A study published in 2000 showed that healthy centenarians (people of over 100 years old) all had high levels of the antioxidants vitamin A and vitamin E. Studies on a number of mammalian species also seem to show a link between the levels of antioxidant vitamins in the blood and life expectancy (see **fig. 1.6.16**).

Eating plenty of fruit and vegetables increases the dietary intake of antioxidants and this is accepted by most scientists as sound advice. However, the benefits of rubbing antioxidants into your skin, or eating excessive amounts in the diet, have still to be proven scientifically.

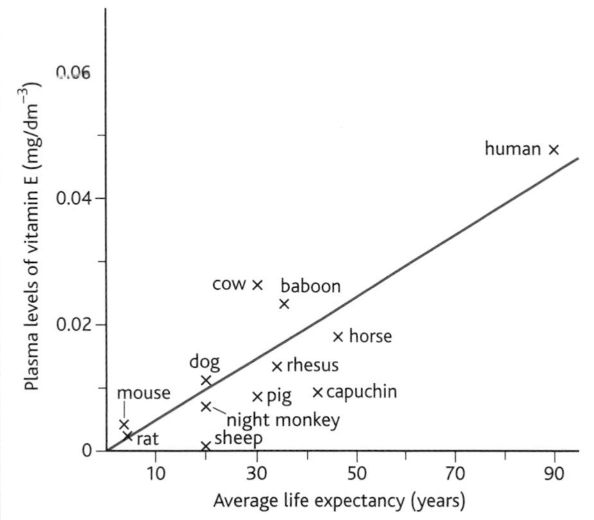

fig. 1.6.16 This graph suggests a link between vitamin E levels in the blood plasma and the life expectancy of the animals studied. This suggests that the antioxidant effect of the vitamin may help prolong life by reducing damage by free radicals. (Rhesus and capuchin are types of monkey).

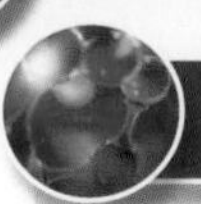

HSW Ethical issues

Cars and society

The combustion of alkanes in the form of fossil fuels drives most of the great economies of the world. One of the major uses of fossil fuels is in the internal combustion engine that powers road vehicles. Public attitudes to the car are changing and more people are now questioning the use of the car than at any time since they were first given the freedom of the roads at the turn of the twentieth century.

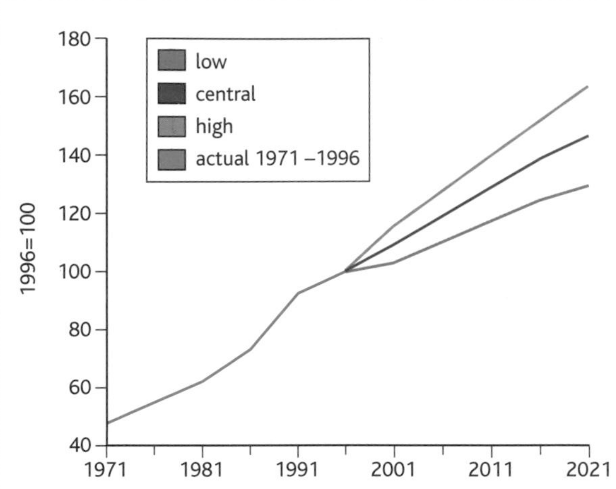

fig. 1.6.17 Helped by the power of advertising, the car you drive gives an instant impression of you and your lifestyle. The graph, produced by the UK Department for Transport, shows how road traffic has increased – and is predicted to continue increasing in the future.

Cars are one of the major polluting agents on planet Earth, but using a car has become so convenient and even necessary that it is difficult for individuals to see the collective damage that each journey is doing. The advantages of restricting motor vehicle use are to society as a whole; the disadvantages are all personal. This balance of personal and public good has to be addressed in all the arguments over how to make the car more environmentally acceptable.

What's the problem?

In a car engine, hydrocarbon combustion is not always complete and so carbon monoxide is formed along with carbon dioxide. Carbon monoxide is not only a greenhouse gas (see below) but is also highly toxic as it combines with the haemoglobin of the blood more effectively than oxygen, preventing the blood from carrying oxygen around the body. Oxides of nitrogen and sulfur are also produced from impurities in the fuel, and these cause acid rain and smog if they are not removed from the exhaust fumes. Catalytic converters convert carbon monoxide and nitrogen oxides into less harmful gases, but cannot remove the carbon dioxide.

The quantity of pollution produced today is too great to be absorbed by the atmosphere without an adverse effect on the environment. How is this effect felt?

Increased levels of carbon dioxide in the atmosphere are thought to be causing **global warming**, with resultant changes in weather patterns (see below). Any process that adds to the level of carbon dioxide or other greenhouse gases is therefore a problem.

The greenhouse effect

Radiated energy from the Sun reaches the Earth. The Earth's atmosphere reflects some of this radiation back into space, but some of it passes through the atmosphere to the Earth's surface. The Earth absorbs this radiation and also radiates its own infrared radiation, which has a different wavelength from that of the Sun. Some of this infrared does not pass out through the atmosphere and back into space. It is re-radiated back down to the Earth again by **greenhouse gases** in the atmosphere, so keeping the temperature of the Earth higher than it would otherwise be. Greenhouse gases include water vapour, carbon dioxide and methane amongst others. This re-radiation as part of the natural balance in the atmosphere is called the **greenhouse effect**.

Increased amounts of greenhouse gases in the atmosphere lead to an increased greenhouse effect, and this is widely believed to be causing global warming. Twentieth-century human activities have produced a massive increase in certain greenhouse gases in the upper atmosphere, including carbon dioxide from the burning of fossil fuels and methane from rotting vegetation, from paddy fields and the flatulence produced by cattle raised for cheap beef. At the same time people have removed vast areas of the world's vegetation in the much-publicised destruction of the rainforests. This adds to the carbon dioxide loading because the trees are often burnt after felling, and the loss of trees reduces the ability of the biosphere to absorb carbon dioxide. What is more, the trees are frequently replaced by cattle which add to the greenhouse gases by

producing methane. The ability of trees to take up carbon dioxide also seems to be reduced as temperatures increase. Carbon dioxide makes the largest single contribution to the increased greenhouse effect, although methane is chemically a more potent gas.

The long-term effects of global warming remain to be seen, although increasingly violent and unpredictable weather patterns are often blamed on global warming.

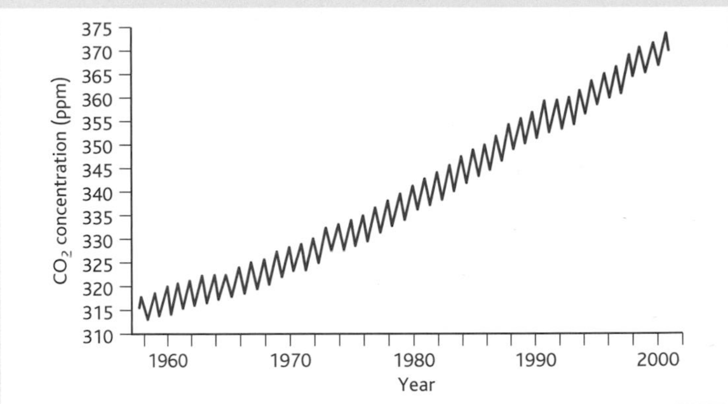

fig. 1.6.18 The data shown in this graph were measured on a mountaintop in Hawaii. The overall trend shows a steady rise in carbon dioxide levels.

	Amount (gigatonnes p.a.)	Carbon storage	Amount (gigatonnes)
Annual emission:	**8.7–9.1**	Atmosphere	750
Fossil fuels	6.9–7.0	Forests	610
Land-use change (deforestation)	1.8–2.0	Surface ocean	1580
Other	0.1	Deep ocean	38100
Annual absorption:	**8.7–9.1**	*Fossil fuels:* *	5000
Remains in atmosphere	4.5	Coal	4000
Absorbed by oceans	2.3	Oil	500
Absorbed by vegetation	1.9–2.3	Natural gas	500

table 1.6.2 Published data on some of the sources of global carbon dioxide, and where it is stored in the Earth (*Source: Kasting 1998*). This data varies widely depending on when it was collected and who publishes it. *Carbon is locked up in fossil fuels and is released on burning.

fig. 1.6.19 There have been many cases of catastrophic flooding and other natural disasters over the last decade or more. Are these the result of global warming – or just natural fluctuations in climate?

Predicted outcomes of global warming include the melting of the polar icecaps causing a rise in sea levels and the loss of low-lying regions of many countries. It is also suggested that relatively small changes in global temperature could bring about major shifts in climate, destroying the present pattern of world food production as many currently fertile countries become barren deserts. Whether these predictions hold true remains to be seen. It is easy to blame any unusual weather conditions on global warming – yet a look back in history often reveals similar events centuries ago. However, the evidence of a real problem seems to be mounting and there is a growing consensus among scientists that the rising levels of greenhouse gases in the atmosphere do pose a real threat to the future well-being of the planet. At present, all we can do is act to reduce or avoid adding to the current levels of greenhouse gases and then wait and see.

Developing new fuels

Modern cars have catalytic converters which absorb most of the polluting gases from the exhaust emissions. They are certainly cleaner than their predecessors. Six modern cars produce less pollution than one 1970s model. But the escalating numbers of vehicles outweigh the improvements made. Over 100 000 cars are produced per day, with an estimated 600 million vehicles on the road worldwide. To make a real impression on pollution levels alternative, sustainable energy sources must be developed.

There are a number of options. Cars which run on methane, cars powered by ethanol produced by the fermentation of sugar cane, hydrogen-burning vehicles and solar-powered cars are all possible alternatives to the traditional petrol- or diesel-powered vehicle. For every alternative fuel, you need to consider the *overall* carbon economy. What carbon emissions are released in the production of the alternative fuel? How much carbon dioxide is released as it is burnt? How easy is it to refuel the vehicle? Is the process really carbon neutral? For example, methane can be produced by the fermentation of plant and animal waste. Carbon dioxide is absorbed during the growth of the plants. Ethanol too is produced by the fermentation of plants such as maize, so the plants act as a **carbon sink** or reservoir. But the energy needed for processing, and the amount of land needed to produce **biofuels** such as ethanol and biodiesel, mean they may not be the answer that was hoped for. People need land to grow food before fuel!

Battery-powered vehicles

The most practical globally available alternative to petrol is the electrically powered car. Electric vehicles are not new – eg electric milk floats have long been seen on our streets. No exhaust gases are produced because these vehicles are powered by an electric battery – but this has its own problems. The vehicle has a limited range and cannot travel very fast once the battery runs down. There are two main types of electric vehicles being designed by the major motor companies – battery-powered and fuel-cell vehicles. The technology for the latter is still in its infancy, but battery-powered cars and cars that can switch between petrol power and electric power are already gaining in popularity.

fig. 1.6.20 This is one of the first 'dual fuel' cars to become popular. The car switches between battery power for urban driving and petrol power for longer journeys.

However, batteries have a high energy density – kilogram for kilogram they provide far less energy than petrol. This means that, with conventional batteries at least, electric cars have to carry extremely heavy batteries to provide them with a reasonable range of travel. Batteries need to be recharged – refuelling an electric car entails not a five-minute stop at the garage but around 12 hours plugged into the mains! And although the cars themselves are pollution-free, the production of mains electricity from the combustion of gas, oil and coal is far from a pollution-free process. Only when nuclear fuel is used to generate the electricity to charge the battery is the environmental impact considerably less in terms of carbon emissions than for a petrol-fuelled car. Real advances have been made in battery technology in recent years, and research is still continuing.

The hydrogen cell

One of the most exciting developments is the **hydrogen-cell** motorbike. The ENV (Emissions Neutral Vehicle) was designed in the UK. It is an almost silent motorbike, which can travel at around 50 mph. It runs on hydrogen and so combustion produces only water and heat. It can travel 100 miles on a single tank of compressed hydrogen costing around £2. The system is based on a proton-exchange membrane (PEM) fuel cell. This works by using the energy produced during the oxidation of hydrogen to generate electricity. The cell contains special membranes for electrolysis to take place. The electricity produced by the fuel cell is stored in batteries and used to drive the motor.

If the supply of chlorine is limited, the net result of this reaction is lots of chloromethane, CH_3Cl and hydrogen chloride, and relatively little ethane. However, if there is a plentiful supply of chlorine then further substitutions of the methane will take place to give di-, tri- and tetrachloromethane as follows:

$CH_4 + Cl_2 \rightarrow CH_3Cl + HCl$; $CH_2Cl_2 + Cl_2 \rightarrow CHCl_3 + HCl$
$CH_3Cl + Cl_2 \rightarrow CH_2Cl_2 + HCl$; $CHCl_3 + Cl_2 \rightarrow CCl_4 + HCl$

Some of these halogenated alkanes (halogenoalkanes) are well-known chemicals. Trichloromethane or chloroform, $CHCl_3$ was one of the first anaesthetics to be used for surgical operations and to ease the pain of childbirth, whilst tetrachloromethane, CCl_4 was widely used as a solvent until its carcinogenic properties were recognised.

Questions

1 Under what conditions do the alkanes react with chlorine?

2 Explain, using chemical equations, the mechanism of the reaction between hexane and chlorine, showing why the reaction conditions are so important.

3 What type of experiments would you like to see to help to confirm the effects of antioxidants such as vitamin E in preventing ageing?

HSW Antioxidants and ageing

A growing body of experimental evidence shows that free radicals exist in the human body where they are responsible for many negative effects. As you saw earlier they are linked to the development of cancer. They have also been linked to heart disease and very specifically to the ageing process. For example, your skin changes very noticeably as you age. The smooth, elastic appearance of young skin is due in part to collagen, a flexible protein. One theory of ageing is that free radicals in the body attack the cross-links between collagen fibres, making them stiffer and less flexible, which in turn affects the appearance of the skin.

Free radicals also seem to affect the biochemistry of cells and may even attack the DNA in the nuclei of cells. This could be how they cause cancers.

To try to counteract these effects, people are increasingly turning to **antioxidants** – compounds that 'mop up' free radicals. Vitamin E is a well-known example of a natural free-radical inhibitor. Many people take antioxidants as dietary supplements and use them in cosmetics. Do they work? There is evidence that people deficient in vitamin E show skin changes similar to those of ageing, so maintaining healthy levels of such vitamins in the diet should have a protective effect. A study published in 2000 showed that healthy centenarians (people of over 100 years old) all had high levels of the antioxidants vitamin A and vitamin E. Studies on a number of mammalian species also seem to show a link between the levels of antioxidant vitamins in the blood and life expectancy (see **fig. 1.6.16**).

Eating plenty of fruit and vegetables increases the dietary intake of antioxidants and this is accepted by most scientists as sound advice. However, the benefits of rubbing antioxidants into your skin, or eating excessive amounts in the diet, have still to be proven scientifically.

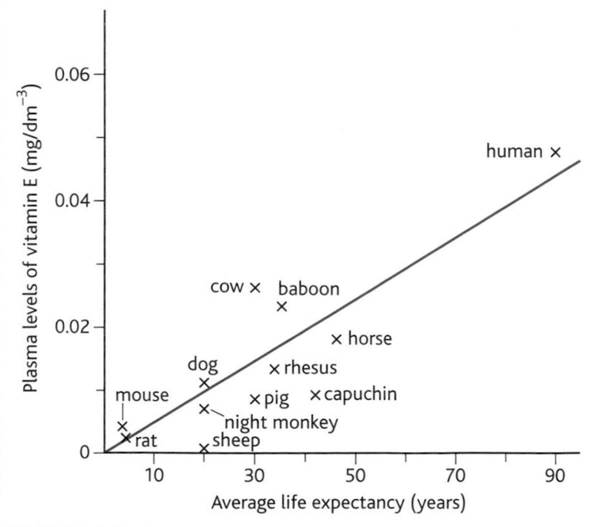

fig. 1.6.16 This graph suggests a link between vitamin E levels in the blood plasma and the life expectancy of the animals studied. This suggests that the antioxidant effect of the vitamin may help prolong life by reducing damage by free radicals. (Rhesus and capuchin are types of monkey).

HSW Ethical issues

Cars and society

The combustion of alkanes in the form of fossil fuels drives most of the great economies of the world. One of the major uses of fossil fuels is in the internal combustion engine that powers road vehicles. Public attitudes to the car are changing and more people are now questioning the use of the car than at any time since they were first given the freedom of the roads at the turn of the twentieth century.

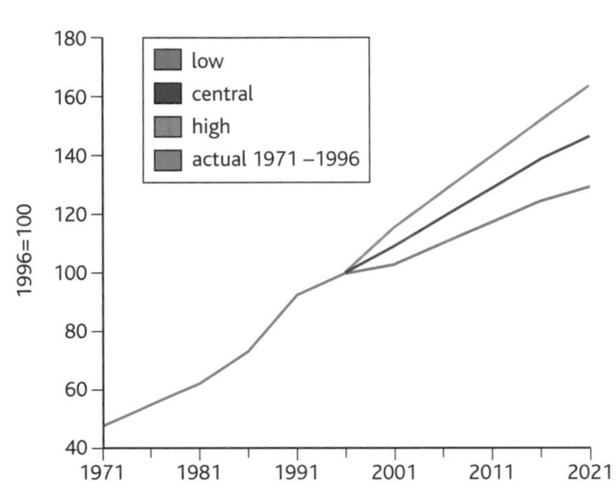

fig. 1.6.17 Helped by the power of advertising, the car you drive gives an instant impression of you and your lifestyle. The graph, produced by the UK Department for Transport, shows how road traffic has increased – and is predicted to continue increasing in the future.

Cars are one of the major polluting agents on planet Earth, but using a car has become so convenient and even necessary that it is difficult for individuals to see the collective damage that each journey is doing. The advantages of restricting motor vehicle use are to society as a whole; the disadvantages are all personal. This balance of personal and public good has to be addressed in all the arguments over how to make the car more environmentally acceptable.

What's the problem?

In a car engine, hydrocarbon combustion is not always complete and so carbon monoxide is formed along with carbon dioxide. Carbon monoxide is not only a greenhouse gas (see below) but is also highly toxic as it combines with the haemoglobin of the blood more effectively than oxygen, preventing the blood from carrying oxygen around the body. Oxides of nitrogen and sulfur are also produced from impurities in the fuel, and these cause acid rain and smog if they are not removed from the exhaust fumes. Catalytic converters convert carbon monoxide and nitrogen oxides into less harmful gases, but cannot remove the carbon dioxide.

The quantity of pollution produced today is too great to be absorbed by the atmosphere without an adverse effect on the environment. How is this effect felt?

Increased levels of carbon dioxide in the atmosphere are thought to be causing **global warming**, with resultant changes in weather patterns (see below). Any process that adds to the level of carbon dioxide or other greenhouse gases is therefore a problem.

The greenhouse effect

Radiated energy from the Sun reaches the Earth. The Earth's atmosphere reflects some of this radiation back into space, but some of it passes through the atmosphere to the Earth's surface. The Earth absorbs this radiation and also radiates its own infrared radiation, which has a different wavelength from that of the Sun. Some of this infrared does not pass out through the atmosphere and back into space. It is re-radiated back down to the Earth again by **greenhouse gases** in the atmosphere, so keeping the temperature of the Earth higher than it would otherwise be. Greenhouse gases include water vapour, carbon dioxide and methane amongst others. This re-radiation as part of the natural balance in the atmosphere is called the **greenhouse effect**.

Increased amounts of greenhouse gases in the atmosphere lead to an increased greenhouse effect, and this is widely believed to be causing global warming. Twentieth-century human activities have produced a massive increase in certain greenhouse gases in the upper atmosphere, including carbon dioxide from the burning of fossil fuels and methane from rotting vegetation, from paddy fields and the flatulence produced by cattle raised for cheap beef. At the same time people have removed vast areas of the world's vegetation in the much-publicised destruction of the rainforests. This adds to the carbon dioxide loading because the trees are often burnt after felling, and the loss of trees reduces the ability of the biosphere to absorb carbon dioxide. What is more, the trees are frequently replaced by cattle which add to the greenhouse gases by

GET A HEAD START TO PASSING GCSE SCIENCE THIS SUMMER

📕 Stress-free Tutoring 👩‍🏫 with a Qualified Teacher

Does your child STRUGGLE in science and need to PASS their science GCSEs?

Perhaps they are predicted a STANDARD pass (grade 4) and need to move up a few grades so they can do the further education courses of their choice?

Or maybe they're currently working at a grade 5/6 (STRONG PASS) and need support to ensure they achieve their predicted grade or higher.

Maybe your child is a KS3 student and recognises that the foundation to passing GCSE science starts NOW!

If any of these examples ring true for you, my tutoring service can help.

Small Groups & Block booking discounts available
Get a FREE trial lesson with Online Tuition.

📞 07846847610 or 📧 passgcsescience@gmail.com for RATES and more info.

This is a promising way of powering vehicles. Hydrogen is relatively safe – it does not explode readily when kept in a pressurised bottle away from the air. However, refuelling is not easy at the moment, because garages are not set up to supply hydrogen. And at present, greenhouse gases are produced in the formation and pressurisation of the hydrogen. In the future, biofuels may be used as the source of the hydrogen which would reduce carbon emissions further. This technology is still developing.

fig. 1.6.21 The hydrogen-powered motorbike is quiet and clean. Will vehicles like this reduce greenhouse emissions enough to save the planet?

Your studies of the alkanes have given you an introduction to the chemistry of the hydrocarbons, and into some of the reaction mechanisms that are important in all the families of organic chemicals. In the next chapter you will look at the chemistry of the unsaturated hydrocarbons and see how this differs from that of the saturated alkanes.

Questions

1. What are the main environmental issues linked to the use of fossil fuels in car engines?
2. Using this book and other resources, investigate the evidence for the greenhouse effect and global warming.
3. Choose a way to represent the data in **table 1.6.2** to show clearly (in gigatonnes):
   - a annual carbon emissions
   - b annual absorption
   - c the main carbon storage reservoirs.
4. From the data presented in **table 1.6.2**,
   - a What are the maximum and minimum percentages of global emissions which come from fossil fuels?
   - b What are the maximum and minimum percentages of global absorptions which is the result of plant activity?
   - c How does this information support or undermine your ideas about the role of deforestation in global warming?
5. Use ICT to bring the story of the hydrogen-cell motorbike up to date. What impact have they made, are they selling well and what is the current state of hydrogen cell technology in the car industry as a whole?

7 The alkenes – a family of unsaturated hydrocarbons

ethane C_2H_6

```
    H   H
    |   |
H — C — C — H
    |   |
    H   H
```

ethene C_2H_4

```
H       H
 \     /
  C = C
 /     \
H       H
```

fig. 1.7.1 Alkenes contain two fewer hydrogen atoms than the corresponding alkane.

The **alkenes**, and ethene in particular, are of immense importance to the chemical industry. They occur naturally only in very small quantities but are obtained from crude oil by the processes of cracking and catalytic reforming described in chapter 1.6. Ethene is the starting point for a great many synthetic processes – it is used for the production of polymers, detergents, solvents and many other chemicals.

Like the alkanes, the alkenes are hydrocarbons (they contain carbon and hydrogen only), but they are **unsaturated**. Unsaturated compounds do not contain the maximum amount of hydrogen possible because they have one or more carbon–carbon double or triple bonds. All alkenes contain at least one carbon–carbon double bond, and for every carbon–carbon double bond in the molecule there will be two fewer hydrogen atoms than in the corresponding saturated molecule. The general formula for an alkene containing just one double bond is C_nH_{2n}.

The carbon–carbon double bond

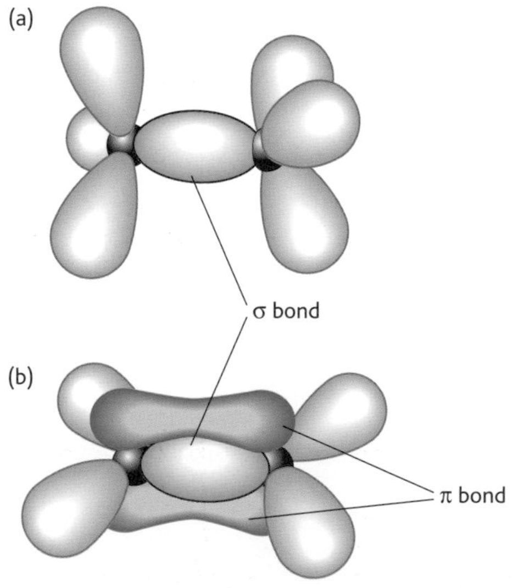

fig. 1.7.2 The distribution of the electrons in (a) a carbon–carbon single bond and (b) a double bond. A double bond consists of both a σ bond and a π bond. This affects the shape of the molecule, which in turn dictates the properties of the compound.

In a carbon–carbon single bond (also known as a **σ bond** or **sigma bond**) the electron cloud is symmetrical about the central axis of the molecule (see **fig. 1.7.2**). Because it lies along the line joining the two carbon atoms, the σ bond can rotate about this axis, so the two ends of the ethane molecule are free to rotate relative to each other.

In the carbon–carbon double bond the geometry is different. The ethene molecule $CH_2{=}CH_2$ is flat. The double bond in ethene involves a σ bond plus a second bond which has its electron density concentrated in two regions on either side of the axis of the bond, above and below the plane of the molecule. This second bond is known as a **π bond** (**pi bond**). The π bond does not allow rotation around the axis, and this has a big effect on the structure, and therefore also the properties of compounds that have a π bond.

The electron density in the π bond is on average further from the nuclei of the two carbon atoms concerned, so the double bond is less than twice as strong as a σ bond. As you will see in this chapter, this arrangement of electron density also explains two other properties of molecules containing carbon–carbon double bonds:

- They are reactive, due to the attraction of positively polarised groups to the electron-rich π bond within the double bond. These electron-seeking groups are known as **electrophiles**. The π bond provides an electron pair to form a new bond with the electron-seeking group.
- Many unsaturated compounds show **geometric isomerism**. Rotation about the bond axis is not possible without breaking the π bond.

Naming the alkenes

As you saw in chapter 1.5 the alkenes are named using the IUPAC system, based on both the number of carbon atoms in the chain and the position of the functional group, which in this case is the carbon–carbon double bond. For example, for an alkene with four carbon atoms, the double bond can occur in three different places, forming but-1-ene, but-2-ene and methylpropene.

The properties of the alkenes are physically very similar to the equivalent alkanes, although the melting and boiling temperatures are a little lower. Also, like the alkanes, the unbranched alkenes form a family with regular trends in properties.

Alkene	Structural formula	Boiling temperature (°C)
Ethene	$H_2C{=}CH_2$	–104
Propene	$H_2C{=}CH{-}CH_3$	–47
But-1-ene	$H_2C{=}CH{-}CH_2{-}CH_3$	–6
Pent-1-ene	$H_2C{=}CH{-}CH_2{-}CH_2{-}CH_3$	30

table 1.7.1 The alkenes show a rise in boiling temperature as the number of carbon atoms in the molecule increases.

HSW Ethene – ripe fruit, big carbon footprint?

Ethene is the smallest alkene. It is also a plant hormone, which has the effect of ripening fruit. This property of ethene is used commercially to make sure fruit arrives in the supermarket in perfect condition. So, for example, bananas are picked in the Caribbean or South America long before they are ripe. They are chilled, stored, transported, stored again and then, once the supermarket is ready for them, they are exposed to ethene to ripen them ready for sale. This means that bananas and many other exotic fruits that do not grow in the UK can be available in our shops all year round.

This is great for consumer choice – but is it good for the planet? Transporting fruit thousands of miles, and keeping it in refrigerated storage over periods of weeks followed by exposure to ethene, all use huge amounts of energy whose generation releases vast amounts of carbon dioxide into the atmosphere. Is this the right use of our resources? Another aspect of this issue is that the livelihoods of many people in the banana-growing countries are dependent on this method of fruit production – if we change our eating habits to reduce our carbon footprint, what happens to them?

fig. 1.7.3 Is the use of ethene to manipulate nature and provide us with bananas whenever we want them a good thing for the planet?

Questions

1 Give the structural formula of the following compounds:
   a but-2-ene b 2-methylbut-2-ene.

2 a How does a carbon–carbon double bond differ from a carbon–carbon single bond?
   b How does the presence of a double bond affect the physical and chemical properties of the alkenes?

3 Outline the pros and cons of using ethene as a fruit-ripening hormone. What social and ethical problems might arise if:
   a the process continues
   b the process was banned?

Geometric isomerism

But-2-ene, C_4H_8 can be separated by distillation into two components that have significantly different boiling temperatures and other physical properties – but that show no significant differences in chemical reactivity. How do we explain this?

The lack of free rotation around the double bond in but-2-ene results in molecules known as **geometric isomers**. Geometric isomers occur when components of the molecule are arranged on different sides of the molecule. The traditional way of naming geometric isomers is known as ***cis–trans* isomerism**. The *cis*- and *trans*-isomers of but-2-ene are shown in **fig. 1.7.4**.

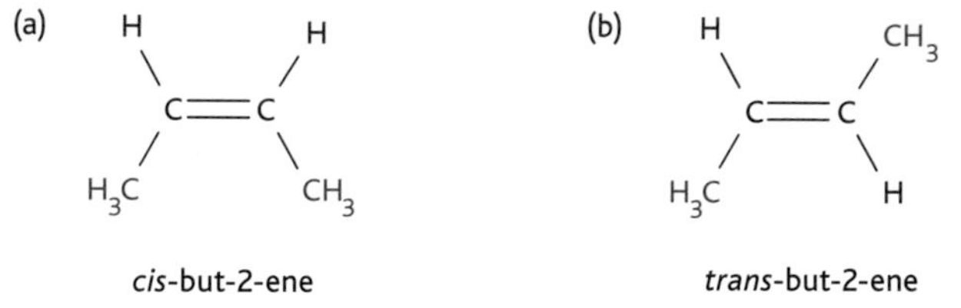

(a)

```
H         H
 \       /
  C === C
 /       \
H3C       CH3
```

cis-but-2-ene

(b)

```
H         CH3
 \       /
  C === C
 /       \
H3C       H
```

trans-but-2-ene

fig. 1.7.4 *Cis*- and *trans*-but-2-ene. Geometric isomers of the same molecule often have different properties.

In the *cis*-isomer both methyl groups are on one side of the bond and both hydrogen atoms are on the other. In *trans*-but-2-ene the two methyl groups are on opposite sides of the double bond (*cis* means 'on the same side' and *trans* means 'on opposite sides'). Geometric isomers frequently show different physical properties. Their melting and boiling temperatures as well as the density of the compound are often affected. Sometimes the chemical reactions of the isomers are influenced too, but often the chemical reactivity of the isomers is the same (eg the two isomers of but–2–ene).

Isomer	Melting temperature (°C)	Boiling temperature (°C)
cis-but-2-ene	–139	4
trans-but-2-ene	–106	1

table 1.7.2 Physical properties of the geometric isomers of but-2-ene.

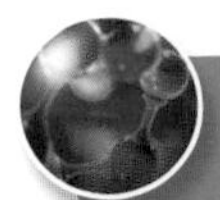

HSW *cis-trans* isomerism, cooking tomatoes and healthy hearts

Biological systems are often very sensitive to different geometric isomers. For example, in recent years it has been discovered that the antioxidant lycopene is very effective in helping to prevent a variety of problems, from heart disease to prostate cancer. It may even help reduce the effects of ageing. Lycopene is found in many red-coloured fruits and particularly high levels are found in tomatoes. Scientists have found that cooked tomatoes seemed to be more biologically active than raw ones, although their lycopene levels are similar. Then it was shown that there are different isomers of lycopene. The dominant isomer found in most red tomatoes is *trans*-lycopene, a straight-chain molecule. The biologically active form found in the human body is *cis*-lycopene, a bent molecule. Scientists are not sure whether the body converts one isomer into another using an enzyme or whether the *cis*-isomer is absorbed into the body more easily.

However, they have discovered that cooking tomatoes can alter the ratio of *trans*:*cis* isomers in the fruit from 19:1 to 11:9. The more *cis*-isomers in the food, the more readily the lycopene is absorbed and the more useful it is in the body. So cook your tomatoes thoroughly before you eat them for maximum benefit!

What's more, specially bred tangerine tomatoes have been produced which contain high levels of *cis*-lycopene. In a trial in the US, 12 adults ate two different test meals each. One had a spaghetti sauce made from tangerine tomatoes, while in the other the sauce was made from ordinary tomatoes. The trial participants were asked to avoid tomato-rich foods for 13 days before the test meals. Blood samples were taken at regular intervals after the meals.

The overall level of lycopene absorption was 2.5 times higher in people who had eaten sauce made from the tangerine tomatoes, showing that the human body is very sensitive to the difference. Commercial varieties of these tangerine, *cis*-lycopene rich, tomatoes are now in the pipeline.

fig. 1.7.5 Tangerine and red tomatoes. In future, just by choosing food containing the right isomer to eat, you may be able to improve your health and preserve your looks all in one delicious meal!

E-*Z* isomerism

The *cis–trans* system for naming geometric isomers has been used for many years. However, it has some limitations because the system does not work for all geometric isomers. The IUPAC system for naming geometric isomers is now the system of ***E–Z* isomerism**. Although you will still come across *cis–trans* isomers, the *E–Z* system is better because it works for all geometric isomers.

In the *E–Z* system, groups attached around a carbon–carbon double bond (or other rigid bond) are ranked based on their atomic number. The atom with the highest number has the highest priority – eg carbon outranks hydrogen. So you look at the position of the higher-priority group at one end of the carbon–carbon double bond and compare it with the higher-priority group at the other end of the double bond. If the two higher-priority groups are on the same side they are ***zusammen*** (together) and this is the ***Z*-isomer**. If the higher-priority groups are on opposite sides of the rigid bond then they are ***entgegen*** (opposite) and this is the ***E*-isomer**.

Look at isomer (a) of but-2-ene in **fig. 1.7.4**. At the left-hand side of the double bond, the two atoms attached to the carbon atom are hydrogen on top, and carbon below. Carbon has the higher atomic number so it has the higher priority, and the higher-priority group is 'down'. At the other end of the double bond the same is true – again the atoms are hydrogen and carbon. Carbon has higher priority – and it is in the 'down' position. Both higher-priority groups are on the same side of the double bond (*zusammen*) so the isomer is *Z*-but-2-ene.

Now look at isomer (b). You can see that here the higher-priority group is 'down' on the left-hand side of the double bond but 'up' on the right-hand side. The groups are on opposite sides of the double bond (*entgegen*) and so this is *E*-but-2-ene.

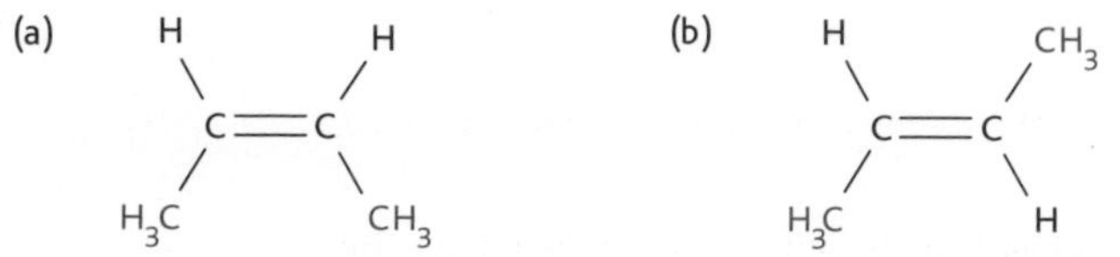

(a) H, H, C=C, H_3C, CH_3

Z-but-2-ene

(b) H, CH_3, C=C, H_3C, H

E-but-2-ene

fig. 1.7.6 **In this case *Z*-but-2-ene is the *cis*-isomer and *E*-but-2-ene is the *trans*-isomer. However, *cis* and *Z* or *trans* and *E* are not always the same.**

For but-2-ene, both the *cis–trans* system and the *E–Z* system work. However, this isn't true in all cases, as the example in **fig. 1.7.7** will show you.

(a) Br, F, C=C, I, Cl

Z-isomer

(b) Br, Cl, C=C, I, F

E-isomer

fig. 1.7.7 **Two different structures of 1-bromo-2-chloro-2-fluoro-1-iodoethene.**

These two chemicals are geometric isomers with different properties and different structures. The *cis–trans* system cannot name these isomers, but the *E–Z* system works well. In isomer (a), on the left-hand side of the double bond, the atoms are bromine and iodine. Iodine has the higher priority because it has the higher atomic number. It is in the 'down' position. At the other end of the bond, chlorine has the higher atomic number and so has higher priority over fluorine. It too is in the 'down' position. So in this isomer the two priority groups are on the same side of the double bond and this is the *Z*–isomer. In isomer (b) iodine is in the 'down' position but chlorine is in the 'up' position. The priority groups are on opposite sides of the double bond so this is the *E*–isomer.

Questions

1 Look at the evidence for the effect of the *cis*-isomer levels in tangerine tomatoes. How would you assess this research in terms of reliability and validity? What further studies would you like to see done before the new tomatoes are used in the mass production of food?

2 Work out the main isomers of pentene. Identify any geometric isomers, and give them their correct *E–Z* nomenclature.

Reactions of the alkenes

The double bond is the obvious main feature of the alkenes, and it is this that makes their reactions rather different from those of the alkanes. The double bond has two main effects on the chemistry of the alkenes:

- Alkenes exhibit *E–Z* geometric isomerism. There is no rotation around the double bond.
- Alkenes undergo **addition reactions** rather than the substitution reactions of the alkanes. An addition reaction is one in which two substances react to form a single product.

As you saw earlier, alkenes are more reactive than alkanes because the energy required to break the double bond is less than twice the energy required to break the single bond. For example, the bond enthalpy for the C–C bond in ethene is 347 kJ mol^{-1}, while the bond enthalpy for the C=C bond in ethene is 612 kJ mol^{-1}. This shows that the bond enthalpy for the π bond is only 265 kJ mol^{-1}. The reactions of the alkenes do not involve free radicals. Instead, heterolytic fission of the double bond occurs, and this is what determines how the molecules behave. The high electron density associated with the double bond means that the alkenes are attacked by both electrophiles (negative-charge-loving species) and oxidising agents.

Electrophiles

An alkene's carbon–carbon double bond is attacked when a species containing an atom with a whole or partial positive charge approaches the electron-rich double bond. Such a species is said to be **electrophilic**, since it is attracted to areas of high electron density. An electrophile is an electron-deficient species that can form a new covalent bond, using an electron pair provided by the carbon compound. The most common electrophilic agent is the proton, H^+.

Addition reactions of the alkenes

Reaction of the alkenes with hydrogen

The alkenes do not react with hydrogen under normal conditions of temperature and pressure. However, in the presence of a finely divided nickel catalyst and at a moderately high temperature (around 200 °C), alkenes undergo an addition reaction with hydrogen to form the corresponding alkane. The reaction of ethene with hydrogen (see fig. 1.7.8) is a typical example.

```
H       H                          H   H
 \     /              Ni           |   |
  C==C     +  H2   -------->   H — C — C — H
 /     \                           |   |
H       H                          H   H
```

fig. 1.7.8 **The addition of hydrogen across a double bond is called hydrogenation, a process widely used in the manufacture of margarine.**

Reactions of the alkenes with the halogens

The alkenes react with the halogens in a very different way from the alkanes. For example, when ethene is bubbled through bromine in the dark, the bromine is decolourised and a colourless liquid which is immiscible with water is formed. This reaction is so typical of the alkenes that it is used in analysis to demonstrate the presence of a double bond. In it, the ethene undergoes an addition reaction with the bromine forming 1,2-dibromoethane, a colourless liquid which does not mix with water. This is a disubstituted halogenoalkane. (See p.133 for more details of the reaction mechanism).

```
H       H                          H   H
 \     /                           |   |
  C==C     +  Br2  -------->   H — C — C — H
 /     \                           |   |
H       H                          Br  Br
                                1,2-dibromoethane

H       H                          H   H
 \     /                           |   |
  C==C     +  Cl2  -------->   H — C — C — H
 /     \                           |   |
H       H                          Cl  Cl
                                1,2-dichloroethane
```

fig. 1.7.9 **The reaction of the alkenes with the halogens produces addition products such as those shown here.**

All alkenes react vigorously in addition reactions with fluorine. The vigour of the addition reactions with the halogens decreases down the halogen group; reactions with iodine are relatively slow. Ethene and fluorine react explosively to form carbon and hydrogen fluoride, while with iodine a much slower reaction takes place to form 1,2-diiodoethane. The reaction of ethene with chlorine shown in **fig. 1.7.9** is particularly important as part of the process in the manufacture of chloroethene (which has the non-systematic name of vinyl chloride). This in turn is used to make poly(chloroethene) (also known as poly(vinyl chloride) or PVC), a widely used polymer – you will be looking at polymers in more detail later in this chapter.

All the reactions of the alkenes with the halogens proceed at room temperature, which suggests that the addition reaction does not involve free radicals. This is borne out by further studies, as you will see when you look at the reaction mechanisms below.

Testing for alkenes – bromine water and the carbon–carbon double bond

Bromine water (an aqueous solution of bromine) is used as a test for the alkenes. If you shake an alkene with bromine water, or bubble a gaseous alkene through bromine water, you will see the solution become colourless. This demonstrates the presence of the carbon–carbon double bond because an addition reaction takes place across the double bond, which is why the bromine water becomes colourless. The major product of the reaction is 2-bromoethanol, because OH^- ions from the water take part in the reaction as well as Br^- ions. However, some 1,2-dibromoethane is formed as well. So there are two different reactions taking place (see **fig. 1.7.10**) but the end result is always that bromine water is decolorised.

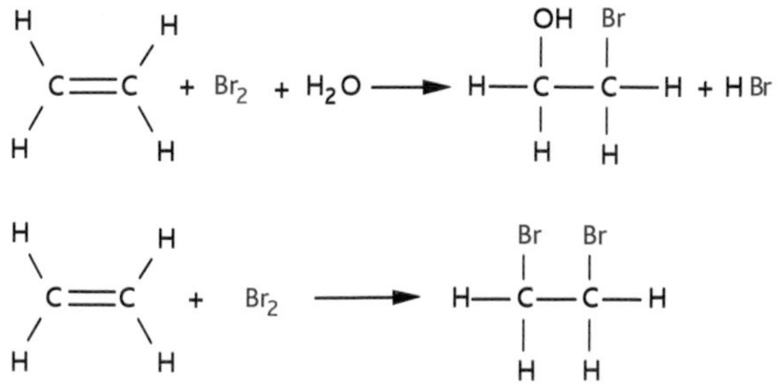

$$CH_2{=}CH_2 + Br_2 + H_2O \longrightarrow CH_2(OH){-}CH_2Br + HBr$$

$$CH_2{=}CH_2 + Br_2 \longrightarrow CH_2Br{-}CH_2Br$$

fig. 1.7.10 The decolorisation of bromine water – a test for alkenes.

Reactions of the alkenes with the hydrogen halides

The double bond in the alkenes reacts readily with the hydrogen halides, producing the corresponding monosubstituted halogenoalkane. For example, ethene reacts as follows:

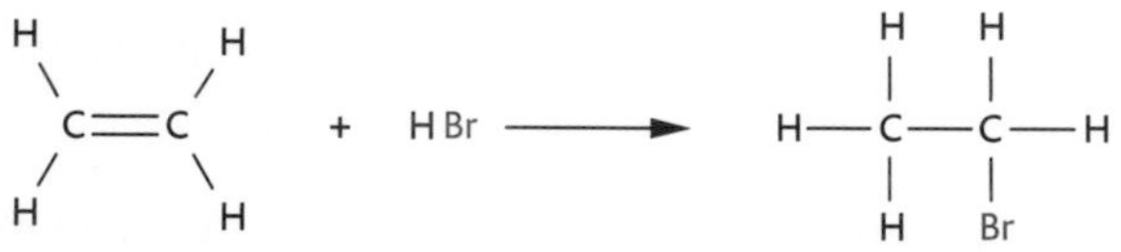

$$CH_2{=}CH_2 + HBr \longrightarrow CH_3{-}CH_2Br$$

This reaction proceeds rapidly at room temperature, forming bromoethane. The addition of hydrogen halides to alkenes, such as propene, which are asymmetric can lead to two possible products, eg:

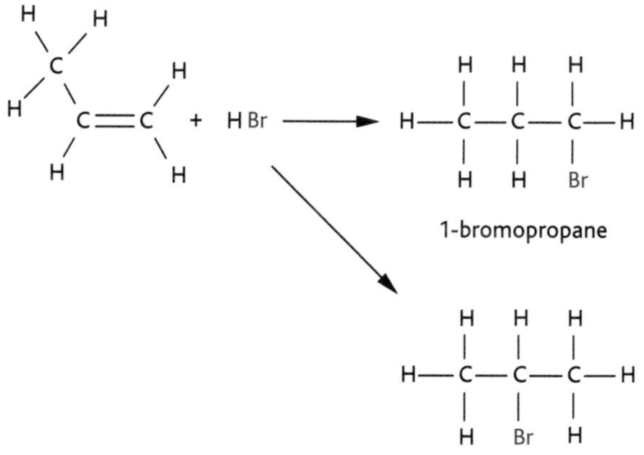

$$CH_3{-}CH{=}CH_2 + HBr \longrightarrow CH_3{-}CH_2{-}CH_2Br \text{ (1-bromopropane)}$$

$$CH_3{-}CH{=}CH_2 + HBr \longrightarrow CH_3{-}CHBr{-}CH_3 \text{ (2-bromopropane)}$$

In this case, the major product formed is 2-bromopropane, with a smaller amount of the alternative 1-bromopropane (see p.132 for more details). The likely products of the addition of hydrogen halides to asymmetric alkenes can be predicted using **Markovnikov's rule**:

- When HX adds across an asymmetric double bond, the major product formed is the molecule in which hydrogen adds to the carbon atom in the double bond with the greater number of hydrogen atoms already attached to it.

This is almost always the most stable alternative. Like all simple rules in science, this one has its exceptions – however, Markovnikov's rule does provide a useful way of deciding the most likely product of this type of reaction. There is more about this on the following pages.

Reaction of the alkenes with acidified potassium manganate(VII)

The reaction of the alkenes with acidified potassium manganate(VII) solution involves both addition across the double bond and oxidation. The products of the reaction are alkanediols, and the manganate(VII) solution is decolorised in the process, turning from purple to colourless. For example, when ethene is added to an acidified solution of potassium manganate(VII) the solution is decolorised and ethane-1,2-diol is formed:

$$H_2C{=}CH_2 \xrightarrow{KMnO_4\ /\ dil\ H_2SO_4} H{-}\underset{H}{\overset{OH}{C}}{-}\underset{H}{\overset{OH}{C}}{-}H$$

ethane-1,2-diol

The manganate(VII) ions are reduced to manganese(II) ions in the reaction.

Although ethane-1,2-diol is an extremely useful chemical in a variety of industrial processes, this is not the way it is made commercially because potassium manganate(VII) is a relatively expensive chemical. It is cheaper and more efficient to produce ethane-1,2-diol from epoxyethane, produced in the catalysed reaction of ethene with oxygen.

The reaction of ethene with potassium manganate(VII) can be used to distinguish the alkenes from the alkanes – the alkanes do not react.

Questions

1 Why are the alkenes more reactive than the alkanes?

2 How does the reaction of the alkenes with the halogens differ from the reaction of the alkanes with the halogens?

3 Give the chemical equation for the reaction when propene is mixed with hydrogen in the presence of a finely divided nickel catalyst at around 200 °C. Explain what happens in the reaction.

How do organic reactions happen?

You have seen how the halogens and the hydrogen halides react differently or don't react with alkanes and alkenes. Observing a wide range of reactions has given chemists information that helps build up a picture of how organic reactions happen – described as **reaction mechanisms**. These mechanisms not only tell you more about how organic reactions occur, they also help you to predict the outcome of similar reactions that you have not met before.

Using curly arrows

As chemists have developed models of reaction mechanisms, they have also established a way of showing what is going on using **curly arrows**. In chapter 1.6, you saw how a curly half-arrow represents the movement of a single electron. A full curly arrow represents the movement of a pair of electrons. In reactions involving alkenes, scientists have hypothesised that the electron pair of the π bond forms a new bond with a positively polarised attacking group (an electrophile). **Figure 1.7.11** shows the breaking and making of bonds in the first stage of the reaction of a hydrogen halide with ethene.

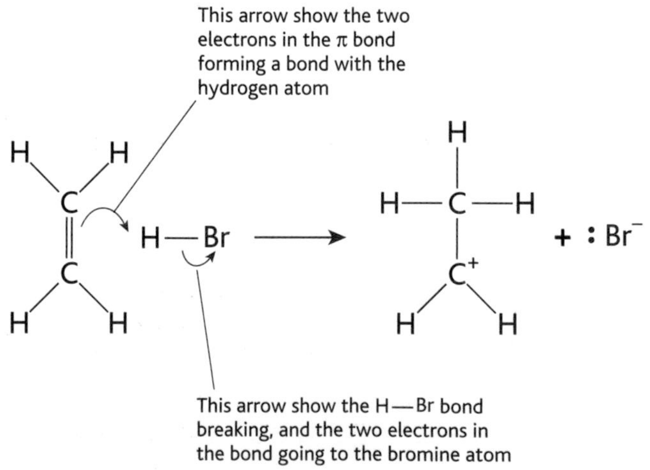

This arrow show the two electrons in the π bond forming a bond with the hydrogen atom

This arrow show the H—Br bond breaking, and the two electrons in the bond going to the bromine atom

fig. 1.7.11 **In the first stage of the reaction of H—Br with an alkene, the π bond breaks, a C—H bond is formed, and the H—Br bond breaks, liberating a Br^- ion.**

The first stage of this reaction involves **electrophilic attack** on the electron-rich π bond. As a result of this attack, one of the carbon atoms in the ethene molecule gains a positive charge, forming an ion called a **carbocation** or sometimes a **carbonium ion**. You will find out more about these ions later in the course.

This positively charged carbon is then open to attack by **nucleophiles** (positive-charge-loving species). In this case the bromide ion formed in the first stage of the reaction attacks the carbocation, forming a new C—Br bond (see **fig. 1.7.12**).

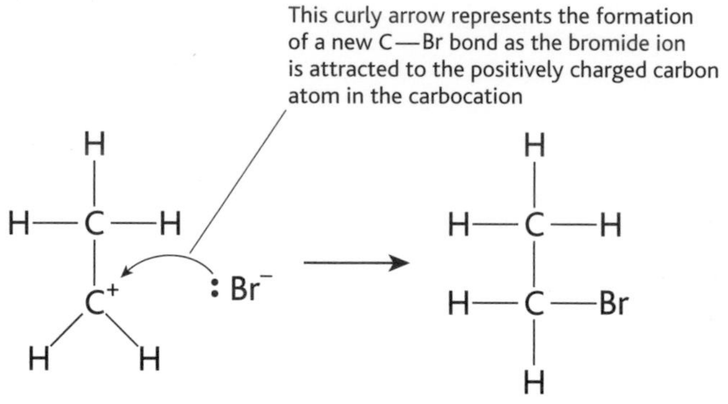

This curly arrow represents the formation of a new C—Br bond as the bromide ion is attracted to the positively charged carbon atom in the carbocation

fig. 1.7.12 **The nucleophilic attack of a carbocation by the bromide ion.**

Ethene is a symmetrical alkene. Propene in contrast is asymmetrical, with three carbon atoms and only one double bond. If you react hydrogen bromide with propene, the electrophilic addition reaction has two possible end products.

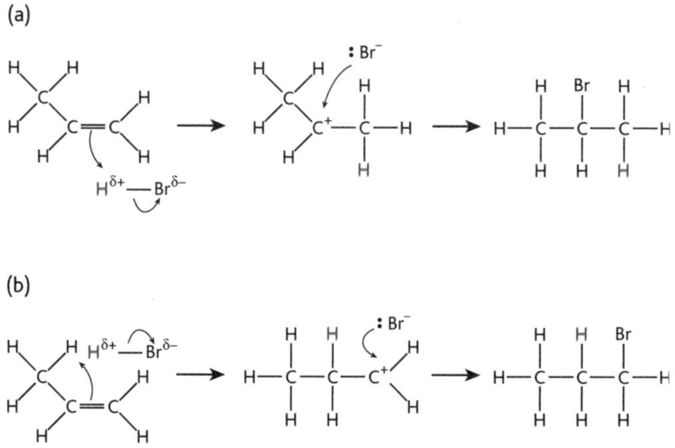

(a)

(b)

fig. 1.7.13 **The electrophilic addition of hydrogen bromide to propene.**

The carbocation in **fig. 1.7.13(a)** is more stable than its alternative in **(b)**, because the two methyl groups donate electron density and stabilise the positive charge. So carbocation (a) tends to be formed in preference. This means that the major product of the reaction is $CH_3CHBrCH_3$, with less $CH_3CH_2CH_2Br$ formed. The reaction largely follows Markovnikov's rule.

The addition of bromine to ethene is another example of an electrophilic addition reaction. The Br—Br molecule is partly polarised by the electron-rich π bond. The mechanism involves an electrophilic attack, followed by nucleophilic attack on the resulting carbocation. This is the mechanism of the reaction of all the halogens with ethene. **Figure 1.7.14** shows the addition reaction of bromine with ethene.

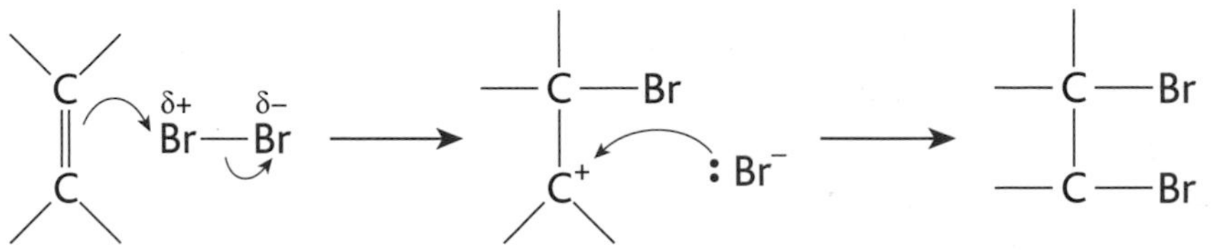

fig. 1.7.14 The mechanism of attack of bromine on a carbon–carbon double bond involves electrophilic attack by a Br_2 molecule with an instantaneous dipole.

HSW Evidence for reaction mechanisms

Chemists have developed models of reaction mechanisms by devising hypotheses based on observations, and then looking for further evidence to support their ideas. For example, scientists hypothesised that organic reactions often involve the formation of several different intermediate species as atoms are added or removed. Unfortunately there often isn't direct evidence for these different intermediates. Much of the evidence that allows scientists to develop hypotheses about reaction mechanisms into models of how they work is indirect. Experiments tend to be designed to confirm that the proposed mechanism is logical. Sometimes a theoretical intermediate can be shown to exist during the process. In some cases reaction rates play an important part in determining the mechanism. If scientists can show that the conversion of one compound to another follows the pathway they predict, this helps to confirm their model.

Scientists have some concrete evidence for the existence of the intermediate carbocation in both the reaction of hydrogen bromide with propene and the reaction of bromine with ethene which you considered above. This evidence comes from the observation that if competing nucleophiles are present in the reaction mixture (eg if chloride, Cl^- ions are present when ethene reacts with bromine), a mixture of products is formed as you can see in **fig. 1.7.15**.

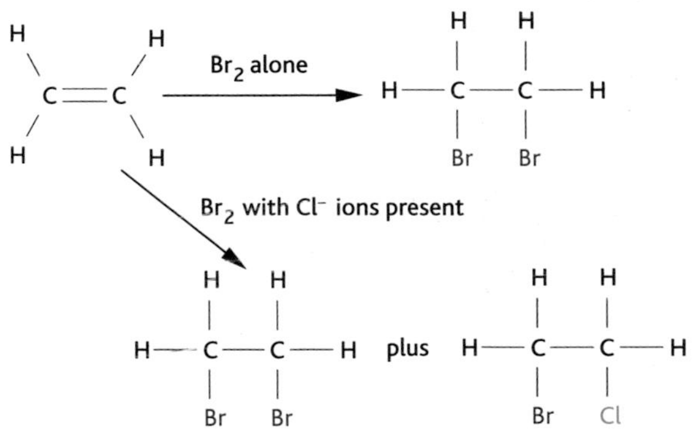

fig. 1.7.15 When ethene and bromine react, 1,2-dibromoethane is formed. In the presence of chloride ions, 1-bromo-2-chloroethane is formed too, as the Cl^- ions compete with the Br^- ions to attack the carbocation.

Questions

1 What are curly arrows used to represent in reaction mechanisms?

2 Propene is bubbled through bromine in the dark. Describe, using chemical equations, the reaction you would expect to take place.

3 What is the mechanism for the reaction described in question 2?

Polymerisation – the most important reaction of the alkenes

A **polymer** is a very large molecule made up of long chains of smaller units joined together. These smaller units are known as **monomers**. Molecules are generally regarded as polymers, rather than just very large molecules, when there are around 50 or more monomer units in the chain. The alkenes and their derivatives, with their reactive double bonds, are some of the main sources of monomers in the production of many commercial polymers. Artificial polymers include what we commonly call plastics, and these polymers play an important role in modern life.

The synthesis of polymers

One widely used method of forming polymers is by addition reactions. Two monomer units, which are usually identical, react together and an addition reaction occurs across a double or triple bond. The polymer is the only product.

monomer + monomer + monomer + … → polymer

Ethene, the simplest of the alkenes, undergoes polymerisation to form **poly(ethene)**, more commonly referred to as polythene. The double bonds break and the carbon atoms of the repeating units link together to form a long chain (see **fig. 1.7.16**).

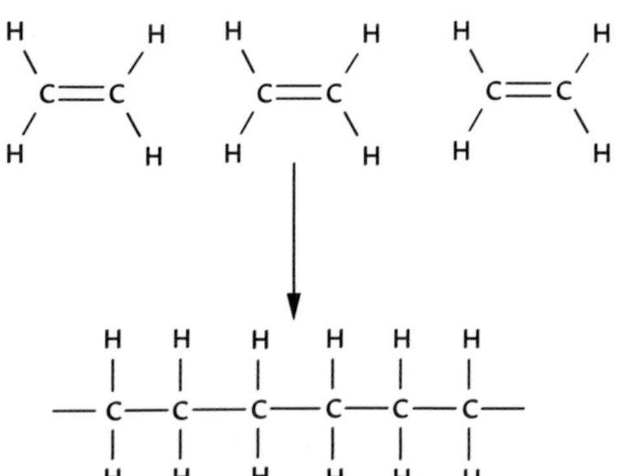

fig. 1.7.16 **The formation of poly(ethene) from its ethene monomers.**

Polythene was first discovered in 1933 by the British chemical company ICI. The first polythene, made by the polymerisation of ethene with traces of oxygen at 200 °C and 1200 atm pressure, had highly branched chains which made it relatively soft and malleable with a fairly low melting point. This material is useful for packaging and forming utensils and it is still the main form of polythene made today. Because of its branching molecules it is known as **low-density polythene** (LDPE).

Karl Ziegler, the German chemist who won a Nobel prize in 1963 for his work on polymers and plastics, developed a different way of producing polythene which uses catalysts and as a result needs only low temperatures (around 60 °C) and atmospheric pressure. The polythene produced by the Ziegler process is called **high-density polythene** (HDPE) – it has few branched chains which make it more rigid and denser than low-density polythene. It also melts at a higher temperature (135 °C as opposed to 105 °C). Polythene of both types softens as it is warmed, but each type has a distinct melting temperature.

Other plastics made from the alkenes include **poly(propene)**, which is also made using the Ziegler process. Poly(propene) has closely and regularly packed chains. It can be moulded, used in films or turned into fibres to make, for example, ropes that do not rot. Chloroethene polymerises to form **poly(chloroethene)** which is better known as PVC from its non-systematic name poly(vinyl chloride). There are many intermolecular forces in PVC as a result of the polar nature of the C—Cl bonds, and so the basic polymer is rather rigid. The addition of other compounds known as plasticisers makes the plastic much softer, and increases its range of uses considerably.

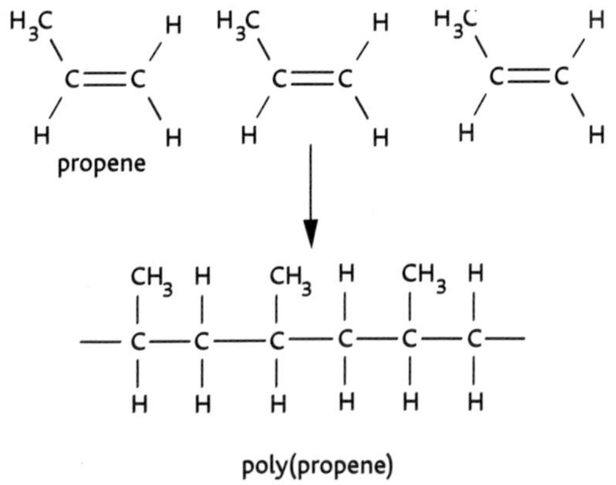

propene

poly(propene)

fig. 1.7.17 **The formation of poly(propene) from its monomer.**

The properties of polymers

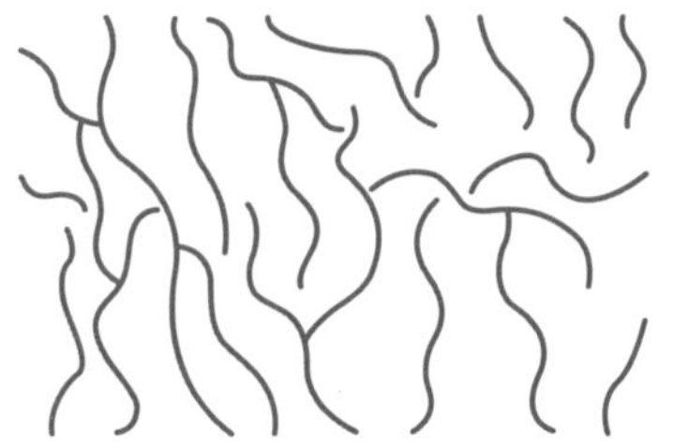

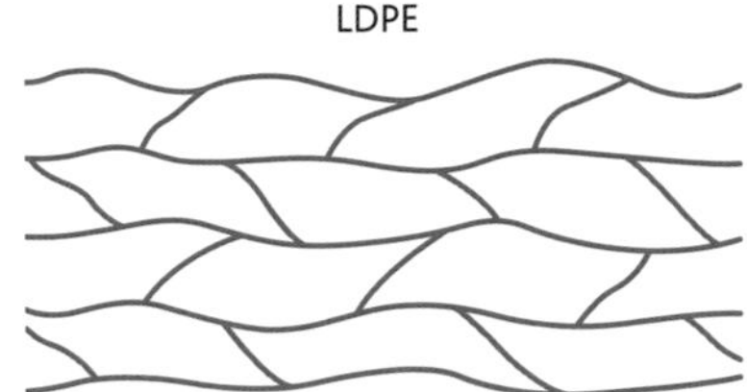

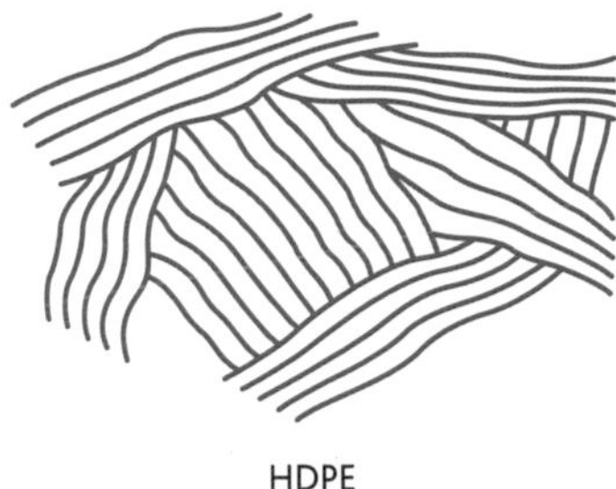

fig. 1.7.18 **The physical properties of a polymer are greatly influenced by the structure of the polymer chains and the forces between them.**

The properties of polymers depend on the chains of monomers from which they are made. They can vary greatly, from very soft and flexible with very low melting temperatures to hard, brittle materials with very high melting temperatures. What features of the polymer chain have most effect on the properties of the polymer?

- *The average length of the polymer chain* – tensile strength and melting temperature increase with the length of the chain until there are about 500 units in the chain. Increase in chain length beyond this has relatively little effect on the properties of a polymer.
- *Branching of the chain* – branched chains cannot pack together as regularly as straight chains, so polymers with highly branched chains tend to have low tensile strengths, low melting temperatures and low density.
- *The presence of intermolecular forces between chains* – these are of immense importance in natural polymers and also affect synthetics. If there are strong intermolecular forces between the chains the polymer will be strong and tend to have a high melting temperature. The strength of the intermolecular forces is largely determined by the side groups on a polymer chain.
- *Cross-links between chains* – these chemical bonds holding the chains together make a polymer very rigid, hard and brittle, usually with a very high melting temperature.

Synthetic polymers made from the reactions of the alkenes, their derivatives and other organic molecules are an excellent example of the way in which chemists can take chemicals from naturally occurring materials such as oil and coal and create novel materials of great usefulness to the human race. The double bonds of the alkenes and their tendency to undergo addition reactions means that they and their derivatives are frequently used in polymerisation reactions. Hundreds of alkene monomers react with each other to form long-chain polymers which have become a fundamental part of our everyday life. Polymer chemists can now develop 'designer polymers' with exactly the properties required for a particular task. Where would we be without plastics?

Questions

1 Why do alkenes often form the basis of the monomer units in polymers and plastics?

2 Show how addition polymerisation takes place using tetrafluoroethene as an example.

$$F_2C{=}CF_2 \quad F_2C{=}CF_2 \quad F_2C{=}CF_2$$

Polymer problems and solutions

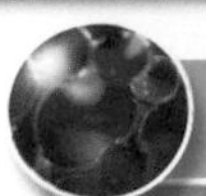

HSW The downside of polymer products – what can be done?

The development of synthetic polymers has led to a great change in the world, with many natural materials having been replaced by their synthetic equivalents. One possible consequence of this is that the development of plastics has saved much natural material from destruction. The ecological effects are not as straightforward as this, however. When looking at the energy balance of these useful products, a number of different factors have to be considered. The benefits, the problems and the possible solutions of the use of synthetic polymer products depend on science – and also the will of society!

The energy costs – polymers are produced from chemicals obtained from fossil fuels. These **hidden energy costs** of polymer products are very high, because large amounts of energy are used both in the production of the polymers and the manufacture of goods made from them. At the moment around 4% of world production of fossil fuels is used each year to generate the electricity used to make polymer products. That is a lot of energy!

The resources used – because many polymer products are made from alkenes from fossil fuels, they use up valuable resources. The limited supply and ever-rising costs of fossil fuels give a strong incentive to develop different polymers based on monomer units from different sources.

Disposal problems – synthetic polymers are not easy to dispose of. They are causing a substantial waste problem. When they burn a variety of toxic gases may be produced which can include hydrogen chloride (from halogen-containing polymers like PVC) and hydrogen cyanide, which are damaging both to human health and to other living organisms. Plastics themselves can also damage animals in the wild.

fig. 1.7.19 **Simply burning plastics often produces poisonous smoke, so it is not the straightforward solution to the disposal problem.**

The carbon footprint of plastics

A **carbon footprint** is defined as a measure of the impact that a particular human activity has on the environment in terms of the amount of greenhouse gases produced. Your carbon footprint is measured in units of carbon dioxide – and the objective is that we should all try to make our carbon footprint as small as possible. The use of fossil fuels in the production of plastics, both to generate energy and as raw materials, releases carbon. Burning waste plastics releases carbon dioxide and toxins into the atmosphere. So using plastics adds to your carbon footprint in a number of ways, particularly if you throw away plastic items rather than recycling or reusing them.

The boom in the development of synthetic polymers in the last 50 years has improved our quality of life in many ways making products lighter, cheaper and often more durable. Plastics are an integral part of our everyday life, from sandwich wrappings to artificial valves working in someone's heart, and they have brought many benefits.

However, the disadvantages highlighted above need to be addressed. The quantities of non-biodegradable polymers around the Earth is growing all the time. But as the problems of plastic production, use and disposal have become more apparent, so some possible solutions are emerging.

Solutions to polymer problems

Renewable energy sources

Making polymer products uses fossil fuels and energy. Using renewable energy sources in their manufacture is therefore a good option. If electricity generated from nuclear, solar or wind power is used instead of fossil fuels, the environmental effect is greatly reduced. Another option is to generate heat by burning methane from biodigesters, and also to combine energy recovery (see p.140) with power generation.

Reducing use of polymer products

There is a move to avoid using too many polymer products, eg to use fabric or paper bags instead of plastic ones in supermarkets. This is a good idea, although resources are used in the production of all bags.

Reusing polymer products is an effective way of reducing their energy costs. For years many plastic products have been disposable, single-use only items. By developing polymer products designed to be used for a much longer time, the energy costs, carbon footprint and resources used can be greatly reduced. Supermarket bags and packing crates that can be reused for years are just two obvious examples. Reusing polymer products also reduces the amount of waste going into landfill sites. On an individual level people can also reuse plastic bags and bottles, and reduce their use of single-use material such as plastic food wrapping.

fig. 1.7.20 **Multiplied by the millions of people who shop at supermarkets around the world, a change from single-use to multiple-use bags can have a real effect on both the energy demands and the resources needed to carry shopping home.**

Recycling

Many plastics are very difficult to dispose of in a way that does not damage the environment. Plastic articles can be dumped on rubbish tips along with other household waste, but many synthetic polymers are not broken down by living organisms in the way that natural compounds are – they are not **biodegradable**. Recycling not only reduces the need for plastics disposal – it also reduces the demand for the production of new plastics.

Recycling thermoplastics

So far only thermoplastics (plastics that soften on heating) can be recycled. This includes much of the waste plastic produced in manufacturing industries – this waste is clean and often does not need to be transported far for processing. The resulting recycled polymers can often be used as if they were new and so the process is often referred to as reprocessing rather than recycling. About 250 000 tonnes of plastic is reprocessed like this each year in the UK.

However, to most people recycling means the processing of used domestic plastic items. In many countries there are now collection points for milk bottles and other drinks bottles made of a plastic called PET [poly(ethyleneterephthalate)]. When recycled this plastic can be used to make fibres in the manufacture of carpets and textiles, and new uses as a waterproofing material are being developed. In the US almost a quarter of all plastic bottles are now recycled – that means almost a quarter of a million tonnes of plastic reused each year! In the UK and across Europe similar recycling of plastic bottles is taking place (see fig. 1.7.21). What is more, many car manufacturers now label plastic components in their vehicles to make reclamation and recycling easier when the car is eventually scrapped.

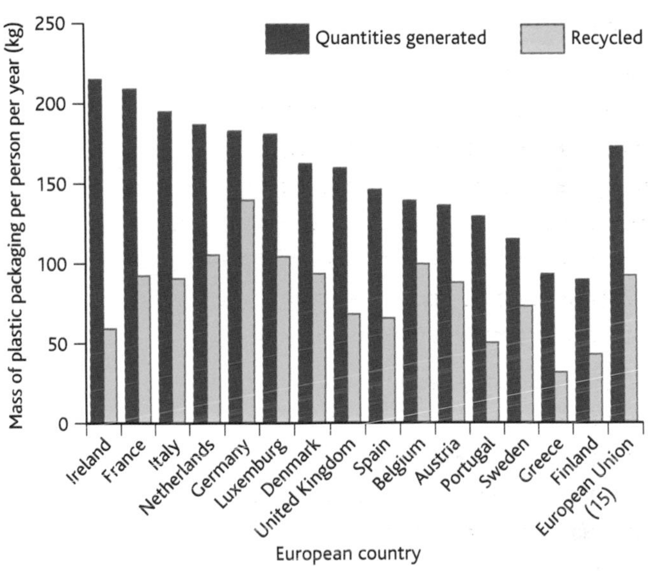

fig. 1.7.21 **Levels of recycling of plastic packaging vary considerably from country to country. Improving this across Europe would greatly reduce the carbon footprint of polymer product use.**

How effective is recycling?

If the plastic that is recovered after recycling is as good as the original, then effectively all the hidden energy from its original production has been saved apart from the energy used in recycling. If the quality of the recycled plastic is less good than the original it cannot be used in the same way the second time around so some of the hidden energy is lost.

When considering the benefits of recycling you need to take into account all the environmental and energy costs. The transport cost of individuals taking the waste plastic to be recycled is reduced when there are kerbside collections or collecting points at supermarkets. Transporting the plastics to the processing plants, the electricity used to break up or melt the plastic and the transport costs of the new goods all have to be taken into account.

A report on the production of carrier bags made from recycled rather than virgin polythene concluded that the use of recycled plastic gave the following environmental benefits:

- **reduction of energy consumption by two-thirds**
- **production of only a third of the sulfur dioxide and half of the dinitrogen oxide**
- **reduction of water usage by nearly 90%**
- **reduction of carbon dioxide generation by two-and-a-half times.**

A different study on the same subject concluded that 1.8 tonnes of oil are saved for every tonne of recycled polythene produced.

fig. 1.7.22 How effective is recycling?

How are plastics recycled?

There are two main ways in which the recycling of plastics can be carried out.

Mechanical recycling involves physical changes to the plastics – they may be melted, shredded or turned into granules before reuse. The plastics have to be sorted carefully before processing. This is usually done by hand, an expensive process in the UK. In fact much plastic waste is shipped to countries such as China where the costs are much cheaper for sorting and processing. The energy costs and environmental impact of the transport have to be taken into account.

People are working hard to develop technologies to allow plastics to be sorted efficiently, cheaply and automatically. At the moment nothing has emerged that is both effective and cheap – and of course technology usually has an energy and environmental cost.

Chemical recycling is also known as **feedstock recycling**. It involves the breakdown of the waste plastic into monomer units which can then be used again in the formation of new polymers or in other reactions in the chemical industry. Scientists are working on a number of ways of doing this which include burning the plastics, thermal cracking and gasification (see p.140 under Energy recovery). The plastics do not need to be so carefully separated as for mechanical recycling processes, and there is more tolerance of impurities.

Chemical recycling is expensive and can require a significant input of energy. It has been estimated that a minimum of 50 000 tonnes of waste plastic a year needs to be processed to be economically viable – but it can be done.

fig. 1.7.23 LINPAC is a global company which produces packaging and also specialises in recycling a variety of polymers. They produce recycled polymers which are comparable in quality to original unused (virgin) polymers. This benefits LINPAC, their customers and the environment.

Biodegradable polymers

Another growing solution to the plastics dilemma is the development of renewable resources known as biopolymers, bioplastics or 'green plastics'. These are polymers which are made by modifying natural polymers such as starch or cellulose. Most biopolymers are derived from plants such as maize or sugar cane. This is a sustainable, renewable source which uses carbon dioxide from the atmosphere during its growing time, reducing the carbon footprint of the subsequent plastic. Bacterial fermentation often plays an important part in breaking down the plant material to produce the monomer units, another low energy, low carbon-emission aspect of the process. What is more, most polymers made from renewable sources can also be attacked and broken down by microbes – they are **biodegradable**. The degradation may take 20–30 years, but eventually the product will be destroyed. The production of biopolymers has many positive points. However, land used to grow plants to make polymers is not being used to grow food. When people are still dying of starvation in the world, should we be using land to grow plastics?

fig. 1.7.24 plastic rubbish turns up everywhere. But the biodegradable plastics we can make from plants also raise ethical issues.

HSW Will *you* use biodegradables?

Two important factors will determine how successful renewable biodegradable plastics become. One is cost – as all the technologies are still very new, they tend to be more expensive than conventional plastics. However, as demand grows the price should fall. The other factor is the performance of the biopolymers. However keen people are to use materials that have a minimal impact on the environment, they will not want lower standards of performance.

Tables 1.7.3 and 1.7.4 compare some data for the biodegradable plastic poly(lactic acid) with the traditional non-biodegradables poly(styrene) and PVC.

Property	Hard poly(lactic acid) sheet	Poly(phenylethene) (polystyrene)	Poly(chloroethene) (PVC)
Thickness (μm)	250	250	250
Tensile strength (MPa)	59	84	50
Elastic modulus (GPa)	2.5	2.8	1.8
Elongation (%)	6	4	160
Tear strength (gf)	110	90	850
Folding endurance strength (times)	560	10	1000

table 1.7.3 Characteristics of hard poly(lactic acid) sheet (PLAS) compared with poly(styrene) and PVC.

Property	Soft poly(lactic acid) sheet	Poly(phenylethene) (polystyrene)	Poly(chloroethene) (PVC)
Thickness (μm)	35	35	35
Tensile strength (MPa)	35	27	14
Elastic modulus (GPa)	1.3	1.0	0.2
Elongation (%)	450	430	340
Tear strength (gf)	20	20	400

table 1.7.4 Characteristics of soft poly(lactic acid) film (PLAF) compared with poly(styrene) and PVC.

Energy recovery

Another relatively new approach to reducing the energy and carbon footprint of polymer products is the concept of **energy recovery** – recovering some of the energy that was put into producing the polymer. One example is burning waste polymer products, eg municipal waste which contains very high levels of plastics and other polymers, and using the energy released to generate electricity. This reduces the use of fossil fuels to generate electricity.

In the process of energy recovery, incinerators are used which operate at very high temperatures to prevent toxin production. These incinerators have special pollution-control systems which make sure that any pollutants such as dioxins are not released into the atmosphere. Energy recovery incinerators are used in the United States, where it has been calculated that one family burning waste in their back garden will produce as much dioxin in a year as a local waste plant burning the rubbish from 50 000 homes!

Polymers contain a great deal of stored energy so they are a high-energy fuel source. The heat produced in the incinerators can be used to drive the generation of electricity and to produce hot water for heating in combined heat and power systems.

There are a number of other ways in which the energy from polymer products can be recovered and reused. These include **pyrolysis** and **gasification**. Both of these processes convert polymer wastes into energy-rich fuels which can be used for electricity generation and as feedstock for the chemical industry.

In pyrolysis, the polymers are broken down by heat in the absence of oxygen. The formation of charcoal from wood is an example which has been used for centuries – now more modern processes are being developed to deal with polymers.

Gasification involves the breakdown of solid hydrocarbons in a limited oxygen atmosphere to make **syngas**. This is mainly a mixture of hydrogen and carbon monoxide, which can be used as feedstock in a wide range of different chemical processes. So far, work on gasification has been based on low-density poly(ethene) film and a mixture of different rigid plastics which are very common in domestic waste.

Many of these new processes are not very efficient at the moment, and they are also often seen as relatively expensive. There are about 100 sites worldwide and they process around 4 million tonnes of waste per year. However, as the need for energy recovery and more sustainable use of resources increases, and the demand to lower the carbon footprint of our lifestyles increases, processes such as these will become more and more important in the way we manage our waste polymer products.

Life cycle analysis

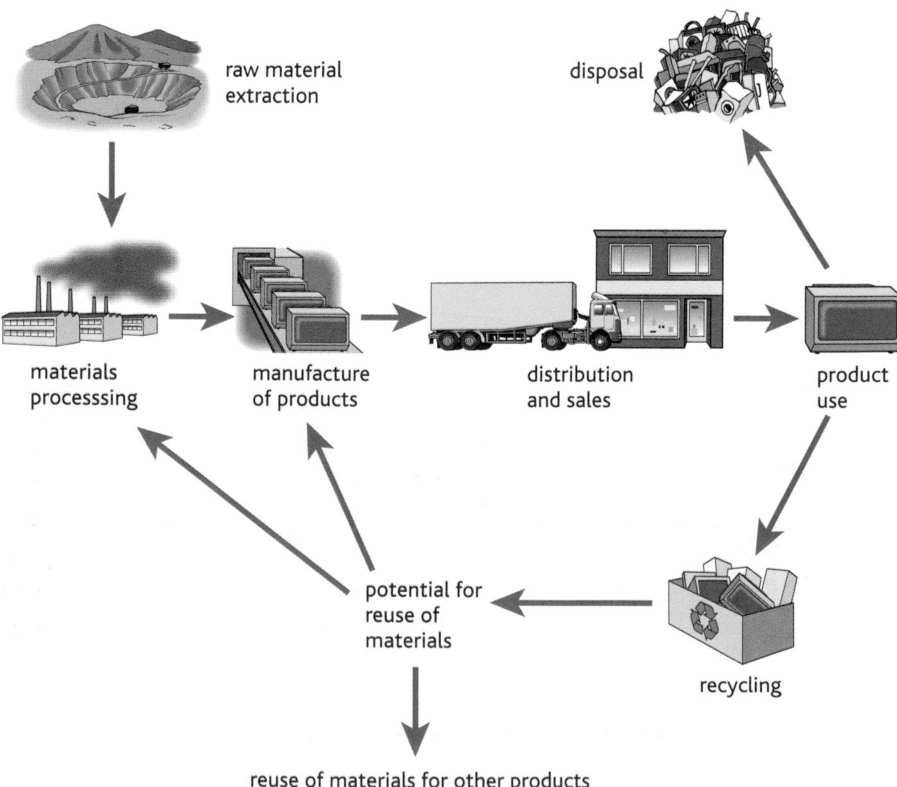

fig. 1.2.25 A life cycle analysis looks at all these facets of the life of a polymer product to quantify its impact on the environment.

Increasingly industry is using a **life cycle analysis** of polymer products to quantify the effect they have on the environment. The product's life cycle is examined from the extraction of the raw materials used in its manufacture to the costs of recycling, reusing or disposing of it at the end of its useful life. An inventory is made of the energy and materials used and any environmental emissions. This type of analysis makes it possible to identify where improvements can be made in a process to reduce the environmental impact, and also to compare two different polymer products and decide which to use on the basis of its life cycle impact on the environment.

For example, table 1.7.5 shows some life cycle analyses carried out in the Netherlands on different materials for making gas pipes. The researchers themselves were surprised at the results! As a result of this study, the Dutch government recommends that PVC is used in gas supply pipes. Analyses like these will play an ever-increasing role in our choice of materials and how they are manufactured and disposed of.

Environmental impact	Material considered for gas supply pipes		
	Cast iron	PE80 (polythene)	PVC
Photochemical oxidant creation (POCP)	216	216	215
Aquatic ecotoxicity (ECA)	26	6.4	3.9
Global warming potential (GWP)	31	22	21
Acidification potential (AP)	13	2.5	1.1
Abiotic depletion potential (ADP)	9.7	1	0.45
Human toxicity (HT)	14	1.9	0.76
Nutrification potential (NP)	2.5	0.61	0.33
Odour threshold limit (OTL)	11	0.39	0.48
Ozone depletion potential (ODP)	1.25	0.17	0.07
Energy content (EC in GJ)	748	303	139

table 1.7.5 **Comparison of different materials used in gas supply pipes.**

Questions

1 What are the main environmental problems which result from our increasing use of polymer products?

2 What is the difference between reusing and recycling polymer products?

3 How do biopolymers differ from compounds such as poly(ethene) and PVC?

4 Use tables 1.7.3 and 1.7.4 to compare the hard and soft types of poly(lactic acid) with poly(styrene) and PVC, and produce graphs to show the tensile strength, the elastic modulus (stretchiness) and the tear strength of the polymers as both hard sheets and soft films. How does the PLA compare with conventional polymers?

5 What is meant by the term 'energy recovery' and how can it contribute to the more sustainable use of resources?

6 For each method of reducing polymer problems outlined in the text, summarise its effect on the carbon footprint of polymer use.

7 **a** Make a bar chart to compare the environmental features shown in table 1.7.5 for the three materials.

b Suggest why the results may have surprised the scientists who carried out the work.

Examzone

You are now ready to try the second Examzone test for unit 1 (Examzone Unit 1 Test 2) on page 250. This will test you on what you have learnt in Unit 1, chapters 5 to 7.

Unit 2 Application of core principles of chemistry

In unit 2 you are going to develop your knowledge of the core principles of chemistry by extending the scope of your studies. You are going to uncover more of the secrets of the atom as you learn how chemical reactions take place. You will also be investigating the techniques of analysis used by chemists and finally, just as real chemists do, you will apply your chemical knowledge to 'green issues'.

Chemical ideas

Many of the ideas you met in unit 1 will be developed further. For example, you will be looking at the bonding between the extremes of covalent and ionic bonding and the nature of the intermediate forces between molecules. The nature of the bonding and intermediate forces has an effect on the properties of substances and in this unit you will be finding out just what these effects are.

You will meet two further groups of the periodic table (groups 2 and 7). This will give you further insight into the nature of the periodic table and how it enables you to predict the properties of the elements and their compounds.

You will have met oxidation and reduction reactions at GCSE. Now you are going to extend those ideas by considering oxidation and reduction in terms of electron transfer. In organic chemistry you will be looking at two more families – the alcohols and the halogenoalkanes. In addition you will develop a deeper understanding of the way in which different organic reactions take place by considering models of possible reaction mechanisms.

In unit 2 you are going to look at rates of reaction and the direction and extent of reactions in a largely qualitative way, discussing ideas and theories but leaving a more quantitative treatment to A2. Throughout the unit you will develop and use chemical equations including molecular and ionic equations. You will also use them to calculate quantities of reactants and products.

How chemists work

All through the unit you will be devising and using simple models of molecules and ions. You will start to use electron-pair repulsion theory and develop an understanding of how the Maxwell-Bolzmann distribution and collision theory explain observations in reaction kinetics. You are also going to look at the way chemists use experimental evidence to support a model or to modify it. For example, chemists have developed a model of the water molecule as a bent molecule rather than a linear one. They support this model by using practical investigations of dipole moments.

Chemistry in action

Chemists frequently use a variety of analysis techniques. You will be introduced to the use of volumetric analysis and spectral analysis. These methods of analysis can be used to identify chemicals in a sample or to measure the quantities of chemicals present, and you will be considering how accurate they are.

Finally there are many 'green issues' which are causing considerable concern at the moment including the greenhouse effect, depletion of the ozone layer, reducing waste, avoiding pollution and the need to develop safe and efficient manufacturing processes. Chemists have a major role to play in finding solutions to all of these problems.

The shape of a molecule will have an influence on its physical and chemical properties. In chapter 2.1, you will look at how nanotube molecules such as this may play an important part in the future of technology.

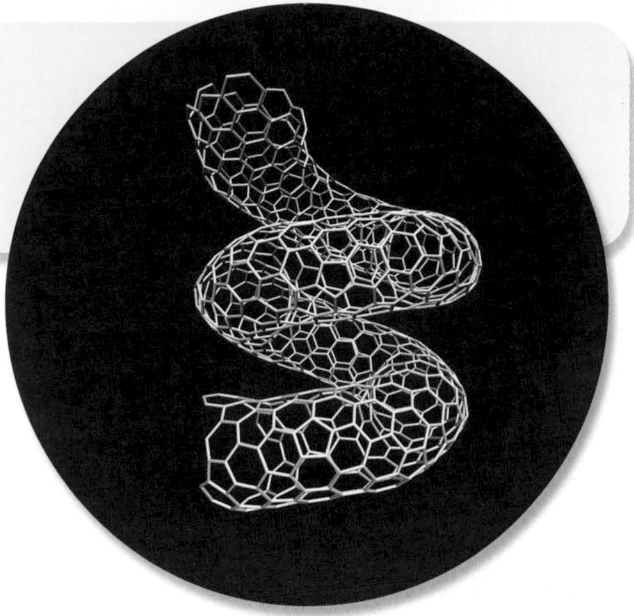

In Chapter 2.4 you will be looking at oxidation and reduction reactions (called redox reactions). Fuel cells, such as this one from a spacecraft, use a redox process to produce electricity.

In Chapter 2.8 you will study the improvements to surgery that have taken place in the past two hundred years. The primitive methods shown in the picture are history due to the developments of halogenoalkanes as anaesthetics.

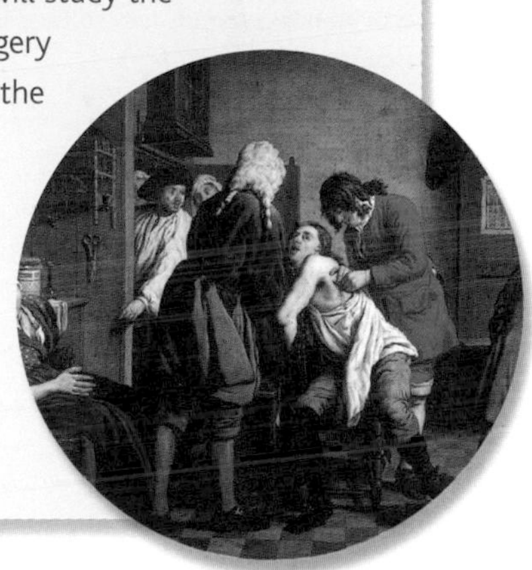

But halogenoalkanes can cause problems as you will see in this unit. The dramatic holes produced in the ozone layer by certain halogenoalkanes increase the likelihood of skin cancer and cataracts on the Earth.

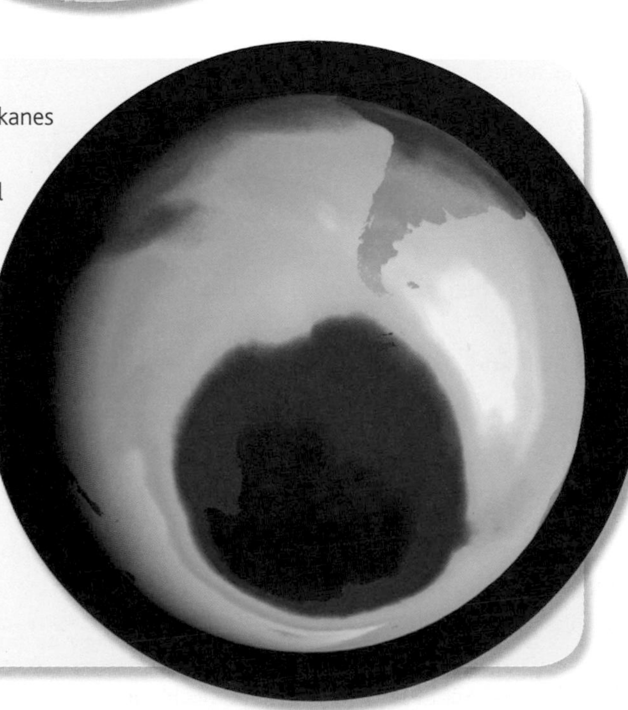

The contrails (frozen water trails) produced by jet aircraft can increase the greenhouse effect because water vapour is an effective greenhouse gas. In the final chapter of the book you will explore some of the many ways that chemists are working to combat climate change and reduce our carbon footprint.

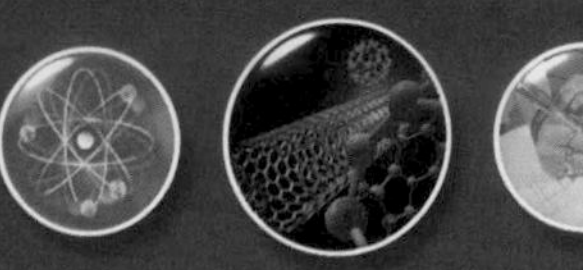

1 Shapes of molecules and ions

fig. 2.1.1 Spraying organophosphates protects crops from insect damage and helps provide more food for the world population. But the molecules are the right shape to interfere with human enzymes too, so they have to be used carefully.

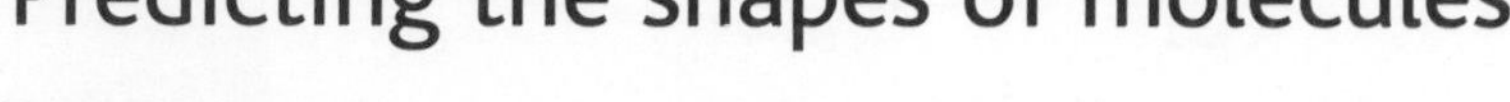

Predicting the shapes of molecules

You have already seen in chapters 1.3 and 1.4 that the type of bonding in a compound is central in determining its properties. What is more, covalent compounds form molecules and ions whose bonds are highly directional, giving them a fixed shape which in turn also determines the physical and chemical properties of the compound. Nowhere is molecular shape more important than in the structure of enzymes, substances which control the rates of chemical reactions in biological systems. For example organophosphates are commonly used insecticides. The molecules are exactly the right shape to interfere with an enzyme that controls nerve impulses in an insect's body, so killing it. The shapes of molecules and ions are important when considering not only chemical structures but also reaction mechanisms.

Shapes of ionic compounds

Although both ionic and covalent bonding depend on electrostatic attraction to hold particles together, ionic bonding is very different from covalent bonding in that it is non-directional. The ions in an ionic compound attract ions with the opposite charge no matter what direction they are in – and, for that matter, repel ions with the same charge in just the same way. Because of this, the structure of ionic compounds is simply the arrangement of ions in a lattice which maximises the attractive forces between oppositely charged ions and minimises the repulsion between similarly charged ions. You saw an example of this arrangement in the model of the sodium chloride ionic lattice (see **fig. 1.4.6** in unit 1).

Shapes of covalent compounds

In contrast to the ionic bond, covalent bonds are highly directional. This leads to molecules with more than two atoms having a very definite shape, in which the three-dimensional relationship between the atoms is constant. This spatial relationship is important because it governs the chemical and physical properties of molecules or ions.

You saw in chapter 1.4 how **electron-pair repulsion theory** can be used to interpret and predict the shapes of molecules (and also ions that contain more than one atom). This is based on the ideas that:

- the shape of a molecule or ion depends on the number of electron pairs around the central atom
- electron pairs repel one another so they stay as far apart as possible.

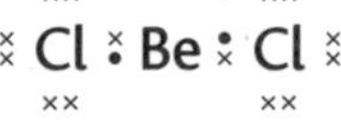

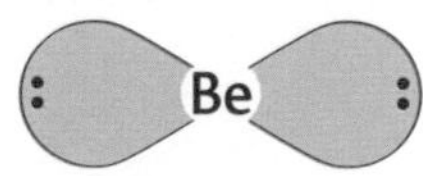

Cl—Be—Cl

fig. 2.1.2 The dot and cross diagram for beryllium chloride allows you to predict the shape of the molecule.

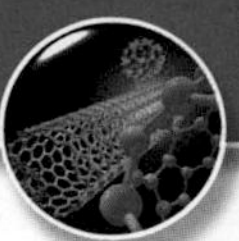

The simplest example of such repulsion occurs in the molecule beryllium chloride, $BeCl_2$, which has the electronic structure shown in **fig. 2.1.2**.

In the $BeCl_2$ molecule, beryllium has only two pairs of electrons around it (it is one of the exceptions to the octet rule you met earlier). To minimise the repulsion between these pairs, they must be arranged so that they are on opposite sides of the beryllium atom. This gives the $BeCl_2$ molecule the shape shown in **fig. 2.1.2** – it is a **linear** molecule. The greatest number of electron pairs that usually occur around a central atom is six. **Table 2.1.1** shows the shapes of molecules with three to six pairs of bonding electrons that this simple model predicts, and also shows the predicted **bond angles** for each arrangement. These shapes are typical and if you know the electronic structure of a compound, you can predict its molecular shape using this table.

Using this electron-pair repulsion theory you can predict the structure of molecules. For example, boron trifluoride, BF_3 has three pairs of electrons around the gallium atom, giving the molecule the triangular, or trigonal planar, shape shown in **table 2.1.1**.

HSW Seven bonding pairs

How can you develop the model of electron-pair repulsion theory to predict the shape of a molecule that has more than six pairs of electrons around the central atom? One example is the compound formed between iodine and fluorine, iodine heptafluoride, IF_7. Here there are seven bonding pairs around the central iodine atom. Electron-pair repulsion theory predicts that five of these bonding pairs are in the same plane. The angle between these bonds is 360/5 = 72°. There is one pair above this plane and the other below, so forming two bonds at right angles to the plane. The bond angles here are 90°. The shape is a **pentagonal bipyramid** (see **fig. 2.1.3**).

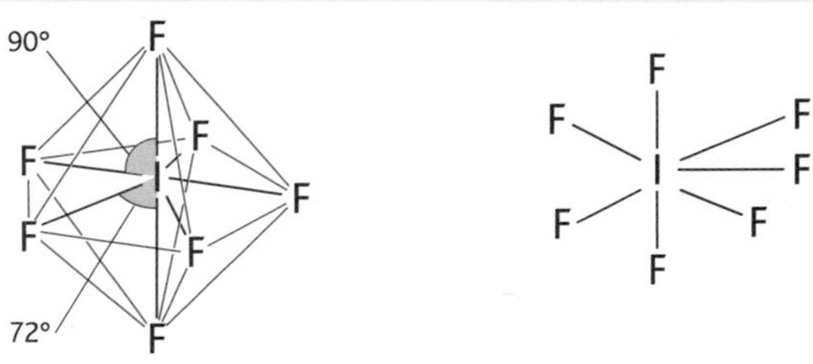

fig. 2.1.3 The shape of the IF_7 molecule can be predicted from electron-pair repulsion theory.

	Number of electron pairs around central atom			
	3	4	5	6
Example	Boron trifluoride, BF_3	Methane, CH_4	Phosphorus(V) chloride, PCl_5	Sulfur(VI) fluoride, SF_6
Dot and cross diagram	B	C	P	S
Shape and bond angle	bond angle = 120°	bond angle = 109.5°	bond angle = 90° and 120°	bond angle = 90°
Name of shape	Trigonal planar	Tetrahedral – a triangular-based pyramid	Trigonal bipyramidal – two tetrahedrons with bases joined	Octahedral – two square-based pyramids with bases joined

table 2.1.1 The shapes of molecules can be predicted from the number of electrons around the central atom. They are based on minimising the repulsion between the electron pairs.

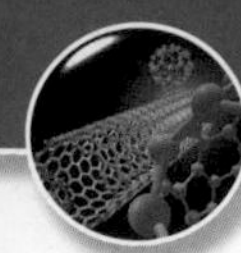

How do lone pairs affect the shapes of molecules?

Not all molecules contain a central atom with electron pairs in bonds around it. For example, ammonia can be represented as shown in **fig. 2.1.4(a)**. The central nitrogen atom has four electron pairs around it, only three of which make up the N—H bonds – the fourth is a non-bonding pair or **lone pair**. As a result, the electron pairs take up a shape which is tetrahedral (see **fig. 2.1.4(b)**), and the shape of the molecule (that is, the nitrogen atom and the three hydrogen atoms) is trigonal **pyramidal** (see **fig. 2.1.4(c)**).

Careful measurements of the H—N—H bond angles in ammonia shows that they are 107°. This is 2.5° less than the predicted 109.5° if the electron pairs were arranged around the nitrogen atom in a tetrahedral shape, all repelling each other equally (as they do in methane). The reason for this difference is that the electron pair in a bond is further from the nucleus of the central atom than the electron pair in a lone pair. This means that the repulsion between a lone pair and a bonding pair is greater than the repulsion between two bonding pairs. This extra repulsion 'squeezes' the bonding pairs in ammonia closer together, accounting for the smaller than expected bond angle.

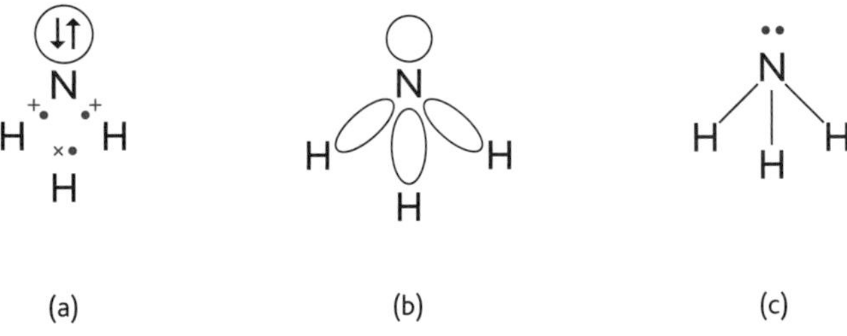

fig. 2.1.4 The electronic arrangement and shape of the ammonia molecule.

As you saw in chapter 1.4, in a water molecule there are two bonding pairs and two lone pairs of electrons around the central oxygen atom. The four pairs of electrons are approximately tetrahedral (see **fig. 2.1.5**). The result is a **V-shaped** molecule.

The H—O—H bond angle is 104.5°, compared with approximately 109.5° in methane or 107° in ammonia. As before, the two non-bonding pairs of electrons give extra repulsion.

fig. 2.1.5 The electronic arrangement and shape of the water molecule.

HSW Xenon tetrafluoride

As you saw in chapter 1.3, chemists originally accepted the idea that noble gas elements do not form compounds. The completed electron shell in a noble gas atom gives the atom stability. In 1933 however, Linus Pauling, about whom you will find out more later in this unit, predicted that heavier noble gases might be able to form compounds. In 1962 Neil Bartlett produced a complex xenon compound which led other chemists to try to make compounds. Later in 1962 Howard Claassen found that xenon would combine with fluorine at high temperatures to form xenon tetrafluoride, XeF_4. This molecule is not tetrahedral in shape like methane, but planar. How can electron-pair repulsion theory explain this?

The central xenon atom forms four covalent bonds and also has two lone pairs of electrons. The four covalent bonds lie at the four corners of a square with the xenon atom in the centre. One pair of non-bonding electrons is above the plane at right angles, and one below (**see fig. 2.1.6**). All bond angles are 90°.

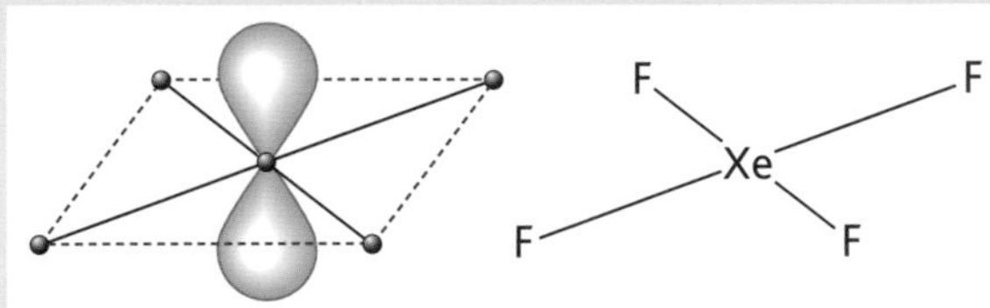

fig. 2.1.6 Xenon tetrafluoride is a planar molecule.

Other noble gas compounds have since been made, eg krypton(IV) fluoride, KrF_4 and xenon(II) fluoride, XeF_2. No compounds have been made with helium or neon.

Multiple bonds

For molecules containing double or triple bonds, the electron pairs making up the bond all exist together between the two atoms they join. A multiple bond can therefore be treated in exactly the same way as a single bond when predicting molecule shapes. **Figure 2.1.7** shows how this leads to a linear shape for carbon dioxide, CO_2 but a V shape for sulfur dioxide, SO_2.

(a)

O=C=O

(b)

fig. 2.1.7 The electronic arrangement and shape of (a) the carbon dioxide molecule, (b) the sulfur dioxide molecule.

The shapes of ions

A molecule that gains or loses one or more electrons forms an ion, and the shape of the ion can also be predicted by electron-pair repulsion theory. For example, an ammonium ion is formed when an ammonia molecule and a hydrogen ion combine. **Figure 2.1.8(a)** shows a dot and cross diagram for the reaction. The ammonium ion contains a central nitrogen atom with four bonding pairs of electrons around it. This is similar to the structure of methane – the ammonium ion is tetrahedral with bond angles of 109.5.

(a)

(b)

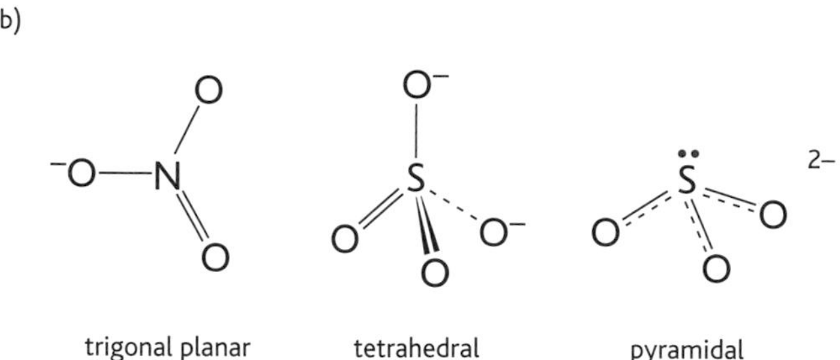

fig. 2.1.8 (a) The tetrahedral ammonium ion. (b) The nitrate(V) ion, the sulfate(VI) ion and the sulfate(IV) ion.

Figure 2.1.8(b) shows the structure of the nitrate(V) ion, the sulfate(VI) ion (sulfate) and the sulfate(IV) ion (sulfite). These are trigonal planar, tetrahedral and pyramidal respectively

Worked example

For each of the following shapes give the formula of a molecule that has the shape: **a** linear **b** trigonal planar **c** tetrahedral **d** octahedral.

Note that the question asks for a molecule, not an ion. The linear molecule must have more than two atoms – HCl would not be acceptable.

There are many examples that could be chosen, eg

a $BeCl_2$ **b** BCl_3 **c** CH_4 **d** SF_6

Aluminium forms an anionic complex with hydrogen with the formula AlH_4^-. Use electron-pair repulsion theory to predict the shape of this ion. Explain your answer. What is the approximate H—Al—H bond angle?

Figure 2.1.9 shows the structure of the AlH_4^- ion. There are four bonding pairs and no non-bonding pairs. For minimum repulsion between electron pairs the shape is **tetrahedral** with bond angle 109.5°.

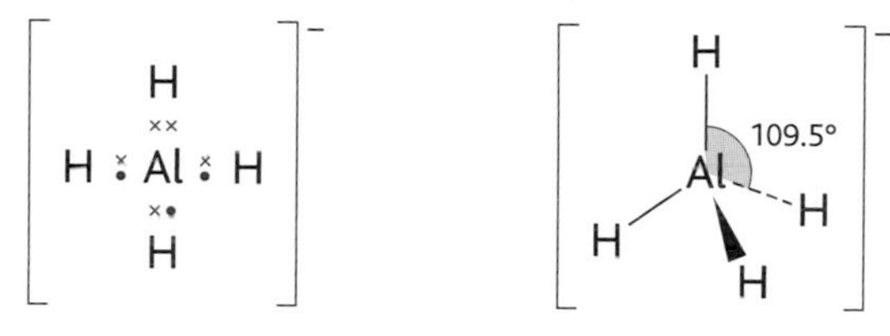

fig. 2.1.9 A dot and cross diagram shows the electron pairs around the central atom, allowing you to work out the shape of the AlH_4^- ion.

Questions

1 What shape would you expect a tetrachloromethane molecule, CCl_4 to have? Explain your answer.

2 How many pairs of electrons are there around the central S atom in a molecule of hydrogen sulfide, H_2S? What shape would you expect this molecule to have?

3 Sulfur trioxide, SO_3 is a planar molecule. Explain how electron-pair repulsion theory predicts this shape.

4 Suggest a shape for a molecule of oxygen difluoride, OF_2. Explain your choice.

5 Explain why ICl_4^- is not tetrahedral.

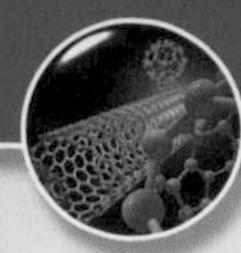

Shapes of simple organic molecules

Alkanes

You have seen that the simplest alkane molecule, methane, has a tetrahedral structure. In an alkane, all the carbon atoms have a tetrahedral arrangement around them.

Figure 2.1.10(a) shows space-filling models of the first three alkanes – methane, ethane and propane. You can see that in each molecule the arrangement around each blue carbon atom is tetrahedral. When drawing molecules to show their three-dimensional shape, the convention shown in **fig. 2.1.10(b)** is used.

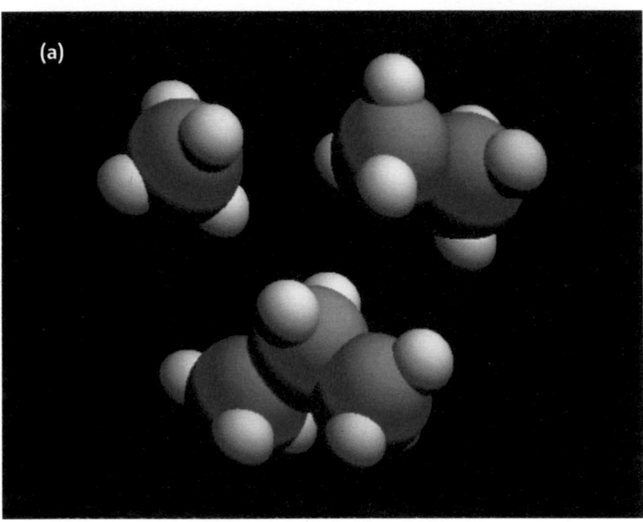

(b)

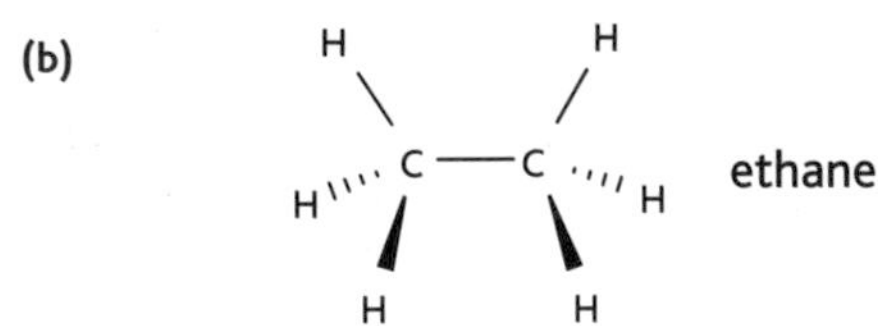

fig. 2.1.10 (a) Space-filling models of methane, ethane and propane. (b) In structural formulae in three dimensions, straight lines show bonds in the plane of the paper. Dashed lines are bonds sticking into the plane of the paper, and wedges show the bonds sticking out of the plane of the paper.

Alkenes

The simplest alkene is ethene, C_2H_4. The two carbon atoms are joined by a carbon–carbon double bond. Around each carbon atom there are three bonds all in the same plane, at an angle of about 120° (see **fig. 2.1.11**).

fig. 2.1.11 The ethene molecule is all in one plane.

Alcohols

The simplest alcohol is methanol. **Figure 2.1.12(a)** shows its structure – there are four bonds around the carbon atom and so the H—C—H bond angle is 109.5° (tetrahedral). Around the oxygen atom there are two pairs of bonding electrons and two non-bonding pairs, so the C—O—H bond angle is about 104°, similar to the angle in the water molecule.

(a)

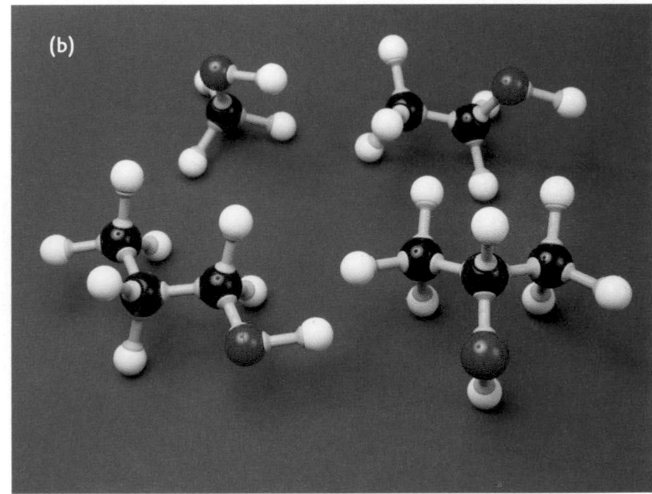

fig. 2.1.12 (a) A methanol molecule. (b) Ball and stick models of four alcohols.

Figure 2.1.12(b) shows ball and stick models of methanol, ethanol, propan-1-ol and propan-2-ol. You can see that the arrangement around the carbon atoms is tetrahedral but the C—O—H atoms are in the same plane in each case.

Carboxylic acids

Methanoic acid is the only carboxylic acid which is planar (see **fig. 2.1.13(a)**). **Figure 2.1.13(b)** shows a computer-generated model of ethanoic acid. In all carboxylic acids the –COOH (carboxyl group) is planar, with the rest of the molecule being based on tetrahedral carbon atoms.

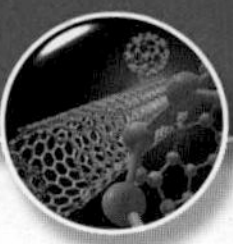

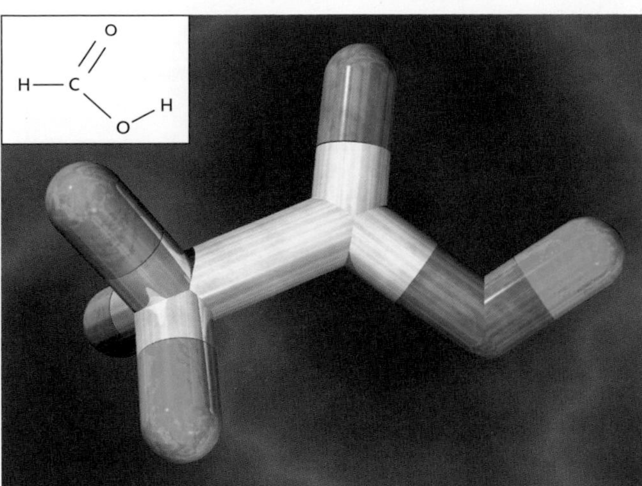

fig. 2.1.13 (a) A diagram of methanoic acid. (b) A computer-generated model of ethanoic acid. Carbon is shown in white, hydrogen in green and oxygen in blue.

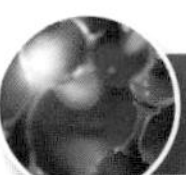

HSW Using molecular models

Chemists frequently use models of molecules to help explain the physical properties of substances and how they react. On these pages you will see three common types of molecular model that you may meet. They all have advantages and disadvantages.

- Space-filling models show the shape of the molecule and which atoms are combined, but they do not give a clear indication of the bonding between atoms.
- Ball-and-stick models are easy to construct and clearly show which atoms are which, the bonds between the atoms and an indication of the relative sizes of the atoms.
- Computer-generated models are easy to produce and easy to rotate on screen, but they do not always give any indication of the relative sizes of the atoms.

Halogenoalkanes

Bromomethane, CH_3Br is a simple halogenoalkane. Like methane it has a central carbon atom with four bonding pairs of electrons. The shape of the molecule is tetrahedral (see **fig. 2.1.14**).

fig. 2.1.14 Halogenoalkanes have similar shapes to alkane molecules.

Aldehydes and ketones

Aldehydes and ketones are two homologous series that both contain a carbonyl group, as you saw in chapter 1.5. Propanal is a typical aldehyde, and propanone the simplest ketone. In both molecules the central carbon atom has three groups bonded to it and this part of the molecule is therefore planar with bond angles of about 120° (see **fig. 2.1.15**).

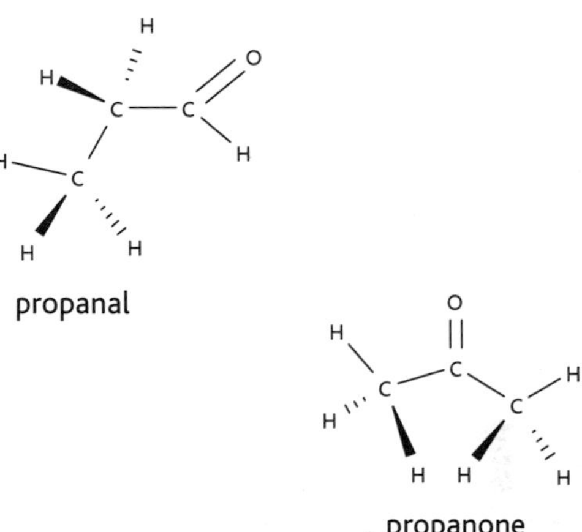

fig. 2.1.15 Structures of propanal and propanone.

Questions

1 Draw a three-dimensional diagram to represent the structure of propene, C_3H_6.

2 Here is a diagram of the alkane octane:

```
    H   H   H   H   H   H   H   H
    |   |   |   |   |   |   |   |
H — C — C — C — C — C — C — C — C — H
    |   |   |   |   |   |   |   |
    H   H   H   H   H   H   H   H
```

It is useful in that it shows how the atoms are bonded together, but it suggests some incorrect information about the shape of an octane molecule. What does it suggest that is incorrect?

3 Write down the names of three common organic compounds that have a planar structure.

4 Methanal, HCHO is the simplest aldehyde. Draw the displayed formula of methanal. Describe the shape of methanal and bond angles in the molecule.

Carbon

The allotropes of carbon

Allotropes are different forms of the same element that exist in the same physical state. The differences in the allotropes are down to different ways of bonding between the atoms, and so different molecular shapes. For example, pure carbon has several allotropes that have different crystal forms. Two of these allotropes are diamond and graphite.

Diamond

Pure diamond is composed entirely of interlocking tetrahedral carbon atoms, each of which is covalently bonded to its four nearest neighbours (see **fig. 2.1.16**).

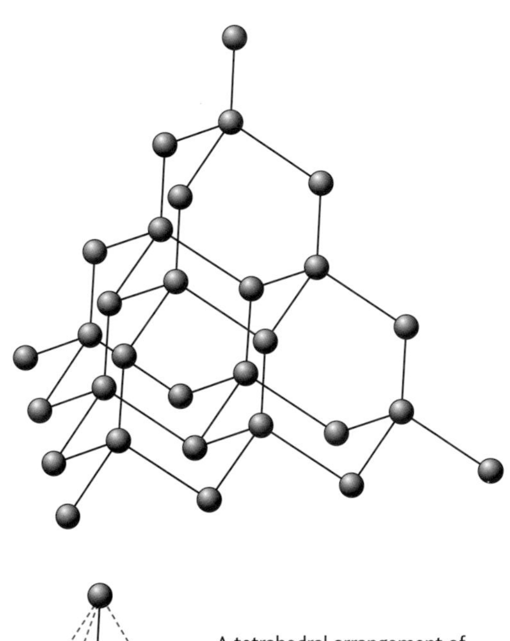

A tetrahedral arrangement of carbon atoms is repeated to give the structure of diamond.

fig. 2.1.16 The combined strength of the many carbon–carbon bonds in the structure of diamond give it both great hardness and a lack of chemical reactivity.

As you saw in chapter 1.4, diamond is uniformly bonded throughout and may be thought of as a giant molecule. The exceptional strength of the carbon–carbon bond and the covalently interlocked crystal structure accounts for the physical properties of diamond. It is the hardest substance known and most of its uses in drilling, cutting and grinding and as bearings are based on this exceptional hardness.

Although pure diamond is colourless and transparent, when contaminated with other minerals it may appear in various colours ranging from pastels to opaque black. The diamond crystal is chemically inert but may be induced to burn in air at high temperatures. It is a poor conductor of heat and an electrical insulator. Until 1955 the only source of diamond was natural deposits of volcanic origin. Since then diamonds have been made artificially from graphite subjected to high pressures and temperatures. Before the 1970s only diamonds of industrial quality could be made in this way – they had the hardness of diamond but did not have the attractive appearance and were not large enough to be used as gemstones. Diamonds of gem quality have now been produced artificially but they are so expensive to create that they are not made in this way for jewellery.

Graphite

Graphite is a black or grey, lustrous substance that easily crumbles or flakes. It has a slippery feel because of its tendency to cleave from the crystal in thin layers. It is chemically inert, although somewhat less so than diamond, and in contrast to diamond conducts both heat and electricity. It occurs as a mineral in nature, usually in a somewhat impure form, and can be produced artificially from amorphous carbon (see below).

Graphite is composed entirely of planes of trigonal carbon atoms joined in a honeycomb pattern. Each carbon atom is bonded to three others at 120°. These planes are arranged in sheets to form three-dimensional crystals. The layers are too far apart for covalent bonds to form between them. Instead weak forces (called **London** (**van der Waals**) **forces**, see chapter 2.3) maintain the arrangement of the layers, which are therefore easily moved relative to each other. Because each atom is formally bonded to only three neighbouring atoms, the remaining outer shell electron (one in each atom) is free to circulate within each plane of atoms, contributing to graphite's ability to conduct electricity (see **fig. 2.1.17**).

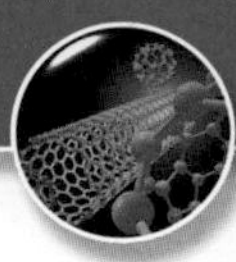

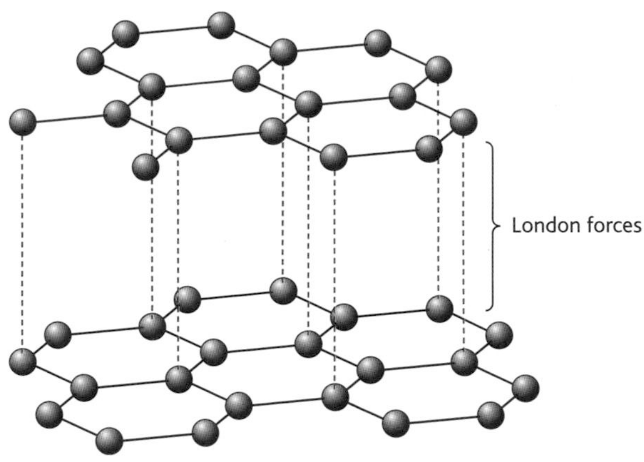

fig. 2.1.17 The structure of graphite. Whilst the bonds between the carbon atoms are strong, those between the layers are not and so they slide over each other easily.

One of the main uses for graphite – as a lubricant – results from the characteristic sliding of one layer over another within the crystal. The 'lead' in pencils is actually graphite and the mark it leaves on the page is the result of layers of graphite being scraped off the crystal. Graphite is also used as an electrical conductor and electrode material (in dry cells, for instance), and in nuclear reactors in order to absorb some of the energy of the neutrons, increasing their ability to cause fission.

Other forms of carbon

Amorphous carbon is less well defined than diamond or graphite. Its physical and chemical properties may vary depending on its method of manufacture and the conditions to which it is later subjected. It is a deep black powder that occurs in nature as a component of coal, and is also seen frequently as soot. It may be obtained artificially from almost any organic substance by heating the substance to very high temperatures in the absence of air. This is how coke is produced from coal, and charcoal from wood. Burning organic vapours with insufficient oxygen also produces amorphous forms such as soot.

Amorphous carbon is the most reactive form of carbon. It burns relatively easily in air, thereby serving as a fuel, and is attacked by strong oxidising agents. Amorphous carbon is not finely divided graphite but appears to have some of its structural features, such as local regions of sheets and layers. Its atomic structure, however, is much more irregular. The most important uses for carbon black are as a stabilising filler for rubber and plastics and as a black pigment in inks and paints. Charcoal and coke are used as clean-burning fuels. Certain types of 'activated charcoal' are useful as absorbents of gases and of impurities from solutions.

Fullerenes

For some years scientists predicted the existence of further allotropes of carbon. A football-shaped molecule was discovered in 1985 by Harry Kroto, Robert Curl and Richard Smalley. In this allotrope 60 carbon atoms are linked to form a more or less spherical molecule. A whole family of spherical allotropes, with differing numbers of atoms, is now known to exist. The first one had 60 carbon atoms and 32 sides (20 hexagons and 12 pentagons). It was nicknamed 'buckyball' and was then formally named **buckminsterfullerene** because it resembles the geodesic domes developed by an American inventor called Robert Buckminster Fuller. The group of spherical carbon molecules are called **fullerenes**.

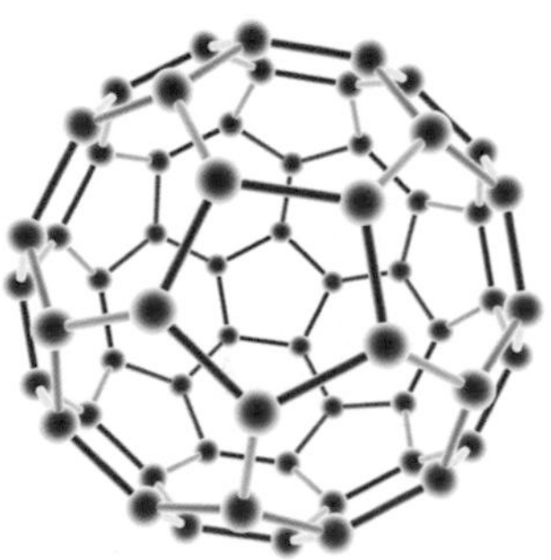

fig. 2.1.18 Harry Kroto (above), Robert Curl and Richard Smalley were awarded the Nobel Prize for chemistry in 1995 for discovering buckminsterfullerene and working out its structure.

Buckminsterfullerene is a black solid that dissolves in petrol to form a deep red solution. Like graphite, it has electrons that are free to move and so it can conduct electricity.

Nanochemistry

In 1991 the Japanese scientist Sumio Iijima reported that in the soot produced while he was making fullerenes he had discovered elongated cage-like structures. Other scientists replicated his work and this led to the development of **nanotubes** and nanotube technology. These closed-cage carbon structures all contain 12 five-membered rings and almost any number of six-membered rings. They make highly complex shapes – doughnut shapes, corkscrews and cones have been produced. **Figure 2.1.19** shows a single-walled corkscrew-shaped carbon nanotube.

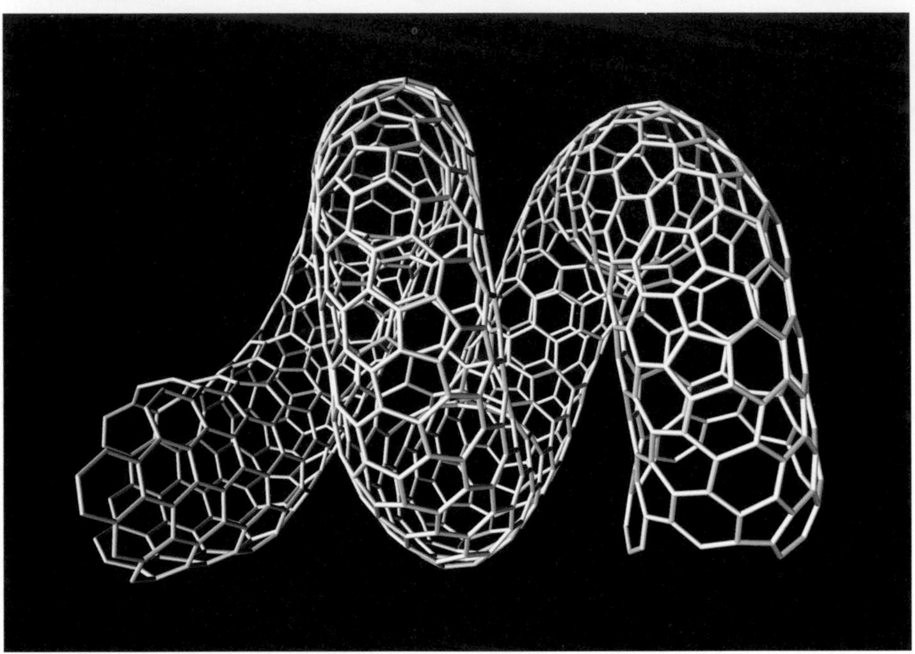

fig. 2.1.19 This is just one of the nanotube molecules that are expected to play an important role in the future of technology.

HSW Applications of nanotechnology

Chemists have produced single-walled nanotubes (SWNTs) and multi-walled nanotubes (MWNTs). Reacting MWNTs with metal oxides produces carbon **nanorods**, some of which are **superconducting** (they have zero resistance to the passage of electric currents). Studies of nanotubes show that they are stiffer than other known materials. If embedded in polymer resins they could produce composite materials which would have good electrical conductivity along with enormous strength and great lightness. They could find uses in the aircraft, space and car industries.

One idea is to develop **nanoprobes** which could circulate around the human blood system and identify cancerous cells. Then surgeons could use lasers to heat up these cells and destroy them. It might sound like science fiction today but some scientists believe it is possible.

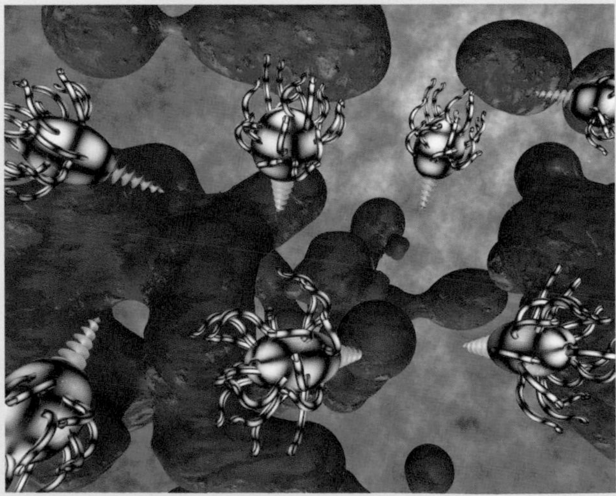

fig. 2.1.20 Nanorobots attacking cancer. This computer-generated image shows nanorobots (grey) attacking a cancerous tumour (red).

HSW Nanoparticles

You are probably aware that mixtures of coal dust and air can explode, although lumps of coal do not explode in air. This is because coal dust has a much larger surface area for the combustion reaction to take place than a lump of coal.

Particles called **nanoparticles** are much smaller than the particles in coal dust, just a few nanometres across. One nanometre (nm) is one billionth of a metre (10^{-9} m). For comparison, typical carbon–carbon bond lengths, or the spacing between these atoms in a molecule, are in the range 0.12–0.15 nm. The smallest cellular lifeforms, the bacteria of the genus *Mycoplasma*, are around 200 nm in length.

Materials reduced to the nanoscale can show very different properties compared with what they exhibit on a macroscale. This offers new opportunities for exciting applications, such as:

- opaque substances become transparent (eg copper)
- inert materials become catalysts (eg platinum)
- stable materials turn combustible (eg aluminum)
- solids turn into liquids at room temperature (eg gold)
- insulators become conductors (eg silicon).

Nanoparticles are now being used in cosmetic preparations. The skin provides a natural barrier to chemicals being absorbed through it. However, if the particles in cosmetics are scaled down to the size of nanoparticles, they become small enough to penetrate the skin. Such particles are smaller than the wavelength of light and therefore too small to see. Very small particles of titanium(IV) oxide are already being used in sunscreens, because they disappear completely into the skin, providing an invisible protective layer. Medications called serums are sold which deliver vitamin E into the skin via nanoparticles.

However, some toxicologists are alarmed by this trend to deliver cosmetics through the skin. The skin forms a barrier with the function of keeping harmful substances out of the body. If nanoparticles can penetrate the skin, might they end up in the bloodstream and the brain? What damage might they do? Will other, less welcome, substances piggy-back on those tiny particles? And what will happen if a number of different nanoparticles, eg from hand cream, sunscreen and foundation, join together?

Major cosmetics companies state that nanoparticles will not penetrate the skin to reach the bloodstream or circulate around the body. They claim that their chemical compounds will not go further than the first layer, the stratum corneum. The reality is that we do not know what the effects of these preparations are and further research is necessary.

Questions

1 Both diamond and graphite have giant structures of carbon atoms. Why do these two allotropes have such different physical properties?

2 Suggest why fullerenes dissolve in petrol but diamond and graphite do not.

3 Explain why we cannot be sure whether using nanoparticles in cosmetics is totally safe.

2 Intermediate bonding and bond polarity

Electronegativity

Ionic to covalent: a continuum

You have already met ionic and covalent bonding in chapter 1.4. These represent extremes in terms of classifying bonds, and compounds frequently have bonding which is partially ionic and partially covalent.

The position of the electrons in a covalent bond

In a hydrogen molecule, one pair of electrons is shared between the two atoms. The positive attraction from each hydrogen nucleus is the same because both nuclei are the same. The pair of electrons in the bond is shared equally. The same is true in a molecule of chlorine (see **fig. 2.2.1**).

H ⁑ H　　　: Cl ⁑ Cl ⁑

hydrogen　　　chlorine

fig. 2.2.1 **In these diatomic molecules both atoms are the same, so the bonding electrons experience the same attraction from each nucleus and are distributed evenly between them.**

However, when the two atoms at each end of a covalent bond are different, the attraction by the nuclei may be different and the pair of electrons is no longer equally shared. For example, in hydrogen chloride the chlorine nucleus is better at attracting the electrons in the covalent bond than the hydrogen nucleus.

H ⁑ Cl ⁑

hydrogen chlorine

fig. 2.2.2 **In this hydrogen chloride molecule the atoms are different, and the bonding electrons experience a greater attraction from the chlorine nucleus than from the hydrogen nucleus.**

We can predict the relative attracting powers of the nuclei of atoms of different elements by using the concept of **electronegativity**. This is a measure of the attraction of an atom in a molecule for a pair of electrons in a covalent bond.

The Pauling scale

There are various scales used to measure electronegativity. The one most commonly used was devised by Linus Pauling, and is the called the Pauling scale of electronegativity. It is a relative scale and so has no units. It runs from 0 to 4. The higher the number, the higher the electronegativity and the more the electrons in the covalent bond are attracted to the nucleus of the atom.

fig. 2.2.3 **Pauling is seen holding a model of water molecules. In 1954 Pauling won the Nobel Prize for chemistry for his work on molecular structure.**

There are trends in the electronegativity of the elements which can be clearly seen in the periodic table. The Pauling electronegativity values make these trends very clear (see **fig. 2.2.4.**).

You can see that:

- Non-metals have higher electronegativities than metals.
- Electronegativity increases across a period of the periodic table.
- Down a group in the main block of the periodic table, the electronegativity decreases.

The most electronegative element is fluorine, F with a value of 4.0 and the least electronegative element is caesium, Cs with a value of 0.7.

H 2.1																
Li 1.0	Be 1.5											B 2.0	C 2.5	N 3.0	O 3.5	F 4.0
Na 0.9	Mg 1.2											Al 1.5	Si 1.8	P 2.1	S 2.5	Cl 3.0
K 0.8	Ca 1.0	Sc 1.3	Ti 1.5	V 1.6	Cr 1.6	Mn 1.5	fe 1.8	Co 1.8	Ni 1.8	Cu 1.9	Zn 1.6	Ga 1.6	Ge 1.8	As 2.0	Se 2.4	Br 2.8
Rb 0.8	Sr 1.0	Y	Zr	Nb	Mo 1.8	Tc	Ru	Rh	Pd	Ag 1.9	Cd 1.7	In	Sn 1.8	Sb 1.9	Te 2.1	I 2.5
Cs 0.7	Ba 0.9	La	Hf	Ta	W	Re	Os	Ir	Pt	Au 2.4	Hg 1.9	Tl	Pb 1.8	Bi 1.9	Po	At

fig. 2.2.4 The electronegativity of some of the elements, measured on the Pauling scale.

Notice there are no electronegativity values given for the noble gases. Atoms of these elements very rarely form covalent bonds and so they have no affinity for electrons.

Values for Pauling's electronegativities may vary from source to source. They were originally calculated by Pauling, who chose hydrogen as a standard which was given an arbitrary value of 2.1. Hydrogen was chosen because it forms a large number of covalent compounds.

HSW Electronegativity and chemical models

You have used several types of model to depict molecules, including ball and stick, space-filling or computer-generated models. However, none of these models takes account of the effects of the electronegativity of the atoms involved. A stick in a ball and stick model represents a pair of electrons in a covalent bond, but it does not show how the pair of electrons is distributed.

In chapter 1.3 you came across electron density maps showing the positions of electrons in orbitals. In chapter 1.4 you saw such a map for sodium chloride in the context of the structure of ionic lattices. These electron density maps can also be used to show the distribution of electrons in two covalent molecules (see fig. 2.2.5). The maps are drawn on a computer using knowledge of the electronegativities of the combining atoms.

In a chlorine molecule, the electron distribution is the same around both atoms. The electron density is greater when it is shaded red and yellow and less when shaded blue and green. The electron charge is evenly distributed between the two atoms. However, in an iodine fluoride molecule, IF, the model shows that the electron density is greater around the smaller fluorine atom.

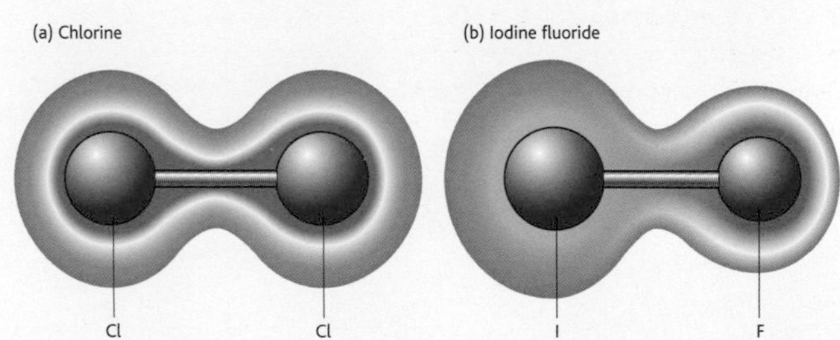

fig. 2.2.5 Electron density maps for (a) chlorine and (b) iodine fluoride.

Questions

1 Fluorine is the most electronegative element. Find the three elements with the next highest electronegativities after fluorine in fig. 2.2.4.

2 Looking at fig. 2.2.4, can you describe any pattern in electronegativities between scandium and zinc?

3 In the electron density map for methane, the electrons are distributed fairly evenly with just a slightly greater density around the carbon atom than the hydrogen atom. Explain what this illustrates about the bonding in methane.

Intermediate bonding

Ionic and covalent bonds

In chapter 1.4 you saw how atoms bond by ionic or covalent bonding.

- In a covalent bond one electron from each atom is made available and is shared between the two atoms.
- In an ionic bond there is a loss of one or more electrons by the metal atom and a gain of one or more electrons by the non-metal atom. The resulting ions are held together by electrostatic charges.

These are two extremes and in practice all degrees of intermediate bonding occur.

Bond character

You can calculate to what degree a bond is ionic and to what degree it is covalent by comparing the difference in electronegativity between the two atoms in the bond.

For example, the electronegativities of carbon and hydrogen are 2.5 and 2.1 respectively. The difference in electronegativity is 0.4. If you read off the percentage of ionic and covalent character from **table 2.2.1** you will see this bond is 96% covalent and 4% ionic.

Electronegativity difference	Percentage ionic character	Percentage covalent character	Electronegativity difference	Percentage ionic character	Percentage covalent character
0.1	0.5	99.5	**1.7**	51	49
0.2	1	99	**1.8**	55	45
0.3	2	98	**1.9**	59	41
0.4	4	96	**2.0**	63	37
0.5	6	94	**2.1**	67	33
0.6	9	91	**2.2**	70	30
0.7	12	88	**2.3**	74	26
0.8	15	85	**2.4**	76	24
0.9	19	81	**2.5**	79	21
1.0	22	78	**2.6**	82	18
1.1	26	74	**2.7**	84	16
1.2	30	70	**2.8**	86	14
1.3	34	66	**2.9**	88	12
1.4	39	61	**3.0**	89	11
1.5	43	57	**3.1**	91	9
1.6	47	53	**3.2**	92	8

table 2.2.1 Using electronegativities to determine the percentage ionic and covalent character of a bond.

Calculating percentages of ionic and covalent bonding

Calculate the percentage of ionic and covalent bonding in:

a lithium fluoride, LiF **b** ammonia, NH_3.

a The electronegativities of lithium and fluorine are 1.0 and 4.0 respectively. The difference in electronegativity is 3.0.

From table 2.2.1, the Li—F bond is 89% ionic and 11% covalent.

b The electronegativities of nitrogen and hydrogen are 3.0 and 2.1 respectively. The difference in electronegativity is 0.9.

From **table 2.2.1**, the N—H bond is 19% ionic and 81% covalent.

Notice that even compounds like lithium fluoride and sodium chloride which we regard as ionic still have appreciable covalent character.

Polar covalent bonds

As you have seen, in a hydrogen molecule, where both atoms in the bond are identical, each atom gets a 'fair share' of the electron pair forming the bond – it is attracted equally to both nuclei. Another way to look at this is to say that the **centres of charge** in the molecule coincide. The idea of centre of charge is directly analogous to the idea of centre of mass (see **fig. 2.2.6**).

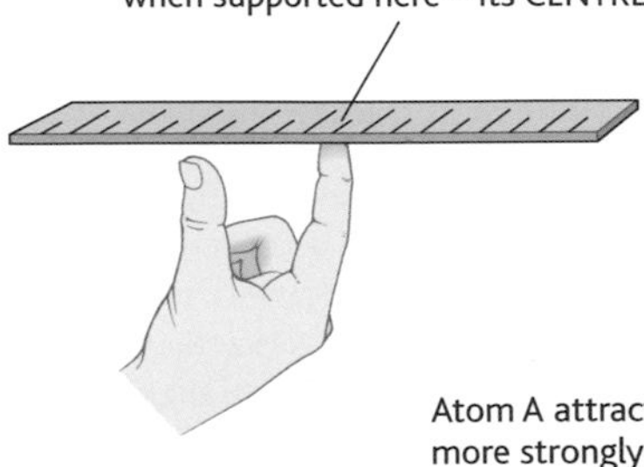

fig. 2.2.6 **The idea of centre of charge is analogous to the idea of centre of mass.**

In a molecule like hydrogen chloride, however, this is not the case. The chlorine atom attracts electrons more strongly than the hydrogen atom, pulling the electron pair in the bond more closely towards it and distorting the electron cloud (see **fig. 2.2.7(b)**). The result of this distortion is that the ends of the molecule have a small charge, shown by the δ+ and δ– in **fig. 2.2.7(c)**. These charges are not full charges of 1+ and 1– like those on an ion, which is why they are written using the lower case Greek delta – in the case of the HCl molecule, the hydrogen carries a charge of +0.17 and the chlorine a charge of –0.17.

(a)

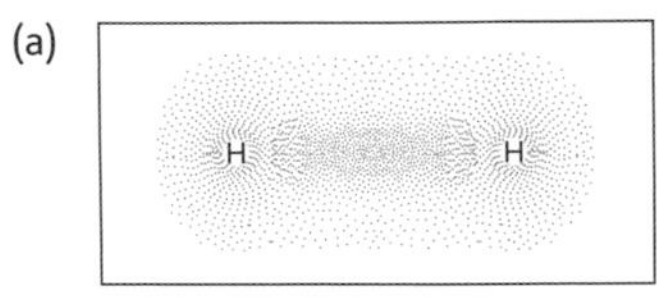

(b)

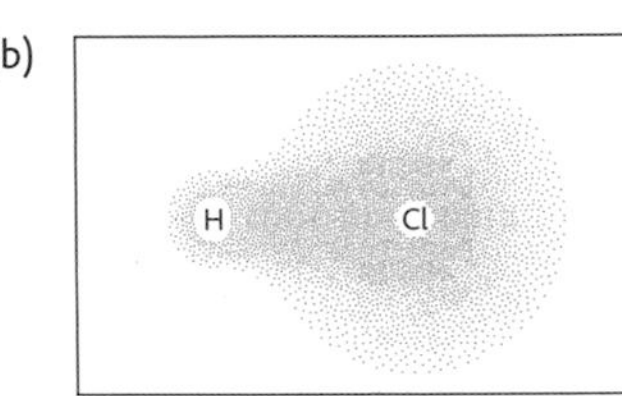

(c) δ+ δ–
H—Cl

fig. 2.2.7 **(a) In the H_2 molecule the electron density is evenly distributed and the centres of positive and negative charge coincide. (b) The uneven distribution of electron density in the HCl molecule gives rise to partial charges at the ends of the molecule. These can be represented on the displayed formula in the way shown in (c).**

In a similar way, as you saw in chapter 1.4, ionic bonds may be distorted by the attraction of the cation for the outer electrons of the anion. **Figure 2.2.8** shows a wholly ionic bond (a), with the attraction of the electron cloud around the anion shown in (b). If the distortion is great, it may even lead to a charge cloud which begins to resemble that of a covalent bond (c).

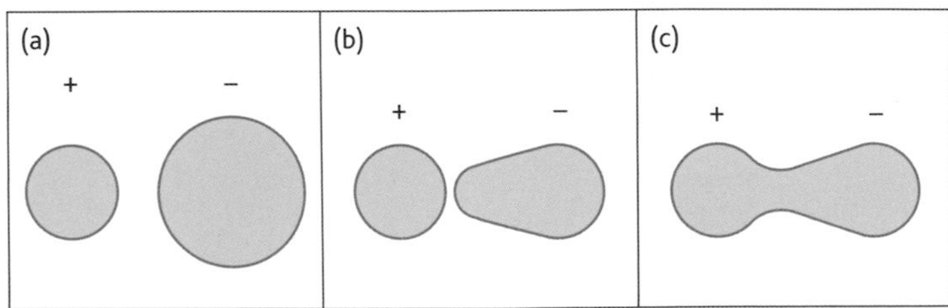

fig. 2.2.8 **The distortion of an ionic bond.**

Questions

1. Which compound in the list below has most ionic character? HF HCl HBr HI
2. Draw a water molecule and show the partial charges using δ+ and δ–.
3. Calculate the percentage of ionic character in lithium iodide. Suggest how the properties of lithium iodide might be different from those of potassium fluoride.

Polarity of bonds and molecules

You have seen that when two different atoms are joined by a covalent bond, the more electronegative atom attracts the electron pair more and as a result has a δ– charge while the other atom has a lesser share of the electron pair and has a δ+ charge. The result is a **polar bond**.

It is possible to have several polar bonds in a molecule. Does this make the whole molecule polar? It depends on the arrangement of the bonds in the molecule.

Are liquids deflected by a charged rod?

A comb or rod is given an electric charge when you rub it on a piece of wool, and the resulting electric field around the comb then attracts some liquids. Liquids with polar molecules, such as water, are deflected, but liquids that show no overall polarity in their molecules are not deflected by the electric field.

Liquids deflected	Liquids undeflected
trichloromethane 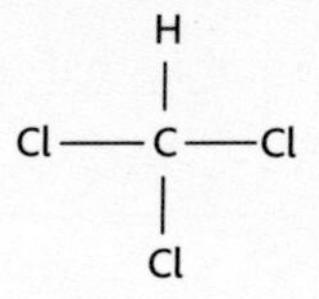	tetrachloromethane 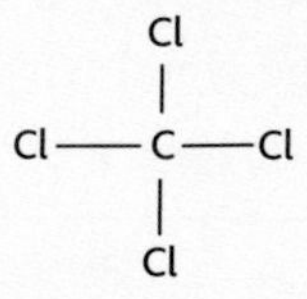
cyclohexene 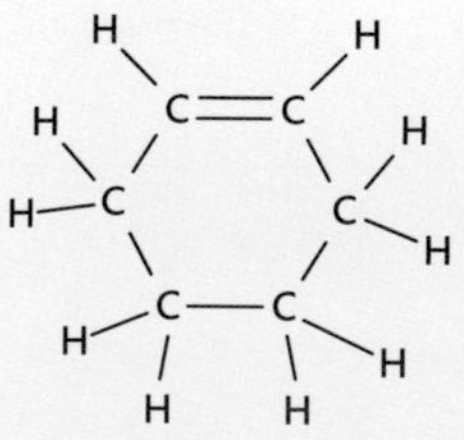	cyclohexane 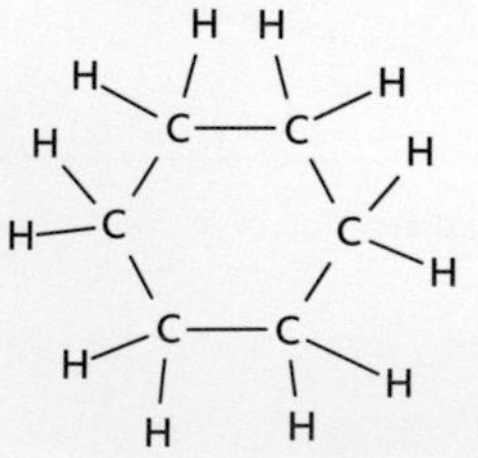

table 2.2.2 The results of testing some liquids with a charged rod.

Why are some molecules polar?

Why are some liquids deflected by an electric field while others are not? Careful examination of the structure of the molecules of liquids that are deflected shows that they are not symmetrical – for example, trichloromethane and cyclohexene. It is this lack of symmetry that leads to a molecule being polar. You know that bonds like the C—Cl bond are polar, because chlorine is more electronegative than carbon

As a result, the electrons in the trichloromethane molecule are attracted towards one end of the molecule, and this distortion of the electron cloud means that the centre of positive charge and the centre of negative charge do not occur in the same place in the molecule. Molecules that contain uneven distributions of charge like this are called **polar molecules**. The charge separation in a polar molecule makes it a **dipole** (there are two types or **poles** of charge in the molecule). This polarity is measured as its **dipole moment** – the amount of charge separation multiplied by the distance between the centres of charge. The unit of dipole moments is the **debye, D**. Measuring the dipole moments of polar molecules provides an important way of checking the predictions made from theories of bonding.

Unlike trichloromethane, the tetrachloromethane molecule is not polar – fig. 2.2.9 compares the two molecules.

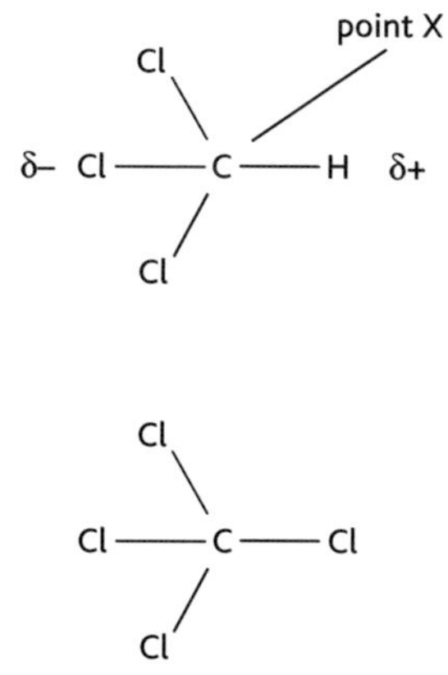

fig. 2.2.9 Chlorine is more electronegative than carbon or hydrogen. This means that point X in the $CHCl_3$ molecule has a net amount of positive charge to the right, and a net amount of negative charge to the left – the molecule is polar. By contrast, the electron-attracting effect of the chlorine atoms in tetrachloromethane cancels out – the centres of positive and negative charge do coincide in this molecule, and it is not polar.

Looking at table 2.2.2 again, you will notice that cyclohexene is deflected by a charged rod but cyclohexane is not. In cyclohexene there is a planar C=C part of the molecule, but in cyclohexane each carbon atom is tetrahedral and any polarities cancel out.

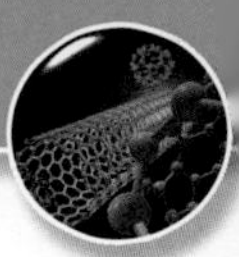

Water

You saw in chapter 2.1 that water is a covalent molecule with a bond angle of 104.5°. In each O—H bond there is a δ+ on the hydrogen atom and a δ– charge on the oxygen atom. This is caused by a movement of electrons towards the oxygen atom in each bond. Because the molecule is not symmetrical, it has an overall dipole moment.

It is interesting to note that if water was a linear molecule (with the H—O—H bond angle 180°), the two dipoles would cancel out and a stream of water would not be deflected by a charged rod.

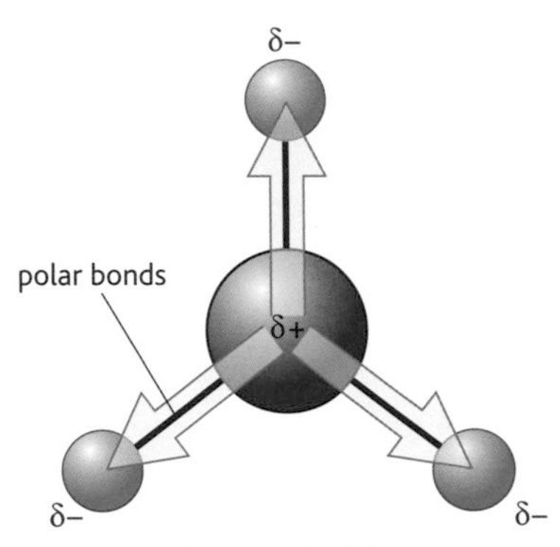

fig. 2.2.10 Boron trifluoride has highly polar bonds, but the molecule is symmetrical so it is not polar. The yellow arrows show the migration of electrons.

Boron trifluoride

Figure 2.2.10 shows how the the net effect of symmetrical polar bonds within boron trifluoride cancel out to give a net polarity of zero.

Predicting whether a substance is polar

Table 2.2.3 will help you to predict whether a substance is polar or not.

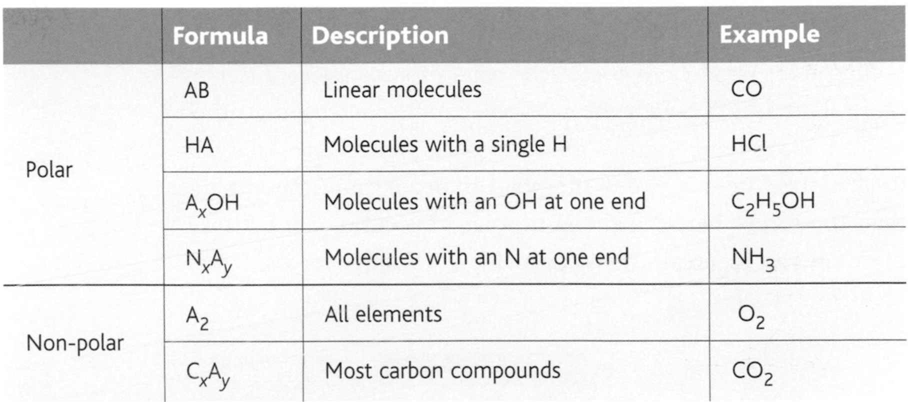

	Formula	Description	Example
Polar	AB	Linear molecules	CO
	HA	Molecules with a single H	HCl
	A_xOH	Molecules with an OH at one end	C_2H_5OH
	N_xA_y	Molecules with an N at one end	NH_3
Non-polar	A_2	All elements	O_2
	C_xA_y	Most carbon compounds	CO_2

table 2.2.3 Polar and non-polar substances.

Questions

1 Is each of the following liquids polar or non-polar?
  - a Liquid nitrogen, N_2
  - b Hydrogen fluoride, HF
  - c Benzene, C_6H_6
  - d Methanol, CH_3OH

2 Explain why a molecule containing polar bonds is not necessarily a polar molecule.

3 Why are chloromethane and dichloromethane polar molecules while tetrachloromethane is not?

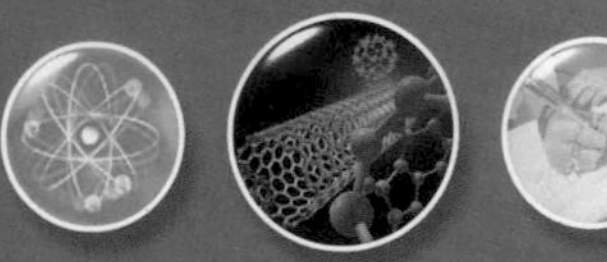

3 Intermolecular forces

Forces between molecules

So far you have seen that the forces between atoms in matter may involve transferring whole electrons to form ions, or sharing them to form molecules. This is not the whole story though, for to fully understand the behaviour of matter we need to know about how molecules interact with each other. These interactions, called **intermolecular forces**, are much weaker than the **intramolecular forces** within molecules.

Polarity and boiling temperature

Figure 2.3.1 compares the boiling temperatures of four substances. Hydrogen chloride and hydrogen sulfide are both polar molecules, while fluorine and argon are non-polar. The boiling temperatures of these two pairs of substances differ by more than 100 °C, with the non-polar molecules having much lower boiling temperatures than the polar molecules. This tells us that the forces between the polar molecules are much bigger than the forces between the non-polar molecules, since it takes more energy to break them apart. How do these forces come about?

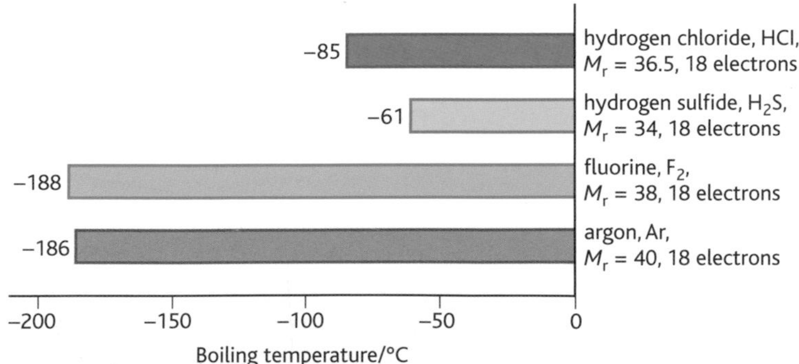

fig. 2.3.1 Polar molecules make for compounds with a much higher boiling temperature than non-polar molecules.

Dipole–dipole interactions

Polar molecules have a permanent dipole – that is to say, a permanent separation of charge. As a result of this, polar molecules are attracted towards one another by forces called **permanent dipole–permanent dipole interactions**, in which the negative end of one molecule is attracted towards the positive end of another (see **fig. 2.3.2**). These interactions decrease quite rapidly as the distance between molecules increases, and they are about 100 times weaker than covalent bonds. This kind of interaction accounts for the forces between molecules of trichloromethane.

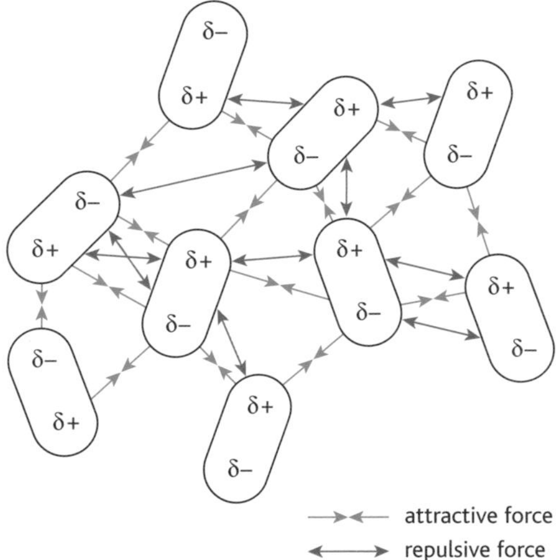

fig. 2.3.2 Permanent dipole–permanent dipole interactions between molecules. Attractive forces between oppositely charged portions of molecules are generally stronger than the repulsive forces between similarly charged portions of molecules, and the molecules experience a net attractive force as a result.

It is quite easy to see how interactions of this sort account for the forces between polar molecules, but what about non-polar molecules? It is possible to liquefy non-polar substances, so there must be forces between their molecules. How do these forces arise?

London forces

The forces of attraction that exist between two non-polar molecules are also explained in terms of charge distribution. In a neutral atom, at any instant in time the electron distribution around the nucleus may result in the centres of positive and negative charge failing to coincide, producing an **instantaneous dipole**. Over time these instantaneous dipoles average out, producing a net dipole of zero. Any other atom next to an atom with an instantaneous dipole will experience an electric field due to the dipole, and so will itself develop an **induced dipole** (see **fig. 2.3.3(a)**). The **instantaneous dipole–induced dipole interactions** between neighbouring molecules provide the force to bring non-polar molecules together to form a liquid when the temperature is sufficiently low. The forces are small however, since the attraction between two molecules lasts only a very short time as the instantaneous dipole is turned on and off as the electron density fluctuates (**fig. 2.3.3(b)**). These interactions are given the name **London forces** (**van der Waals forces**), after the German–American physicist, Fritz London.

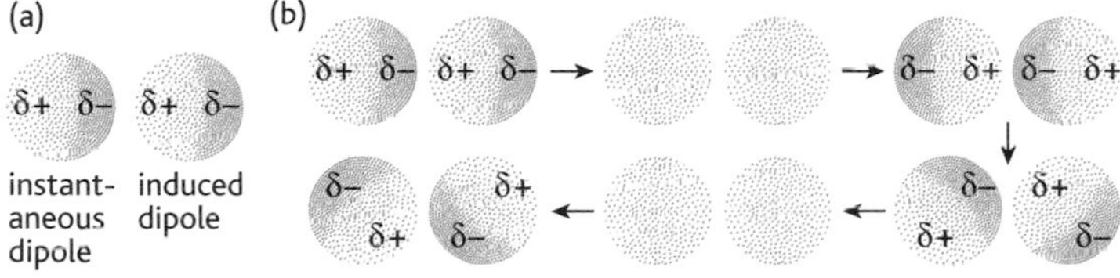

fig. 2.3.3 (a) An instantaneous dipole in one atom gives rise to an induced dipole in a nearby atom, thus causing a net attractive force between the two atoms. (b) The instantaneous dipole varies over time, sometimes having a value of zero. As a result, the attractive force between the atoms fluctuates, and is sometimes zero.

In addition to instantaneous dipole–induced dipole interactions, London forces also include **instantaneous dipole–instantaneous dipole forces**, sometimes called **London dispersion forces**. For example, the Cl_2–Cl_2 interaction in a sample of chlorine gas is an example of a London dispersion force (see **fig. 2.3.4**).

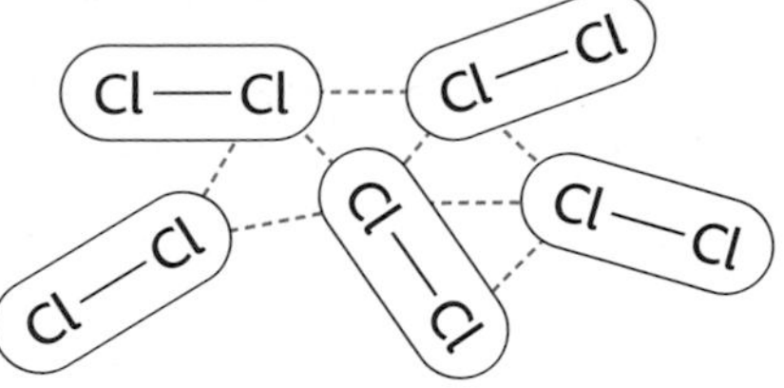

fig. 2.3.4 London forces between chlorine molecules.

London dispersion forces exist between all atoms. London forces are the only forces between noble gas atoms when the gases condense at low temperatures.

The size of the London forces in a substance depends on the size of the electron clouds of the particles that are interacting. Atoms with large electron clouds tend not to hold on to their outer electrons so tightly, so large electron clouds tend to be easily deformed. This favours the existence of an instantaneous dipole (and also favours the creation of an induced dipole), so London forces tend to increase with the size of the electron cloud. This can be clearly seen in the boiling temperatures of the noble gases (see **table 2.3.1**). London forces between molecules are affected by the size of the molecules for the same reason – so ethane (C_2H_6) is a gas at room temperature, while hexane (C_6H_{14}) is a liquid.

	Noble gas					
	He	Ne	Ar	Kr	Xe	Rn
Boiling temperature (K)	4	27	87	121	166	211

table 2.3.1 The boiling temperatures of the noble gases. The boiling temperature rises with atomic size, because the London forces get correspondingly larger.

Hydrogen bonds

The boiling temperatures of the noble gases increase down the group due to the increasing size of the electron cloud. How does the behaviour of other groups compare?

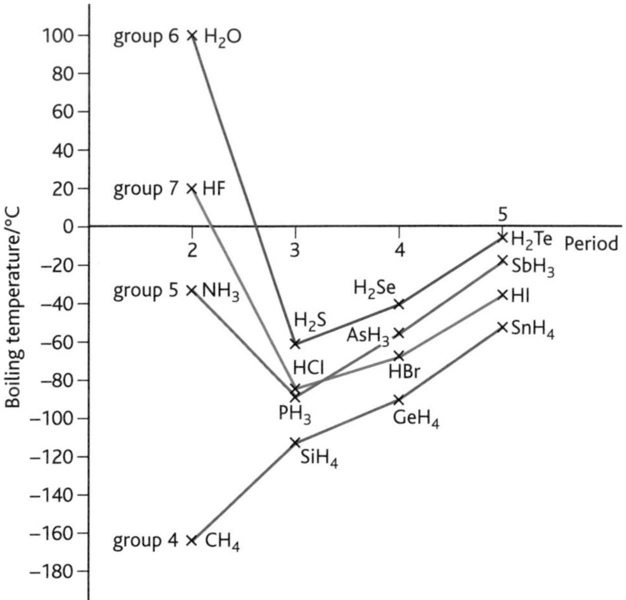

fig. 2.3.5 The variation in boiling temperature of the hydrides of groups 4, 5, 6 and 7.

Figure 2.3.5 shows the boiling temperatures of the hydrides formed by the elements in groups 4, 5, 6 and 7 of the periodic table. While the hydrides of group 4 behave in a very similar way to the noble gases, the hydrides of the other groups do not – at least not as far as the lightest element in each group is concerned. This suggests that the intermolecular forces in the hydrides of the lightest elements are much stronger than expected on the basis of the behaviour of the other elements in each group. The strength of these intermolecular forces can be illustrated by comparing the molar enthalpies of vaporisation of the hydrides of the elements of group 6, as shown in **table 2.3.2**.

The **molar enthalpy of vaporisation** of a liquid is the enthalpy change when one mole of the liquid changes into one mole of its gas at the boiling temperature – this is a direct measure of the intermolecular forces in the liquid. Notice how the energy required to turn one mole of liquid water at 100 °C into steam at 100 °C is more than twice the energy needed to turn one mole of liquid hydrogen sulfide into gaseous hydrogen sulfide at its boiling temperature.

	Hydride			
	H_2O	H_2S	H_2Se	H_2Te
Standard molar enthalpy of vaporisation (kJ mol^{-1})	41.1	18.7	19.9	23.8

table 2.3.2 The molar enthalpies of vaporisation of the hydrides of group 6.

To explain these particularly strong interactions, a different type of intermolecular force comes into play. This interaction is called a **hydrogen bond**, and it exists between a molecule that contains a hydrogen atom attached to an electronegative element and another molecule containing a lone pair of electrons on an electronegative element.

You have seen that a big difference in electronegativity is a good indicator of a polar bond. There is a big difference between the electronegativity of hydrogen (2.1) and the two elements oxygen (3.5) and fluorine (4.0). This results in the H—O and H—F bonds being very polar, with a substantial amount of the electron density in each bond being drawn away from the hydrogen atom, leaving it with a $\delta+$ charge. This polar bond causes the exceptionally strong interactions that lead to the higher than expected boiling temperatures of hydrogen fluoride and water.

Hydrogen bonds form because of the special nature of the hydrogen atom, which has no inner shells of electrons. As a result, the nucleus of the hydrogen atom in these molecules is left unusually exposed by the shift in electron density within the bond, making it easily accessible for strong permanent dipole–permanent dipole interactions to occur. This happens when the lone pair of electrons of an atom (on oxygen in the case of water, on fluorine in the case of hydrogen fluoride) in another molecule is attracted to the positive 'rump' of the hydrogen, leading to the formation of a hydrogen bond. This is shown in **fig. 2.3.6**.

Ammonia is another common example of a compound that shows hydrogen bonding. Nitrogen is more electronegative than hydrogen, and there is a lone pair available for hydrogen bonding on the nitrogen atom.

Relative strengths of intermolecular forces

Covalent bond strengths are typically between 200 and 500 kJ mol^{-1}, although they can be outside this range. Hydrogen bonds are weak in comparison – about 5–40 kJ mol^{-1}. London forces are weaker still. Hydrogen bonds and London forces are not strong enough to influence the chemical behaviour of most substances, although both can affect the physical properties of substances.

HSW Models of bonding and physical properties

Ethanol, CH_3CH_2OH and the ether methoxymethane, CH_3OCH_3 have the same molecular formula and therefore the same molecular mass. They have very different boiling temperatures however. The boiling temperature of ethanol is 78 °C and that of methoxymethane is –25 °C. In both compounds there are strong covalent bonds within the molecules and the differences must be due to intermolecular forces. Our model of intermolecular forces relies on the possibility of hydrogen bonding between a hydrogen atom attached to an electronegative element (oxygen in this case) and a lone pair on an electronegative element in an adjacent molecule. Both of these are present in ethanol. In methoxymethane there is a lone pair on the oxygen but there is no hydrogen attached to it so hydrogen bonding is not possible. Our bonding model explains the difference in boiling temperature.

Questions

1 Predict the type of intermolecular forces between molecules of:

a chloromethane, CH_3Cl

b hydrogen

c methanol, CH_3OH.

2 Arrange the following in order of increasing strength:

a covalent bonds

b hydrogen bonds

c London (van der Waals) forces

d dipole–dipole interactions.

3 Why is it possible to have hydrogen bonding in water or hydrogen fluoride but not in hydrogen or methane?

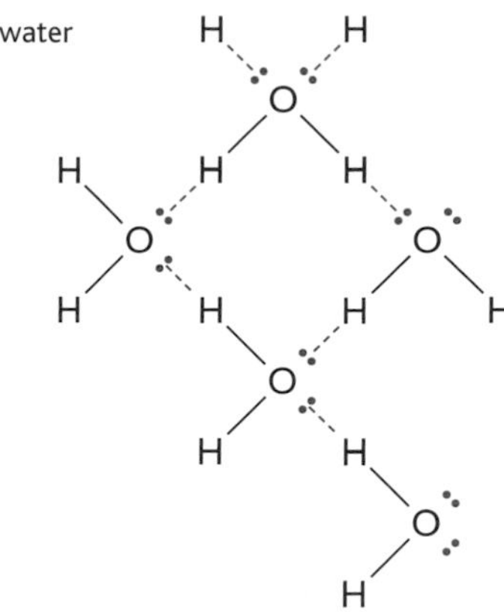

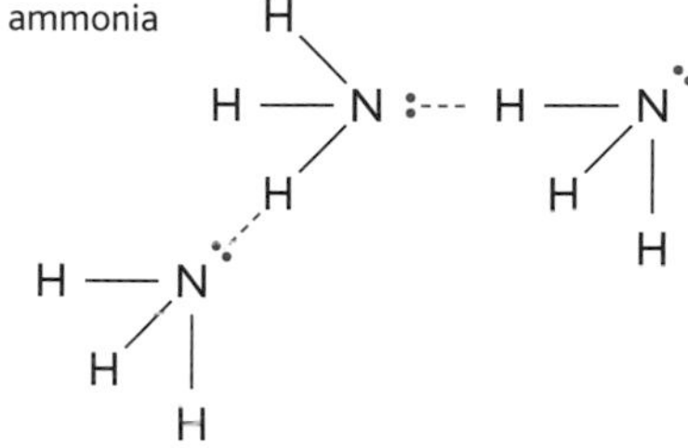

fig. 2.3.6 Hydrogen bonding in water, hydrogen fluoride and ammonia.

fig. 2.3.7 Without water, life on Earth would not be possible. It is hydrogen bonding which gives water its high boiling temperature, and so makes sure the surface of the planet is water which does not simply boil away into the atmosphere!

Trends in physical properties

Introduction

The ideas about intermolecular forces that you have just met are very useful for explaining differences in the physical properties of different substances. These intermolecular forces in turn depend on the detailed molecular structure and polarity of the substances.

Trends in alkanes

In chapter 1.6 you studied the alkanes, a family of saturated hydrocarbons.

Table 2.3.3 compares the melting and boiling temperatures of the first 10 straight-chain alkanes.

Alkane	Molecular formula	Structural formula	Melting temperature (°C)	Boiling temperature (°C)
Methane	CH_4	H–C(H)(H)–H	–182	–164
Ethane	C_2H_6	H–C(H)(H)–C(H)(H)–H	–183	–88
Propane	C_3H_8	H–C(H)(H)–C(H)(H)–C(H)(H)–H	–190	–42
Butane	C_4H_{10}	H–C(H)(H)–C(H)(H)–C(H)(H)–C(H)(H)–H	–138	–0.5
Pentane	C_5H_{12}	$CH_3CH_2CH_2CH_2CH_3$	–130	36
Hexane	C_6H_{14}	$CH_3CH_2CH_2CH_2CH_2CH_3$	–95	69
Heptane	C_7H_{16}	$CH_3CH_2CH_2CH_2CH_2CH_2CH_3$	–90	99
Octane	C_8H_{18}	$CH_3CH_2CH_2CH_2CH_2CH_2CH_2CH_3$	–57	126
Nonane	C_9H_{20}	$CH_3CH_2CH_2CH_2CH_2CH_2CH_2CH_2CH_3$	–51	151
Decane	$C_{10}H_{22}$	$CH_3CH_2CH_2CH_2CH_2CH_2CH_2CH_2CH_2CH_3$	–30	174

table 2.3.3 **Some physical properties of the first 10 straight-chain alkanes.**

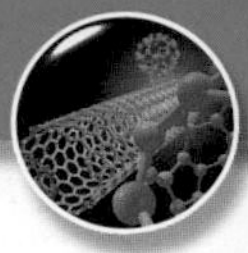

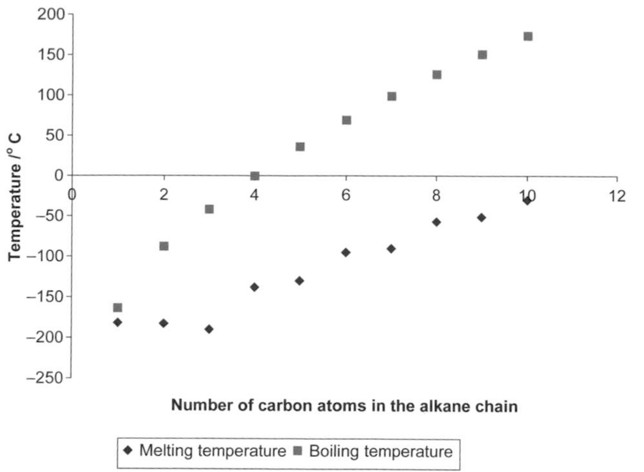

fig. 2.3.8 The temperature of the first 10 straight-chain alkanes plotted against number of carbon atoms in the chain.

The melting and boiling temperatures of the alkanes increase as chain length increases (see **fig. 2.3.8**). The intermolecular forces in the alkanes are weak – they are induced dipole–induced dipole forces. However, despite being weak they are important because they hold together the particles in organic liquids and solids. These forces become larger as the length of the molecules increases because there are more places where these forces can operate. **Figure 2.3.9** shows two long alkane molecules and the weak intermolecular forces between induced dipoles, which are partially overcome when the alkane melts and completely broken when it boils. (Remember that the strong covalent bonds within the molecules are not broken during melting or boiling.)

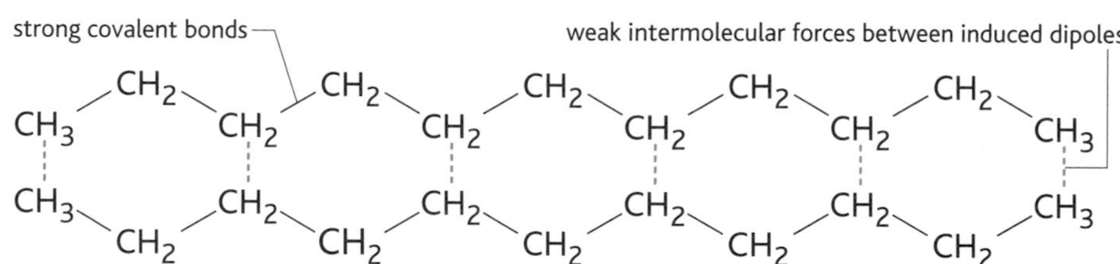

fig. 2.3.9 The longer the molecule, the more places there are where weak intermolecular forces can form.

Volatility is a measure of how easily molecules escape from a liquid. A volatile liquid loses molecules from the surface very easily so it will have a low boiling temperature. Petrol is a volatile liquid.

In very general terms, branching of the molecule lowers the boiling temperature, because the side chains interfere with the packing together of molecules and reduce the formation of weak intermolecular forces between induced dipoles. However, variations in the parent chain and the side chains make it difficult to predict the extent of intermolecular forces that will form.

fig. 2.3.10 The alkanes are easily used as fuels – but to get ethanol burning on your Christmas pudding you need to warm it first so that it starts to vaporise. Even cooks have to take account of the effect of the hydrogen bonds between the molecules of the alcohol!

In **table 2.3.4** the melting and boiling temperatures of three alkanes are shown. They all have the same relative formula mass and the same number of carbon and hydrogen atoms. You can see that the alkanes become more volatile (the boiling temperature decreases) as the amount of branching increases.

	Alkane		
	Pentane	**2-methylbutane**	**2,2-dimethylpropane**
Displayed formula	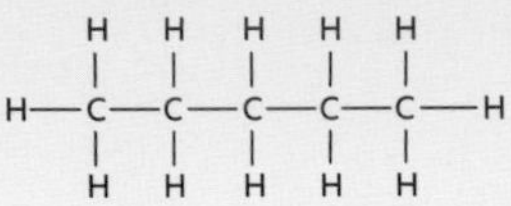 H H H H H H—C—C—C—C—C—H H H H H H	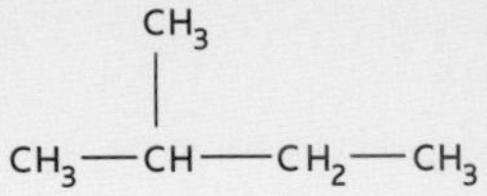 CH_3 CH_3—CH—CH_2—CH_3	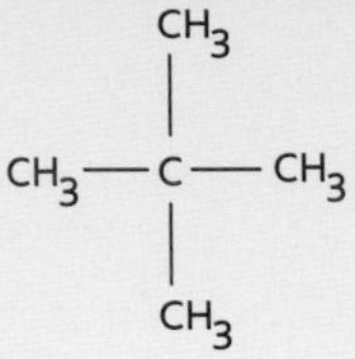 CH_3 CH_3—C—CH_3 CH_3
Melting temperature (°C)	−130	−160	−16
Boiling temperature (°C)	36	28	10

table 2.3.4 Melting and boiling temperatures of straight and branched-chain alkanes.

Comparing alkanes and alcohols

Table 2.3.5 compares the boiling temperatures of some alkanes and alcohols.

	Organic compound		
	Propan-1-ol	**Propane**	**Butane**
Displayed formula	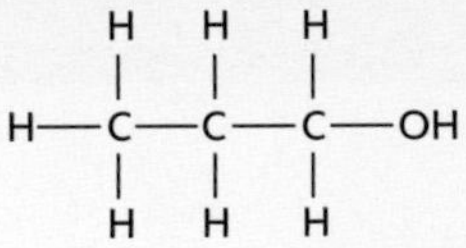 H H H H—C—C—C—OH H H H	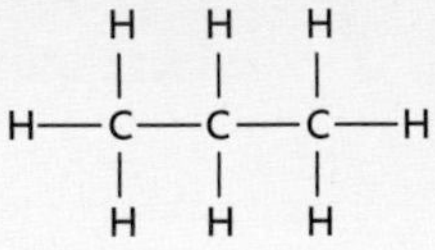 H H H H—C—C—C—H H H H	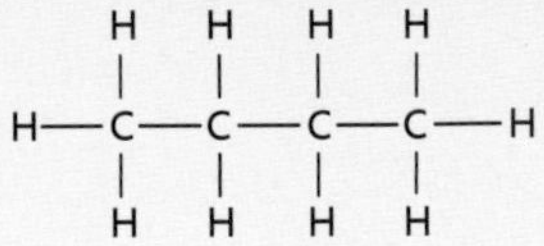 H H H H H—C—C—C—C—H H H H H
Relative formula mass	60	44	58
Boiling temperature (°C)	97	−42	−0.5

table 2.3.5 Boiling temperatures of an alcohol and two alkanes.

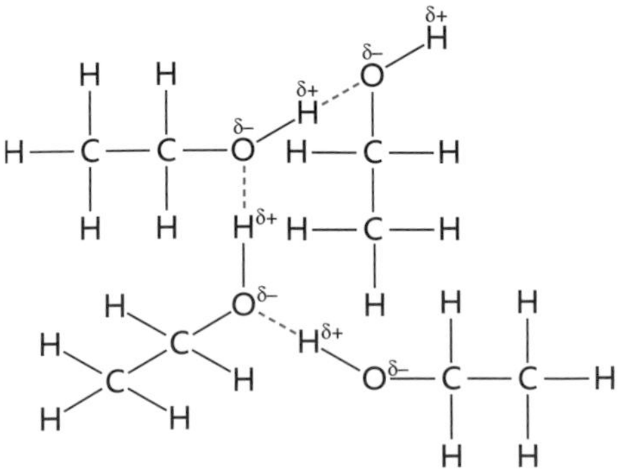

fig. 2.3.11 Hydrogen bonding between molecules in ethanol.

Propan-1-ol and propane both contain three carbon atoms in the chain. However, the boiling temperature of propan-1-ol is much higher than that of propane. Also, propan-1-ol and butane have very similar relative formula masses. Again, the boiling temperature of propan-1-ol is much higher than the boiling temperature of butane. What causes these differences?

In propane and butane the intermolecular forces are weak forces between induced dipoles. These are also present in propan-1-ol. However, propan-1-ol can also form hydrogen bonds between its molecules. These form because an electronegative oxygen atom is bonded to a hydrogen atom, and this positively polarised hydrogen attracts an oxygen atom in another molecule (see **fig. 2.3.11**).

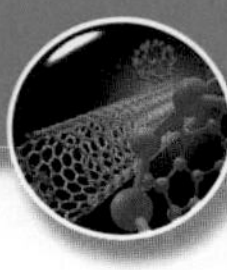

Trends in boiling temperatures of the hydrogen halides

Figure 2.3.12 shows a graph of the boiling temperatures of the hydrogen halides against the period in which the halogen is placed.

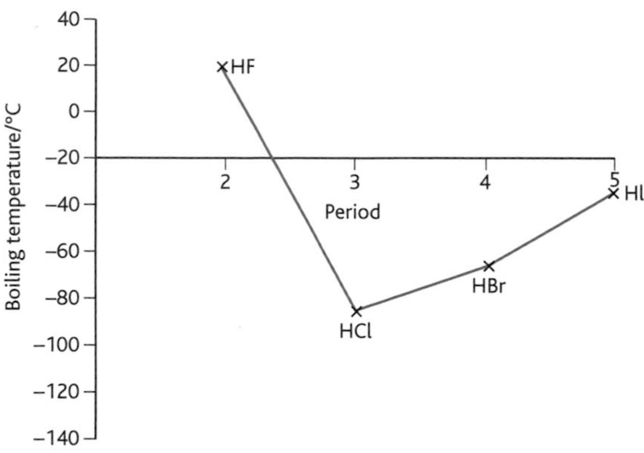

fig. 2.3.12 Graph of the boiling temperatures of the hydrogen halides against period.

You might expect a gradual increase in boiling temperature from hydrogen fluoride to hydrogen iodide because of the increasing size and mass of the molecules leading to increased London forces. This is the case from hydrogen bromide to hydrogen iodide but the boiling temperature of hydrogen chloride is higher than you might expect, and that of hydrogen fluoride is very much higher.

Element	H	F	Cl	Br	I
Electronegativity	2.1	4.0	3.0	2.8	2.5

table 2.3.6 The electronegativities of the halogens and hydrogen.

Table 2.3.6 shows the electronegativities of hydrogen, fluorine, chlorine, bromine and iodine. You can see that the H—Cl bond will be much more polar than the H—Br or H—I bond so dipole–dipole interactions will be greater and this will explain the higher boiling temperature. In hydrogen fluoride, because of the very high electronegativity of fluorine, there will be considerable hydrogen bonding as well. This accounts for its much higher boiling temperature.

Questions

1 Plot a graph of boiling temperature against number of carbon atoms for the first five members of the straight-chain alkanes. What is the effect on boiling temperature of increasing the length of the carbon chain? Explain why this is so.

2 **Figure 2.3.13** shows the boiling temperatures of the hydrides of elements in group 4 and group 5 of the periodic table.

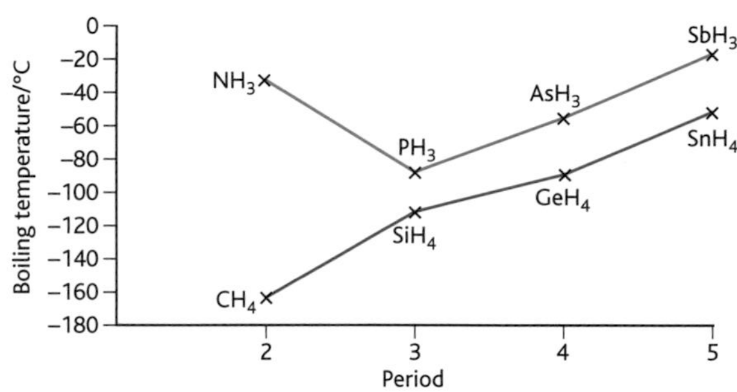

fig. 2.3.13 Boiling temperatures of the group 4 and 5 hydrides.

a Describe the trend in boiling temperatures from CH_4 to SnH_4. Explain this trend in terms of intermolecular forces.

b Describe the trend in boiling temperatures from NH_3 to SbH_3. Explain this trend in terms of intermolecular forces and electronegativities.

Solubility

Introduction

You are familiar with the idea of solubility – a soluble substance will dissolve in a solvent to form a solution. You will also know that water is a very good solvent because it dissolves a wide range of substances. However, it doesn't dissolve some substances – eg chlorophyll, the green pigment in grass, is not soluble in water.

Other solvents, sometimes called **non-aqueous solvents**, will dissolve many of the substances that water does not dissolve. Examples of non-aqueous solvents include hexane and paraffin.

The word **solubility** strictly means the mass of a solute that dissolves in 100 g of solvent at a particular temperature. Here however, the word will be used **qualitatively** rather than **quantitatively**.

Investigating solubility

You can investigate which substances dissolve in a range of solvents. It is advisable to work with small quantities of each substance and small volumes of solvent. Working on a small scale saves materials and also is quicker to do. Check Hazcards to ensure you are using solvents safely.

Patterns in solubility

There are some general rules about solubilities that you may have found out by experiment:

- Highly polar solids such as ionic salts, eg sodium chloride dissolve in water (a polar solvent) but not in hexane (a non-polar solvent).
- Polar organic substances such as sucrose dissolve in water but not in hexane.
- Non-polar solids such as candle wax do not dissolve in water but dissolve in hexane.
- Non-polar liquids such as petrol and diesel mix completely. They are said to be **miscible**.
- Polar liquids such as ethanol and water are miscible.
- A polar liquid and a non-polar liquid such as water and petrol are **immiscible** and form separate layers, with one floating on another (see **fig. 2.3.16**).

These observations may be summarised by the statement 'Like dissolves like'.

Dissolving an ionic solid in water

What happens when an ionic solid dissolves in water? Why does it seem to disappear? The process of dissolving an ionic solid is shown in **fig. 2.3.14**.

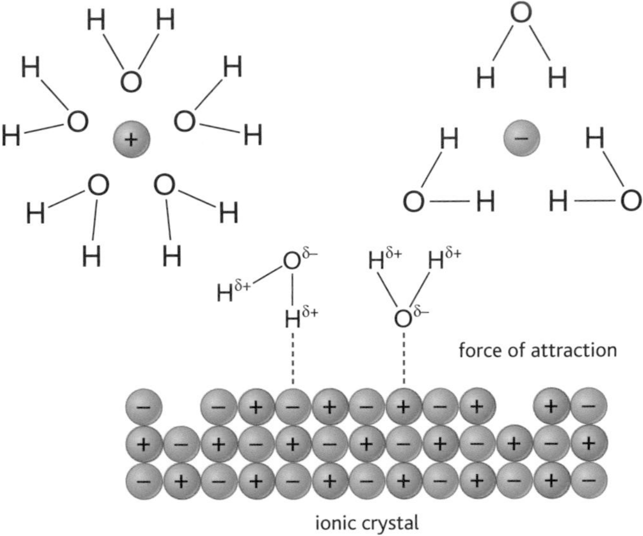

fig. 2.3.14 The process of dissolving an ionic solid in water.

Sodium chloride, for example, dissolves in cold water and there is no apparent change in energy. In the ionic solid the positive and negative ions are held together by strong electrostatic forces. Energy is required to break down the ionic lattice and this is called the **lattice energy** (see chapter 1.4). The water molecules are polarised so that hydrogen atoms have a δ+ charge and oxygen atoms have a δ– charge. Attractive forces between the water molecules and the ions in the lattice break up the lattice and the individual ions become spread throughout the water.

Water molecules arrange themselves around the dispersed ions. With a positive ion such as Na^+, the negative oxygen of the water molecule is closer to the ion. With a negative ion such as Cl^-, the positive hydrogen of the water molecule is closer to the ion.

This process of water molecules arranging themselves around an ion is called **hydration** and the energy released when this happens is **hydration enthalpy**. There is a balance between the lattice energy and this hydration enthalpy.

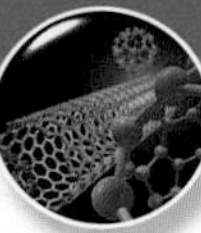

In the case of sodium chloride, the lattice energy is –770 kJ mol^{-1} and the hydration enthalpy is 770 kJ mol^{-1}. The hydration enthalpy is enough to balance the energy needed to break up the lattice and so sodium chloride dissolves in water.

In the case of sodium hydroxide, the lattice energy is much less than its hydration enthalpy. So when sodium hydroxide is mixed with water the surplus energy from hydration is released and the solution gets warmer – dissolving sodium hydroxide is an exothermic change.

For lithium fluoride, the lattice energy is high at –1031 kJ mol^{-1} and the hydration enthalpy is 1025 kJ mol^{-1}. This is insufficient to break down the lattice and lithium fluoride is insoluble in water.

The solubility of alcohols in water

Alcohols are soluble in water because they have polar –OH groups that can hydrogen-bond to the –OH groups in water molecules (see **fig. 2.3.15**).

fig. 2.3.15 Methanol dissolves in water because of hydrogen bonding between the methanol and water molecules.

The solubility of alcohols in water decreases with increasing carbon chain length of the alcohol. This is because a smaller proportion of the molecule is polar. Carboxylic acids dissolve in water in a similar way to alcohols.

Why are non-polar substances insoluble in water?

Non-polar substances such as iodine or hexane do not dissolve in water. **Figure 2.3.16** shows what happens when hexane is added to water. The hexane floats on the water in a separate layer. The forces between the hexane molecules are much weaker than the hydrogen bonds between the water molecules. The hexane molecules are unable to disrupt the water structure and so hexane remains as a separate layer.

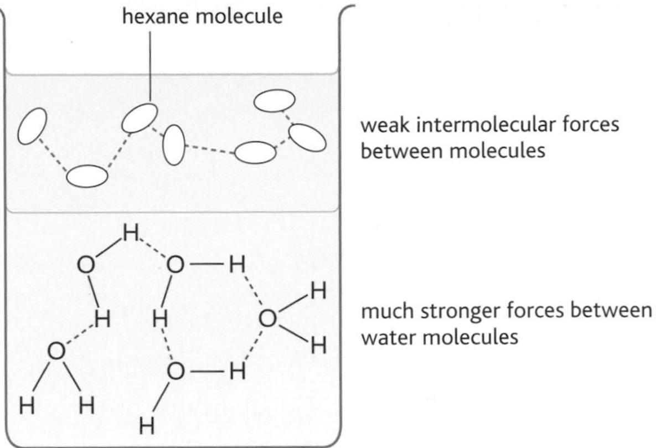

fig. 2.3.16 Hexane is immiscible with water – it cannot disrupt the many strong hydrogen bonds between the water molecules.

Even compounds such as halogenoalkanes which have polar molecules cannot disrupt the water structure, and they remain insoluble. Although the carbon–halogen bond is polar the bulk of the molecule – the hydrocarbon chain – is not. There is insufficient polar character to disrupt the hydrogen-bonded water structure.

Mixing two organic liquids

Hexane and octane are two non-polar liquids that mix completely. Both pure liquids contain weak London forces. When they are mixed these weak forces extend throughout the mixture (see **fig. 2.3.17**).

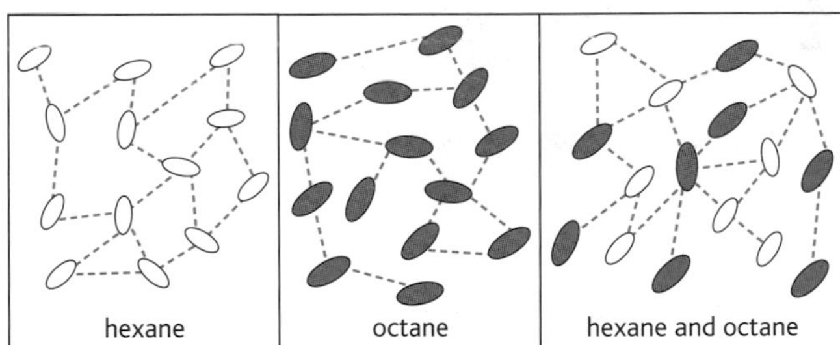

fig. 2.3.17 Hexane and octane mix readily – the weak London forces that form in the pure liquids also form in the mixture in exactly the same way.

Questions

1 For calcium chloride, the enthalpy of hydration is 2378 kJ mol^{-1} and the lattice energy is –2258 kJ mol^{-1}. From these data, what is likely to happen when calcium chloride is added to water?

2 Draw a diagram to show why methanol and ethanol mix well together.

3 Suggest as fully as you can why sugar dissolves in water.

4 Redox

Oxidation number

You will have met examples of **oxidation** and **reduction** reactions before, because combustion is an obvious oxidation reaction. Usually both processes occur together in a **redox reaction**.

Keeping track of electrons

Keeping track of electrons during chemical reactions is important, because the charges on both sides of a chemical equation must balance. One way of doing this is to use **oxidation numbers**, an idea that has a number of applications in chemistry.

For example, in the reaction of chlorine with sodium, you know that one electron is transferred from each sodium atom to a chlorine atom. We write the symbol equation for the reaction as:

$$\underset{0}{2Na(s)} + \underset{0}{Cl_2(g)} \rightarrow \underset{+1\ -1}{2NaCl(s)}$$

The small number underneath the symbol for each element represents its oxidation number. By convention, the oxidation number of an uncombined element is 0. The loss of an electron from sodium is shown by its oxidation number increasing from 0 to +1. In contrast, chlorine gains an electron, which is shown by its oxidation number changing from 0 to −1. Notice how the sum of the oxidation numbers on each side of the equation is the same (in this case it is 0) – this shows that we have not lost or gained any electrons, and our electron book-keeping is correct.

The equation below shows the reaction of lithium with oxygen. On first sight the sum of the oxidation numbers on each side does not appear to be the same. However, they can be balanced if you also take into account the numbers of atoms on each side of the equation:

$$\underset{0}{2Li(s)} + \underset{0}{\tfrac{1}{2}O_2(g)} \rightarrow \underset{+1\ -2}{Li_2O(s)}$$

There are two lithium ions on the right-hand side of the equation, so the total oxidation number associated with lithium is (2 × +1) = +2. Now the book-keeping is correct!

You will be familiar with ionic equations from chapter 1.1. These can be broken down into **half-equations**, which have the advantage of showing how electrons are transferred. For example, for the reaction of sodium with chlorine above:

$$Na(s) \rightarrow Na^+(aq) + e^-$$

$$Cl_2(g) + 2e^- \rightarrow 2Cl^-(aq)$$

Note that for ionic half-equations, the oxidation numbers on each side of the equation do not balance:

$$\underset{0}{Na(s)} \rightarrow \underset{+1}{Na^+(aq)} + e^-$$

Once more, the increase in oxidation number of sodium by 1 shows that it has lost one electron. However, combining both ionic half-equations for a reaction will produce oxidation numbers that balance.

Oxidation numbers in covalent compounds

The examples above show ionic compounds, but chemists also find it useful to extend the oxidation number concept to reactions which involve covalent compounds. For example, think about the reaction of hydrogen and oxygen:

$$2H_2(g) + O_2(g) \rightarrow 2H_2O(l)$$

This reaction looks very different from the reaction of potassium with oxygen, since one involves the formation of K^+ ions and O^{2-} ions while the other produces a molecule held together by shared electrons. But there is a similarity between the two reactions.

In chapter 2.2 you saw that the electrons in the O—H bond are not equally shared, because oxygen is more electronegative than hydrogen (it has a greater tendency to attract electrons than hydrogen in a chemical bond). During the reaction above, the electron density around the atoms changes – a hydrogen atom starts with zero charge in H_2 and finishes with a partial positive charge in H_2O, while an oxygen atom starts with zero charge in O_2 and finishes with a partial negative charge in H_2O. The reaction therefore involves a shift in electron density – which is not so very different from the change that happens in the reaction of potassium and oxygen.

Oxidation numbers and electron density

fig. 2.4.1 Fuel cells use a redox process – the reaction of hydrogen gas and oxygen – to produce electricity. This is a fuel cell from a spacecraft. As well as electricity, the fuel cell produces water.

Oxidation numbers provide a way of following shifts in electron density. We first need to assign oxidation numbers to the elements in the molecules involved. In doing this, we continue to assign an oxidation number of 0 to uncombined elements. Furthermore

- we assign oxidation numbers to elements in covalent compounds so that the oxidation number of an atom is the charge it would have if the electrons in each of its bonds belonged to the more electronegative element.

In this way, in water the hydrogen atoms are treated as though they had lost an electron and the oxygen atom is treated as though it had gained two electrons. The oxidation number of hydrogen in water is therefore +1, while that of oxygen is –2, and the reaction of hydrogen with oxygen can be written with its oxidation numbers included:

$$\underset{0}{2H_2(g)} + \underset{0}{O_2(g)} \rightarrow \underset{+1\ -2}{2H_2O(l)}$$

Once again, notice how the oxidation numbers balance once you take into account the number of atoms involved on each side of the equation. To summarise, here are some simple rules for assigning and using oxidation numbers which will help you keep track of what is happening in reactions.

1 The oxidation number of any uncombined element is 0.
2 The oxidation number of an uncombined ion is the same as its charge.
3 The sum of all the oxidation numbers in a molecule is zero (or in a complicated ion is equal to the charge on the particle).
4 Fluorine always has an oxidation number of –1.
5 Hydrogen always has an oxidation number of +1, except in metal hydrides (when its oxidation number is –1).
6 Oxygen always has an oxidation number of –2, except in peroxides (when it has an oxidation number of –1), and when it combines with fluorine (when it is positive).
7 Chlorine always has an oxidation number of –1, except with oxygen and fluorine when it is positive.
8 Group 1 elements always have oxidation number +1, group 2 always have +2 and group 3 always have +3.

Using these we can assign oxidation numbers to the elements in compounds as shown.

NH_3 Ammonia	Treat as $(N^{3-})(H^+)_3$	so	oxidation number of N in NH_3 = –3 and oxidation number of H in NH_3 = +1.
PCl_5 Phosphorus(V) chloride	Treat as $(P^{5+})(Cl^-)_5$	so	oxidation number of P in PCl_5 = +5 and oxidation number of Cl in PCl_5 = –1.
MnO_4^- Manganate(VII) ions	Treat as $[(Mn^{7+})(O^{2-})_4]^-$	so	oxidation number of Mn in MnO_4^- = +7 and oxidation number of O in MnO_4^- = –2.
OF_2 Oxygen difluoride	Treat as $(O^{2+})(F^-)_2$	so	oxidation number of O in OF_2 = +2 and oxidation number of F in OF_2 = –1.
H_2O_2 Hydrogen peroxide	Treat as $(H^+)_2(O_2)^{2-}$	so	oxidation number of H in H_2O_2 = +1 and oxidation number of O in H_2O_2 = –1.

HSW Does oxidation number exist?

Many students have difficulty with oxidation number because it is an abstract idea. Chemists use this tool because it is convenient and gives the right answer. We know, for example, that in ammonia there can never be N^{3-} or H^+ ions, but treating it as if they exist gives the right answer.

Redox: loss and gain of electrons

Reduction and oxidation – a shift in electron density

The concept of oxidation number is useful in studying the many chemical reactions that involve a shift in electron density – redox reactions. The many important redox processes include burning, respiration and rusting, as well as generating electricity in batteries. Originally the definition of reduction and oxidation was very narrow, being used by chemists to describe reactions that involve hydrogen and oxygen. Nowadays the term is applied to any process involving a transfer of electrons, in which the oxidation number of an element changes.

Oxidation can be defined as any process in which electrons are lost, and **reduction** is any process in which electrons are gained. It might help you to remember this using the mnemonic OIL RIG, which stands for oxidation is loss of electrons, and reduction is gain of electrons.

If you think about the reaction that takes place when chlorine is bubbled through potassium iodide solution:

$$Cl_2(g) + 2KI(aq) \rightarrow 2KCl(aq) + I_2(aq)$$

you would not call this a redox reaction if you used the simple definition of oxidation in terms of loss or gain of oxygen or hydrogen. However, the ionic equation for this reaction is:

$$Cl_2(g) + 2I^-(aq) \rightarrow 2Cl^-(aq) + I_2(aq)$$

Each chlorine atom has gained an electron and each iodide ion has lost an electron. The iodide is therefore oxidised and the chlorine is reduced.

In this reaction iodide ions act as the **reducing agent** – they reduce another substance but are themselves oxidised. Chlorine acts as the **oxidising agent** – it oxidises another substance but is itself reduced.

You can write this equation as two ionic half-equations.

$$2I^-(aq) \rightarrow I_2(s) + 2e^-$$
$$Cl_2(g) + 2e^- \rightarrow 2Cl^-(aq)$$

Adding these half-equations together in such a way that the electrons on each side are cancelled out from the equation gives the overall equation.

Redox reactions: metals with oxygen

Probably one of the first types of redox reaction that most of us come across is the reaction of metals with the oxygen in the air. For sodium:

$$4Na(s) + O_2(g) \rightarrow 2Na_2O(s)$$

In this reaction electrons are removed from sodium and transferred to oxygen – we say that sodium is oxidised and oxygen is reduced. These processes can be shown more clearly by writing two half-equations for the reaction:

$$4Na(s) \rightarrow 4Na^+(s) + 4e^-$$
$$O_2(g) + 4e^- \rightarrow 2O^{2-}(s)$$

Notice how this way of writing the reaction emphasises the fact that the four electrons required by the two oxygen atoms in order to achieve a noble gas configuration are supplied by the four sodium atoms. In the process, the sodium atoms achieve a noble gas configuration too.

This makes it clear that Na loses electrons and is oxidised to Na^+, while O_2 gains electrons and is reduced to O^{2-}. Sodium acts as a reducing agent, while oxygen behaves as an oxidising agent.

Redox reactions: metals with water

The reactions of metals with water will be familiar to you. **Figure 2.4.2** shows sodium reacting with water. You may never have thought of this reaction as a redox reaction, but if you split it into half-equations you can see that it is.

fig. 2.4.2 Sodium reacting with water. An indicator has been added to the water bath which shows the production of the alkali sodium hydroxide by the reaction.

The symbol equation is:

$$2Na(s) + 2H_2O(l) \rightarrow 2NaOH(aq) + H_2(g)$$

The ionic equation is:

$$2Na(s) + 2H_2O(l) \rightarrow 2Na^+(aq) + 2OH^-(aq) + H_2(g)$$

This can be written as two ionic half-equations:

$$Na(s) \rightarrow Na^+(aq) + e^-$$

$$2H_2O(l) + 2e^- \rightarrow 2OH^-(aq) + H_2(g)$$

Before adding these together you need to multiply the first equation by 2 so that there are the same number of electrons in each half-equation and they cancel out.

You can see that sodium loses electrons and so is oxidised, and water gains electrons and so is reduced.

Redox reactions: metals with acid

Again the reactions of metals with acids are familiar reactions that you have probably never thought of as redox reactions. **Figure 2.4.3** shows magnesium reacting with hydrochloric acid.

fig. 2.4.3 Magnesium and dilute hydrochloric acid.

The symbol equation is:

$$Mg(s) + 2HCl(aq) \rightarrow MgCl_2(aq) + H_2(g)$$

This equation can be written as the sum of two ionic half-equations.

$$Mg(s) \rightarrow Mg^{2+}(aq) + 2e^-$$

$$2H^+(aq) + 2e^- \rightarrow H_2(g)$$

Magnesium loses electrons and so is oxidised, and hydrogen ions gain electrons and so are reduced.

Redox reactions: displacement

Iron reacts with copper(II) sulfate solution to form copper and iron(II) sulfate. You will be familiar with this as a **displacement reaction**.

The symbol equation for the reaction is:

$$CuSO_4(aq) + Fe(s) \rightarrow Cu(s) + FeSO_4(aq)$$

Writing two ionic half equations:

$$Cu^{2+}(aq) + 2e^- \rightarrow Cu(s)$$

$$Fe(s) \rightarrow Fe^{2+}(aq) + 2e^-$$

Iron loses electrons so is oxidised, and copper gains electrons so is reduced. Displacement reactions involve a shift of electron density and so are redox reactions.

Questions

1 What is the oxidation number of hydrogen in H_2 and nitrogen in N_2?

2 Outline why the concept of oxidation number is useful even though it is not an accurate representation of the charges in a covalent molecule.

3 Work out the oxidation number of each element in these compounds:
a PH_3 **b** $FeCl_3$ **c** Na_2SO_3 **d** MnO_4^{2-}.

4 In the reaction of sodium with water, which substance is oxidised and which is reduced? Which substance is the oxidising agent and which is the reducing agent?

5 Copper reacts with silver nitrate solution to form silver. Write a symbol equation and then two ionic half-equations. Explain what is happening in terms of oxidation and reduction.

6 Zinc reacts with dilute sulfuric acid solution to form zinc sulfate and hydrogen. Write a symbol equation and then two ionic half equations. Explain what is happening in terms of oxidation and reduction.

7 Does the equation below represent a redox reaction? Explain your answer.
$NaOH(aq) + HCl(aq) \rightarrow NaCl(aq) + H_2O(l)$

Changes in oxidation number

Using oxidation number in naming compounds

Oxidation numbers are a very useful tool for naming compounds. Some elements can exist with more than one oxidation number, so there might be some confusion if we don't use their oxidation number in their names.

For compounds of just two elements this is not usually a problem, so we have:

- PCl_3 phosphorus(III) chloride or phosphorus **tri**chloride
- PCl_5 phosphorus(V) chloride or phosphorus **penta**chloride
- CO carbon **mon**oxide
- CO_2 carbon **di**oxide
- OF_2 oxygen **di**fluoride
- O_2F_2 **di**oxygen **di**fluoride.

In more complex compounds, the name of the element which has a variable oxidation number is written, with its oxidation number in Roman numerals after it:

- $KMnO_4$ potassium manganate(VII)
- K_2CrO_4 potassium chromate(VI)

As you saw in chapter 1.5, the systematic names of chemical compounds are agreed by IUPAC (the International Union of Pure and Applied Chemistry). Generally speaking you should use the systematic names of compounds, although you will often find the common names used too. Some chemicals when named using the IUPAC rules are cumbersome and people often use the traditional or trivial names for these chemicals. Some of examples are given in **table 2.4.1**.

Substance	IUPAC systematic name	Traditional name
HNO_2	Nitric(III) acid	Nitrous acid
$HN0_3$	Nitric(V) acid	Nitric acid
NO_2^-	Nitrate(III)	Nitrite
NO_3^-	Nitrate(V)	Nitrate
H_2SO_3	Sulfuric(IV) acid	Sulfurous acid
H_2SO_4	Sulfuric(VI) acid	Sulfuric acid
SO_3^{2-}	Sulfate(IV)	Sulfite
SO_4^{2-}	Sulfate(VI)	Sulfate

table 2.4.1 Systematic names use oxidation numbers to be unambiguous. Sometimes the traditional names are still used.

Redox in terms of changes in oxidation number

You have seen that if a species gains or loses an electron, then reduction or oxidation has taken place. It follows that a change in oxidation number also indicates a redox reaction, since changes in oxidation numbers represent shifts in electron density. An increase in oxidation number corresponds to oxidation and a decrease in oxidation number corresponds to reduction.

For example, in the reaction of chlorine with hydrogen sulfide:

$$Cl_2(g) + H_2S(g) \rightarrow 2HCl(g) + S(s)$$

the oxidation number of chlorine changes from 0 (in the element) to –1 in HCl. Chlorine is therefore reduced because its oxidation number is reduced from 0 to –1.

The oxidation number of sulfur changes from –2 in H_2S to 0 in S. Sulfur is oxidised because its oxidation number has increased.

As another example, think about reaction of lead(II) oxide with hydrogen. The equation is:

$$PbO(s) + H_2(g) \rightarrow Pb(s) + H_2O(l)$$

The oxidation number of lead changes from +2 in PbO to 0 in Pb, so lead is reduced. The oxidation number of hydrogen changes from 0 in H_2 to +1 in H_2O, so hydrogen is oxidised.

The thermite reaction is a useful process involving the reaction of iron(III) oxide and aluminium (see **fig. 2.4.4**).

fig. 2.4.4 The thermite reaction has many industrial applications. It is an exothermic reaction that produces molten iron, useful in welding.

The equation for the reaction is:

$$Fe_2O_3(s) + 2Al(s) \rightarrow 2Fe(l) + Al_2O_3(s)$$

In this reaction the oxidation number of iron changes from +3 to 0, so iron is reduced. The oxidation number of aluminium changes from 0 to +3, so aluminium is oxidised.

Disproportionation

A **disproportionation reaction** takes place when a substance is both oxidised and reduced in the same reaction. An example is the breakdown of hydrogen peroxide into water and oxygen:

$$2H_2O_2(aq) \rightarrow 2H_2O(l) + O_2(g)$$

Writing this as two ionic half-equations, you can see how the reactant is both oxidised and reduced:

$$\underset{+1\,-1}{H_2O_2}(aq) \rightarrow \underset{0}{O_2}(g) + 2\underset{+1}{H^+}(aq) + 2e^- \quad \text{oxidised}$$

$$\underset{+1\,-1}{H_2O_2}(aq) + 2\underset{+1}{H^+}(aq) + 2e^- \rightarrow 2\underset{+1\,-2}{H_2O}(l) \quad \text{reduced}$$

The oxygen in the hydrogen peroxide disproportionates.

Another example of disproportionation is the reaction of copper(I) oxide with dilute sulfuric acid to form copper(0) and copper(II) sulfate:

$$\underset{+1\;\;-2}{Cu_2O}(s) + \underset{+1\,+6\,-2}{H_2SO_4}(aq) \rightarrow \underset{0}{Cu}(s) + \underset{+2\,+6\,-2}{CuSO_4}(aq) + \underset{+1\,-2}{H_2O}(l)$$

Copper is both oxidised from +1 to +2 and reduced from +1 to 0 in the same reaction.

Other examples of disproportionation that you will meet include the reaction of chlorine with water, and the reaction of chlorine with hot sodium hydroxide solution.

HSW Using oxidation number

Oxidation number is useful to work out changes that take place during a reaction at an atomic level.

Questions

1 In a chemical reaction, MnO_4^{2-} is changed first to MnO_4^- and then to MnO_2. Are these changes oxidation or reduction? Explain your answer in terms of changes in oxidation number.

2 Write the chemical formula of iron(III) sulfate(VI).

3 In the following reactions, work out the oxidation numbers of each element in bold. Then decide whether this element is oxidised, reduced or neither oxidised nor reduced.

a $KI(aq) + AgNO_3(aq) \rightarrow AgI(s) + KNO_3(aq)$

b $CuSO_4(aq) + \mathbf{Zn}(s) \rightarrow Cu(s) + ZnSO_4(aq)$

c $SnCl_2(aq) + \mathbf{HgCl_2}(aq) \rightarrow Hg(l) + SnCl_4(aq)$

d $2\mathbf{CuSO_4}(aq) + 4KCN(aq) \rightarrow 2CuCN(s) + C_2N_2(g) + 2K_2SO_4(aq)$

e $10\mathbf{FeSO_4}(aq) + 2KMnO_4(aq) + 8H_2SO_4(aq) \rightarrow 5Fe_2(SO_4)_3(aq) + K_2SO_4(aq) + 2MnSO_4(aq) + 8H_2O(l)$

Some useful ionic half-equations

Some elements show a whole range of oxidation states. A colourful reaction you might see is the reduction of vanadium(V) through successive oxidation numbers to vanadium(II) (see **fig. 2.4.5**). Zinc and dilute hydrochloric acid are added to a solution of ammonium vanadate(V) (yellow). The zinc and hydrochloric acid produce hydrogen, which reduces vanadium(V) first to vanadium(IV) (blue) then vanadium(III) (green) and finally vanadium(II) (purple).

fig. 2.4.5 Vanadium shows a range of oxidation states.

Common oxidising agents

Below is a list of ionic half-equations for some common oxidising agents. Oxidising agents react by oxidising something else, and are themselves reduced. So the electrons are always on the left-hand side of the equation.

Oxygen

$O_2(g) + 4e^- \rightarrow 2O^{2-}(s)$

Chlorine

$Cl_2(g) + 2e^- \rightarrow 2Cl^-(s)$

Bromine

$Br_2(g) + 2e^- \rightarrow 2Br^-(s)$

Iodine

$I_2(g) + 2e^- \rightarrow 2I^-(s)$

Manganate(VII) in acid solution

$MnO_4^-(aq) + 8H^+(aq) + 5e^- \rightarrow Mn^{2+}(aq) + 4H_2O(l)$

Dichromate(VI) in acid solution

$Cr_2O_7^{2-}(aq) + 14H^+(aq) + 6e^- \rightarrow 2Cr^{3+}(aq) + 7H_2O(l)$

Iron(III) salts

$Fe^{3+}(aq) + e^- \rightarrow Fe^{2+}(aq)$

Hydrogen ions

$2H^+(aq) + 2e^- \rightarrow H_2(g)$

Hydrogen peroxide in absence of another oxidising agent

$H_2O_2(aq) + 2H^+(aq) + 2e^- \rightarrow 2H_2O(l)$

Concentrated sulfuric (VI) acid

$2H_2SO_4(l) + 2e^- \rightarrow SO_4^{2-}(aq) + 2H_2O(l) + SO_2(g)$

Common reducing agents

Below is a list of ionic half-equations for common reducing agents. Reducing agents react by reducing something else, and are themselves oxidised. So the electrons are always on the right-hand side.

Metals

$M(s) \rightarrow M^{n+}(aq) + ne^-$

For example:

$Zn(s) \rightarrow Zn^{2+}(aq) + 2e^-$

Iron(II) salts

$Fe^{2+}(aq) \rightarrow Fe^{3+}(aq) + e^-$

Acidified potassium iodide

$2I^-(aq) \rightarrow I_2(aq) + 2e^-$

Thiosulfate

$2S_2O_3^{2-}(aq) \rightarrow S_4O_6^{2-}(aq) + 2e^-$

Ethanedioic acid and ethanedioates

$C_2O_4^{2-}(aq) \rightarrow 2CO_2(g) + 2e^-$

Sulfuric(IV) acid

$SO_3^{2-}(aq) + H_2O(l) \rightarrow SO_4^{2-}(aq) + 2H^+(aq) + 2e^-$

Hydrogen peroxide in the presence of an acid and absence of a strong oxidising agent

$H_2O_2(aq) \rightarrow O_2(g) + 2H^+(aq) + 2e^-$

Hydrogen

$2H_2 + O_2 \rightarrow 2H_2O$

Iron(II) salts

$Fe^{2+}(aq) \rightarrow Fe^{3+}(aq) + e^-$

Using half equations to write ionic equations

Write a balanced equation for the following reactions by combining the relevant half-equations.

1 **Reaction of manganate(VII) in acid solution with iron(II) ions.**

$MnO_4^-(aq) + 8H^+(aq) + 5e^- \rightarrow Mn^{2+}(aq) + 4H_2O(l)$

$Fe^{2+}(aq) \rightarrow Fe^{3+}(aq) + e^-$

Now you need to add these half-equations together so that the electrons cancel out. To do this multiply the second equation through by 5:

$MnO_4^-(aq) + 8H^+(aq) + 5e^- \rightarrow Mn^{2+}(aq) + 4H_2O(l)$

$5Fe^{2+}(aq) \rightarrow 5Fe^{3+}(aq) + 5e^-$

Adding these together gives the balanced ionic equation:

$\mathbf{MnO_4^-(aq) + 8H^+(aq) + 5Fe^{2+}(aq) \rightarrow Mn^{2+}(aq) + 4H_2O(l) + 5Fe^{3+}(aq)}$

2 **Reaction of manganate(VII) in acid solution with acidified potassium iodide.**

$MnO_4^-(aq) + 8H^+(aq) + 5e^- \rightarrow Mn^{2+}(aq) + 4H_2O(l)$

$2I^-(aq) \rightarrow I_2(aq) + 2e^-$

For the electrons to cancel out on addition, the first equation must be multiplied through by 2 and the second by 5:

$2MnO_4^-(aq) + 16H^+(aq) + 10e^- \rightarrow 2Mn^{2+}(aq) + 8H_2O(l)$

$10I^-(aq) \rightarrow 5I_2(aq) + 10e^-$

Adding these together gives the balanced ionic equation:

$\mathbf{2MnO_4^-(aq) + 16H^+(aq) + 10I^-(aq) \rightarrow 2Mn^{2+}(aq) + 8H_2O(l) + 5I_2(aq)}$

3 **Reaction of acidified dichromate(VII) with hydrogen peroxide solution.**

Fig. 2.4.6 shows the change of colour as dichromate(VI) reacts to produce chromium(III) ions.

In selecting the ionic half-equations, make sure you choose the correct equation for hydrogen peroxide acting as a reducing agent.

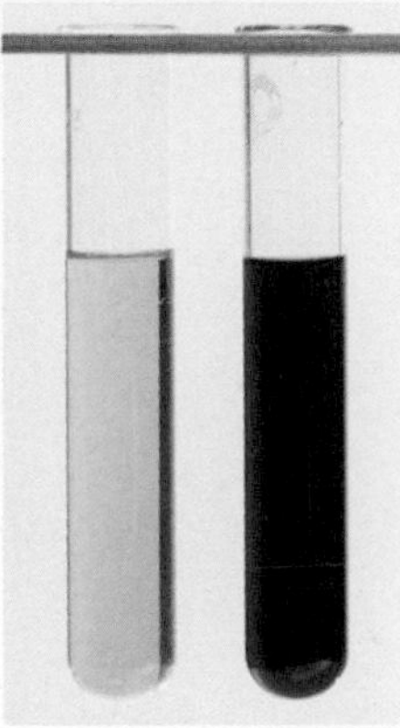

fig. 2.4.6 **Potassium dichromate(VI) solution acts as an oxidising agent. It is reduced to form chromium(III) solution.**

$Cr_2O_7^{2-}(aq) + 14H^+(aq) + 6e^- \rightarrow 2Cr^{3+}(aq) + 7H_2O(l)$

$H_2O_2(aq) \rightarrow O_2(g) + 2H^+(aq) + 2e$

Multiply the second equation by 3 and add them together:

$Cr_2O_7^{2-}(aq) + 14H^+(aq) + 3H_2O_2(aq) \rightarrow 2Cr^{3+}(aq) + 7H_2O(l) + 3O_2(g) + 6H^+(aq)$

You can simplify the equation by subtracting $6H^+(aq)$ from each side:

$\mathbf{Cr_2O_7^{2-}(aq) + 8H^+(aq) + 3H_2O_2(aq) \rightarrow 2Cr^{3+}(aq) + 7H_2O(l) + 3O_2(g)}$

Producing a balanced ionic equation

To write a balanced ionic equation from ionic half-equations, first choose the correct ionic half-equations.

If you have an ionic equation, it is easy to turn it into a molecular equation if you need to. For example, for the reaction of potassium manganate(VII) solution with iron(II) sulfate solution in the presence of dilute sulfuric acid you add in the spectator ions:

$$2KMnO_4(aq) + 8H_2SO_4(aq) + 10FeSO_4(aq) \rightarrow 2MnSO_4(aq) + 8H_2O(l) + K_2SO_4(aq) + 5Fe_2(SO_4)_3(aq)$$

Questions

1 Write an ionic equation for the reaction of thiosulfate ions and iodine.

2 Write an ionic equation for the reaction of bromine and iron(II) ions.

3 Write an ionic equation for the reaction of copper and concentrated sulfuric acid.

4 Turn the ionic equation for the reaction of potassium manganate(VII) with potassium iodide solution in the presence of dilute sulfuric acid into a molecular equation.

5 The periodic table – groups 2 and 7

fig. 2.5.1 Group 2 metals. From the left: beryllium, magnesium, calcium, strontium and barium.

Group 2: the alkaline earth metals

You have probably studied the properties of the alkali metal elements in group 1 of the periodic table at GCSE. You will have seen clear patterns in the physical and chemical properties of these elements. In this chapter you are going to study two further groups – group 2, a family of metals, and group 7, a family of non-metals.

Ionisation energy

As you saw in chapter 1.3, the **ionisation energy** depends on the electronic structure of an element, and it has a marked effect on its physical and chemical properties. The first ionisation energy is the energy change when one mole of atoms of an element in the gaseous state loses one electron each to form one mole of positively charged ions:

$$M(g) \rightarrow M^{+}(g) + e^{-}$$

The second ionisation energy is the energy change when one mole of gaseous single positively charged ions loses one mole of electrons to form one mole of double positively charged ions.

Remember that the value of the first ionisation energy depends on:

- the distance between the outer electron and the nucleus (the atomic radius)
- the effective nuclear charge
- the shielding produced by inner electron shells.

Shielding is caused by the repulsion between the inner, filled shells and the electron being removed in the outer shell.

Ionisation energies of group 2 elements

The elements in group 2 of the periodic table are called the **alkaline earth metals**. They are named after their oxides, called the alkaline earths. Their old-fashioned names were beryllia, magnesia, lime, strontia and baryta. 'Earth' is an old term used by early chemists for non-metallic substances that are insoluble in water and do not split up by heating. Antoine Lavoisier was the first person, in 1789, to realise that these substances were not elements and suggested that they might be metal oxides. Humphry Davy became the first to obtain samples of the metals by electrolysis of their molten earths at the beginning of the nineteenth century. Figure 2.5.1 shows samples of the group 2 metals.

Table 2.5.1 shows the first, second and third ionisation energies and atomic radii of the elements in group 2.

Element	Atomic radius (nm)	Ionisation energy (kJ mol^{-1})		
		First	Second	Third
Beryllium	0.112	900	1757	14849
Magnesium	0.160	738	1451	7733
Calcium	0.197	590	1145	4912
Strontium	0.215	550	1064	4210
Barium	0.224	503	965	3390

table 2.5.1 The ionisation energies of the group 2 elements. These elements form M^{2+} ions readily.

You can see that for all the group 2 elements there is a big increase between the second and the third ionisation energies. These elements readily lose two electrons to form a 2+ ion, but the large third ionisation energies make it impossible for them to lose three electrons and form a 3+ ion.

Table 2.5.2 shows the electronic configurations of the elements in group 2. These elements in the s block have two electrons in the outer shell, so losing two electrons gives them a noble gas configuration. This is consistent with their tendency to form M^{2+} ions readily.

Element	Symbol	Electronic configuration
Beryllium	Be	2,2
Magnesium	Mg	2,8,2
Calcium	Ca	2,8,8,2
Strontium	Sr	2,8,18,8,2
Barium	Ba	2,8,18,18,8,2

table 2.5.2 Here electronic configurations are shown in simple form, with $1s^22s^2$ being shown simply as 2,2 and so on.

You will notice from table 2.5.1 that the ionisation energy decreases down the group – it becomes progressively easier to remove an electron as you move down the group. Why is this?

There is an increasing effective nuclear charge as you move down the group. This might suggest that it should become more difficult to remove an electron down the group. However, you can see in table 2.5.1 that there is an increase in atomic radius down the group so that the outermost electrons are further from the nucleus and so they experience the charge less strongly. Also, there is more shielding from inner subshells as you go down the group – beryllium has just one full inner shell, magnesium has two, and so on up to barium with five. These two factors together are more significant that the increasing effective nuclear charge, and so the ionisation energy decreases down the group.

You will notice a similar trend if you consider the sum of the first and second ionisation energies. You will see later how this links with the reactivity of alkaline earth metals.

HSW Using a simple model of the atom to explain trends

The explanation for trends in ionisation energy given above is based on a simple model of the atom devised by Niels Bohr in 1913. In his atom the nucleus of protons (neutrons were not discovered until 1932) were surrounded by electrons in clearly defined shells. Each shell could hold a maximum number of electrons. We know today that this is a simplified model of the atom but it is sufficient to explain the reactions of the group 2 metals.

Questions

1 Write an equation for a metal M which represents the third ionisation energy.

2 An Mg^{2+} ion has a radius of 0.072 nm, compared with 0.160 nm, the radius of the magnesium atom. Explain why the magnesium ion is smaller than the magnesium atom. (Looking back to chapter 1.3 may help you.)

3 Explain the trends in successive ionisation energies of the group 2 elements, referring to their electronic structures.

Flame tests and characteristic reactions of the group 2 metals

Chemists often want to do a quick and simple test to identify the metal ions (cations) in a sample. A **flame test** is useful for this purpose, particularly for identifying metals from the s block – groups 1 and 2.

Carrying out a flame test

A few drops of concentrated hydrochloric acid are added to the sample on a watch glass. A clean platinum or nichrome wire is dipped into the mixture and then the wire is held in a hot Bunsen burner flame. The colour of the flame allows you to identify the cation in the mixture (see **fig. 2.5.2**).

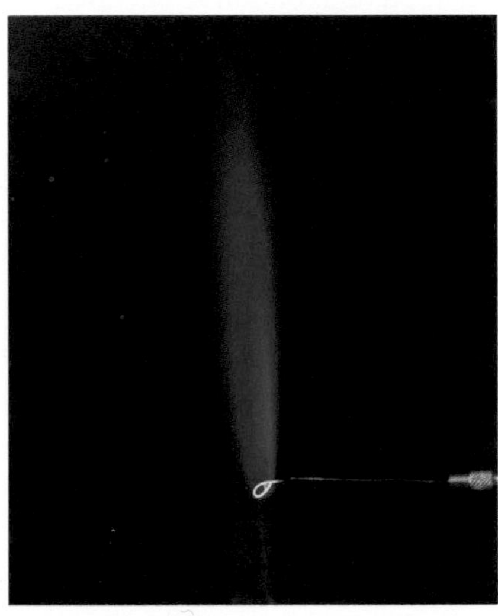

fig. 2.5.2 **The distinctive colours of the group 1 and 2 metals in a flame test is a useful tool in the analysis of unknown compounds in the laboratory. This shows the red flame produced by lithium compounds.**

Table 2.5.3 summarises the colour of the flame produced with different group 1 and 2 cations.

Cation	Flame colour
Lithium	Red
Sodium	Yellow
Potassium	Lilac
Magnesium	No colour
Calcium	(Yellow-) red
Strontium	Red
Barium	(Pale) green

table 2.5.3 **Flame test colours**

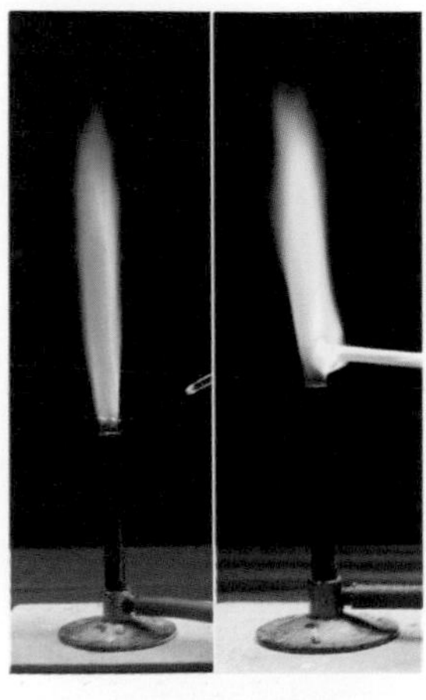

fig. 2.5.3 **Flame tests for sodium and barium.**

Flame tests can only be a guide – many cations have no effect on the flame and sodium, which has an intense yellow flame, can mask the presence of other cations.

What causes coloured flames in flame tests?

In a flame test, the sample is converted to a chloride because chlorides are more volatile than other salts.

You know from chapter 1.3 that electrons occupy certain discrete energy levels in an atom or an ion. When an electron is promoted from its usual energy level to a higher energy level, the atom becomes excited. When the electron drops back to its ground state, the atom emits a photon of light in the visible range (wavelength range about 400–700 nm, see **fig. 2.5.4**).

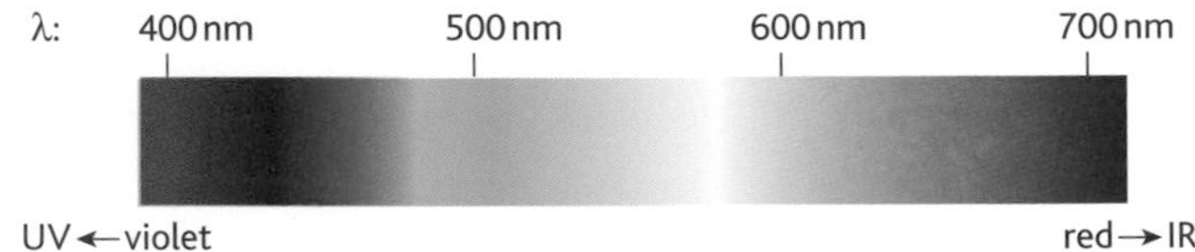

fig. 2.5.4 **The wavelengths of the spectrum of visible light.**

Each element has its own characteristic energy levels. You can see that red light has an approximate wavelength of 800 nm, and this is the wavelength of the energy given out when a lithium ion returns to its ground state from the excited state. Similarly, a sodium ion gives out energy of wavelength 587 nm, in the yellow range. Atoms or ions that give out energy in regions of the spectrum outside the visible region do not show a characteristic colour in a flame test. For some the difference between the energy levels is too large for the ion to become excited in a Bunsen burner flame.

A more accurate test – the flame photometer

A **flame photometer** is an instrument that can be used to measure quantitatively the wavelengths of light given out by ions such as sodium, potassium and barium in a sample by analysing the flame produced (see **fig. 2.5.5**).

A flame photometer is used routinely to find the levels of sodium and potassium ions in blood samples. Millions of blood samples are analysed every day and without this technique the analysis would be extremely slow and very expensive. The technique is also used to monitor the concentrations of sodium, potassium and calcium during the manufacture of wines.

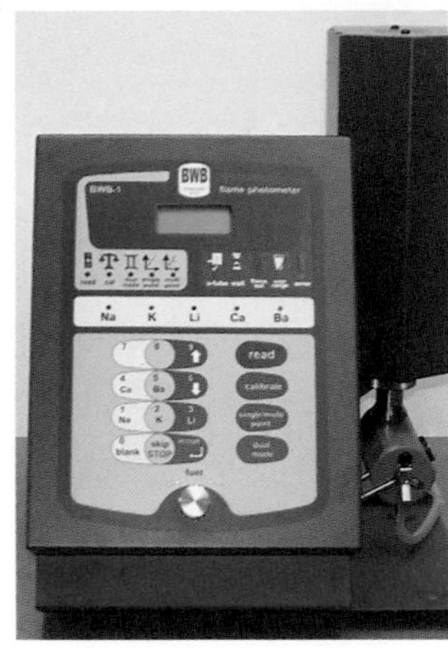

fig. 2.5.5 A flame photometer.

Reactions of the group 2 metals

As well as flame tests, there are other characteristic chemical properties of the group 2 metals and here you will look at their reactions with oxygen, with chlorine and with water. Once again trends in their reactivity down the group are obvious.

With oxygen

The group 2 metals burn in air or oxygen to form solid metal oxides. They usually burn brightly in oxygen (see **fig. 2.5.6**).

(a)

(b)

fig. 2.5.6 **(a) Burning magnesium in air. (b) Burning calcium in oxygen.**

The general equation shows the reaction:

$$2M(s) + O_2(g) \rightarrow 2MO(s)$$

For example:

$$2Mg(s) + O_2(g) \rightarrow 2MgO(s)$$

$$2Ca(s) + O_2(g) \rightarrow 2CaO(s)$$

The reactivity of group 2 metals with oxygen increases down the group. Barium, being the most reactive, is often stored under oil like the alkali metals.

With chlorine

The group 2 metals burn in chlorine gas to form solid metal chlorides:

$$M(s) + Cl_2(g) \rightarrow MCl_2(s)$$

For example:

$$Mg(s) + Cl_2(g) \rightarrow MgCl_2(s)$$

$$Ca(s) + Cl_2(g) \rightarrow CaCl_2(s)$$

Again the reactivity increases down the group, although this is not as clearly seen as in the reaction with oxygen.

With water

The group 2 metals react with water to form oxides or hydroxides. The reactivity increases down the group, as you can see in **table 2.5.4**.

Element	Reaction with water or steam
Beryllium	No reaction due to a thick oxide layer on the surface of the metal
Magnesium	Very, very slow reaction with cold water – rapid reaction with steam
Calcium	Reaction with cold water to produce hydrogen – reactivity increases down the group
Strontium	
Barium	

table 2.5.4 **The reaction of the group 2 metals with water.**

Magnesium reacts with steam to form magnesium oxide and hydrogen:

$$Mg(s) + H_2O(g) \rightarrow MgO(s) + H_2(g)$$

Calcium, strontium and barium react to form the hydroxide and hydrogen, for example:

$$Ca(s) + 2H_2O(g) \rightarrow Ca(OH)_2(aq) + H_2(g)$$

In each case the hydroxide forms an alkaline solution.

Questions

1. Suggest why calcium compounds produce a coloured flame but magnesium compounds do not.
2. How do you clean a wire for a flame test? Explain why this is important.
3. Describe and explain the trends in reactivity of the group 2 metals with water.
4. Barium is the most difficult group 2 metal to extract by electrolysis. Suggest why this is.

Reactions of the group 2 oxides and hydroxides

Group 2 oxides and hydroxides are not found naturally but the elements are usually found as carbonates which decompose on heating to form the oxide. This reacts with water to form the hydroxide. For example, calcium carbonate occurs in different forms, including limestone, chalk and marble. Heating these strongly decomposes them to form calcium oxide (sometimes called quicklime). Adding water to calcium oxide produces calcium hydroxide (sometimes called slaked lime). You will be familiar with a solution of calcium hydroxide in water in the laboratory as limewater.

Reactions of the group 2 oxides

With water

Beryllium oxide does not react with water, while magnesium oxide reacts only slightly.

When cold water is added to calcium oxide, the mixture swells, a fizzing sound is heard and a considerable amount of energy and water vapour are released.

$$CaO(s) + H_2O(l) \rightarrow Ca(OH)_2(aq)$$

This process is sometimes called **slaking** lime and the product is calcium hydroxide. The word 'slake' is Anglo-Saxon and means to satisfy a thirst for water. In this reaction it is clear that calcium oxide has a liking for water.

Calcium hydroxide has many uses including in water treatment, neutralising acidic soils and making whitewash, mortar and plaster.

fig. 2.5.7 Traditional building techniques use a lot of lime in lime washes, lime putty and other materials. There is still a market for these products today – this man is making lime putty.

Strontium oxide and barium oxide react in a similar way to calcium oxide.

With dilute hydrochloric acid and nitric acids

Group 2 metal oxides react with dilute hydrochloric acid to form a chloride salt and water only, eg:

$$MgO(s) + 2HCl(aq) \rightarrow MgCl_2(aq) + H_2O(l)$$

Similarly they react with dilute nitric acid to form a nitrate salt and water only, eg:

$$CaO(s) + 2HNO_3(aq) \rightarrow Ca(NO_3)_2(aq) + H_2O(l)$$

Reactions of the group 2 hydroxides

With dilute hydrochloric acid and nitric acids

Group 2 metal hydroxides react in a similar way to oxides, eg:

$$Mg(OH)_2(s) + 2HCl(aq) \rightarrow MgCl_2(aq) + 2H_2O$$

$$Ca(OH)_2(s) + 2HNO_3(aq) \rightarrow Ca(NO_3)_2(aq) + 2H_2O(l)$$

Solubility of group 2 compounds

Chemists often need to be aware of the relative solubilities of different compounds. For example, sometimes doctors need to see inside the gut in detail – but soft tissues do not show up well on X–rays. However, barium compounds show up really well. If a patient swallows a 'barium meal' containing a barium compound, or is given a barium enema, the shape of that region of the gut will show up clearly on any X–rays taken. However, soluble barium compounds are poisonous to people. Fortunately barium sulphate is insoluble in water, so it is ideal for the job. It shows up well on X–rays but doesn't poison the patient!

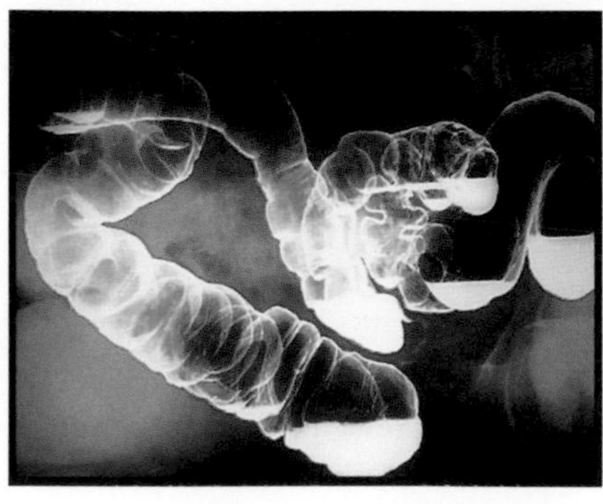

fig. 2.5.8 An X–ray taken after a barium meal, showing the gastrointestinal tract.

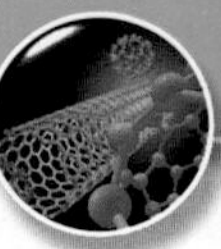

This is just one example where knowing the solubility of different compounds is vital in the way they can be used.

There are a few general trends in solubility in the compounds of the group 2 metals. All the group 2 metal nitrates are soluble, and so are the chlorides. Group 2 salts in which the anion has a charge of –2, such as the sulfates, are largely insoluble with the exception of a few magnesium and calcium salts.

In addition to these trends, there is a tendency for the solubility of the salts to decrease down the group as the atomic number and ionic size of the metals increase. Table 2.5.5 shows the solubilities of group 2 metal sulfates and hydroxides. The solubility of the sulfates decreases down the group while the solubility of hydroxides increases.

Element	Solubility of sulfate (moles/100 g water)	Solubility of hydroxide (moles/100 g water)
Magnesium	3600×10^{-4}	0.2×10^{-4}
Calcium	11×10^{-4}	16×10^{-4}
Strontium	0.62×10^{-4}	330×10^{-4}
Barium	0.0009×10^{-4}	240×10^{-4}

table 2.5.5 Trends in the solubilities of the group 2 metal sulfates and hydroxides.

The insolubility of barium sulfate is used in the laboratory test for sulfate ions. To test an unknown sample to see if it is a sulfate, dilute hydrochloric acid is added to the sample to destroy any carbonate. Then barium chloride or barium nitrate solution is added. A white precipitate of barium sulfate confirms that the salt is a sulfate (see fig. 2.5.9).

fig. 2.5.9 The result of a positive sulfate test.

$$Ba^{2+}(aq) + SO_4^{2-}(aq) \rightarrow BaSO_4(s)$$

The relative insolubility of calcium hydroxide can be demonstrated by adding sodium hydroxide solution to calcium nitrate solution. A white precipitate of calcium hydroxide is formed (see fig. 2.5.10).

fig. 2.5.10 Precipitating calcium hydroxide shows it is more insoluble than sodium hydroxide, calcium nitrate or sodium nitrate.

By comparison, the group 1 metals produce hydroxides, sulfates, chlorides, nitrates and carbonates that are soluble in water. Because group 2 compounds are more likely to contain 2+ and 2– ions, they are likely to have higher lattice energies (eg NaCl –780 kJ mol^{-1} compared with $MgCl_2$ –2526 kJ mol^{-1}). Although the hydration enthalpies of group 2 ions are generally larger, this does not compensate for the higher lattice energies (see chapter 2.3).

Questions

1 Write symbol equations for:
   a the reaction of barium with water
   b the reaction of barium oxide with dilute hydrochloric acid
   c the reaction of barium hydroxide with nitric acid.

2 In table 2.5.5 the units used are moles/100 g of water. Suggest why this is better than using g/100 g of water.

3 With barium oxide and barium hydroxide, the usual reactions of a metal oxide and hydroxide with dilute sulfuric acid do not occur in practice. Referring to table 2.5.5, suggest why this is.

Volumetric analysis

In chapter 1.1 you saw that there are different ways of expressing the concentration of a solution, and you looked at how solutions can be made up to an accurate concentration. But how can you find the concentration of an unknown solution? The most usual way in the lab is by carrying out **volumetric analysis**, and this is explained here.

The concentration of calcium ions in a solution can be measured quantitatively using a flame photometer. However magnesium, which does not colour the flame, cannot be measured in this way. A solution of EDTA (ethylenediaminetetraacetic acid) can be used to measure the concentration of calcium and magnesium in water samples (testing for hardness of water) or in rock samples.

Principles of volumetric analysis

In volumetric analysis you can find the accurate volumes and concentrations of reacting solutions.

The common method of volumetric analysis is called **titration**. In this method, the concentration of an unkown solution can be found by reacting it with a standard solution. A **standard solution** is one whose concentration is known and does not change with time.

If the volumes of two solutions which exactly react with each other are measured, eg an acid and an alkali, the concentration of one can be calculated if the concentration of the other is known.

As you saw in chapter 1.1, the volume of a solution is measured in cubic decimetres, dm^3 and the concentration of a solution in moles per cubic decimetre, mol dm^{-3}.

Finding the molar mass of an organic acid

You can carry out an acid–base titration to find the molar mass of an organic acid. By dissolving a known mass of the acid in water, you can titrate this against a known solution of an alkali. An indicator tells you when the reaction is complete, and the volume of alkali can be used to calculate the number of moles that have reacted.

Precision, accuracy and reliability

Figure 2.5.11 shows the apparatus commonly used for volumetric analysis.

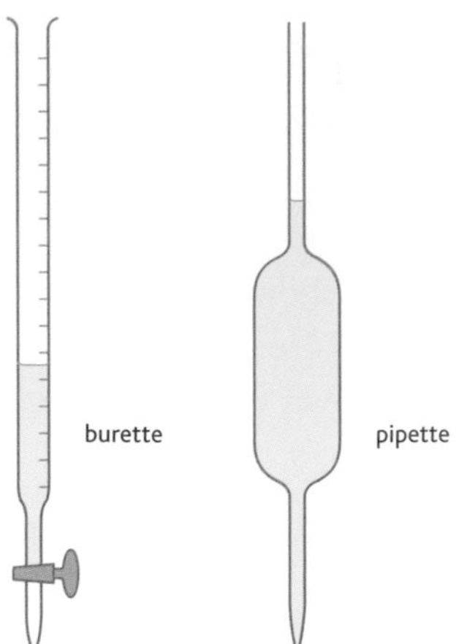

fig. 2.5.11 A burette and a pipette, commonly used for volumetric analysis.

The pipette delivers a fixed volume of solution, usually 25.0 or 10 cm^3. It is a precision instrument capable of delivering an exact volume to within 0.05 cm^3. Similarly the burette is a precision instrument that can deliver any volume measured to the nearest 0.05 cm^3. However, a pipette might be wrongly calibrated and it might actually deliver 25.05 cm^3 each time. It is then **precise** but not **accurate**. It delivers the same volume each time, but not the volume it claims to have delivered.

To check the accuracy of a pipette you can transfer its contents to a weighed beaker and find the mass of the beaker and contents. From this and the density of the contents, you can work out the exact volume delivered.

The **reliability** of the method depends upon errors particular to the experiment and the skill of the experimenter. For example, if you do not line up your eye with the top of the liquid (the meniscus) and with the graduation on the apparatus, you will read the wrong result. Repeating an experiment does not increase the reliability of the experiment or the accuracy of the result. It will, however, give you confidence in the result if readings are the same when the experiment is repeated.

In an experiment 3.02 g of a monobasic organic acid (an acid that will donate just one hydrogen ion (proton) when reacting with an alkali) was dissolved in water and the solution made up to 250 cm^3.

This solution was placed in a burette. 25.0 cm^3 of sodium hydroxide solution (0.10 mol dm^{-3}) solution were pipetted into a conical flask and a couple of drops of phenolphthalein indicator added. Acid solution was run in from the burette until the indicator showed neutralisation. The procedure was repeated three times more, and the results are shown in table 2.5.6.

	Experiment			
	1	2	3	4
Final volume on burette (cm^3)	25.50	25.35	25.55	25.45
Initial volume on burette (cm^3)	0.00	0.10	0.25	0.20
Volume of acid added (cm^3)	25.50	25.25	25.30	25.25

table 2.5.6 Sample results for the titration of an unknown organic acid against 25.0 cm^3 of sodium hydroxide solution.

Questions

1 A solution of sodium hydroxide contains 1.00 g of sodium hydroxide in 250 cm^3 of solution. $M(\text{NaOH}) = 40$ g mol^{-1}. Calculate the concentration of sodium hydroxide in: **a** g dm^{-3} **b** mol dm^{-3}.

2 In the experiment above a student's average titration reading was 0.3 cm^3 higher than the result given above. Calculate the value for the concentration of the acid they obtained. What is the percentage difference?

3 Evaluate the method shown above by considering its limitations, accuracy of measurement and types of error.

In the first experiment, the experimenter had no idea when the indicator was going to change (the **end point**) so did not close the tap on the burette immediately. After that, the titration could be carried out much more accurately as acid could be run in until nearly the end point, and then drop by drop to find the exact point of neutralisation. For this reason the first result was not used when analysing the results.

The average volume of acid that completely reacted with 25 cm^3 of sodium hydroxide is therefore calculated as:

$$\text{Average volume} = \frac{25.25 + 25.30 + 25.25}{3} = 25.25 \text{ cm}^3 \text{ (to the nearest 0.05 cm}^3\text{)}$$

Note that:

- The initial reading on the burette does not have to be exactly 0.00 cm^3 each time.
- All readings are taken to *the nearest 0.05 cm^3*.
- When calculating the average, despite the calculator showing 25.266 666 cm^{-3}, the average recorded should be rounded to the nearest 0.05 cm^3 because that is how accurate your apparatus is.

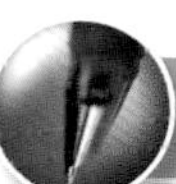

Calculating the molar mass from the titration

Using table 2.5.6, calculate the concentration of the acid solution in mol dm^{-3} and find the relative molecular mass of the unknown organic acid.

The reaction between this unknown organic acid and sodium hydroxide can be represented by the equation:

$$\text{RCOOH(aq)} + \text{NaOH(aq)} \rightarrow \text{RCOONa(aq)} + \text{H}_2\text{O(l)}$$

25.0 cm^3 of 0.10 mol dm^{-3} sodium hydroxide solution reacted with 25.25 cm^3 of acid solution.

Amount of NaOH = 0.025 dm^3 × 0.1 mol dm^{-3} = 0.0025 mol.

From the equation, 1 mol of NaOH reacts with 1 mol of the acid, so 25.25 cm^3 of the acid solution contains 0.0025 mol of the acid. So the concentration of the acid is

$$\frac{0.0025 \text{ mol}}{0.02525 \text{ dm}^3} = 0.099 \text{ mol dm}^3$$

250 cm^3 of the acid solution contains 3.02 g of the acid, so 1 dm^3 contains 12.08 g, and this is 0.099 mol. So 1 mol of it is

$$\frac{12.08}{0.099} \text{ g}$$

So its relative molecular mass is 122.

Group 1 and 2 nitrates and carbonates

s-block elements

The s-block elements in groups 1 and 2 show some similarities in their properties and reactions, but also some differences.

Thermal stability of the salts of the s-block elements

Heating the s-block metal carbonates and nitrates

You can investigate the action of heat on a variety of carbonates and nitrates. Look for changes in state or colour and test any gas evolved.

When the s-block elements react they form salts which all contain positive metal ions – either M^+ or M^{2+}. There are several key factors in predicting the stability of an ionic compound, not the least of which is its lattice energy. This is closely determined by the charge and size of the ions in the lattice (see chapter 1.4).

Because of this, in general, the stability of ionic compounds increases as:

- cationic radius decreases
- the charge on the ions increases.

In thinking about the stability of the salts of the s-block elements you need to think not just in terms of their stability in relation to their elements but also relative to other substances that may be formed. In particular, the carbonate, CO_3^{2-} anion may decompose to produce carbon dioxide and the oxide, O^{2-} anion, while the nitrate, NO_3^- anion may also decompose to give the oxide, O^{2-} anion, or the larger nitrite, NO_2^- anion. The thermal stability of the carbonates and nitrates of the group 1 and 2 elements is greatly dependent on these changes in size and charge.

Table 2.5.7 shows the radius and the charge on the cations in groups 1 and 2.

Group 1			Group 2		
Element	Ionic radius (nm)	Charge on cation	Element	Ionic radius (nm)	Charge on cation
Lithium	0.074	+1	Beryllium	0.027	+2
Sodium	0.102	+1	Magnesium	0.072	+2
Potassium	0.138	+1	Calcium	0.100	+2
Rubidium	0.149	+1	Strontium	0.113	+2
Caesium	0.170	+1	Barium	0.136	+2

table 2.5.7 the stability of ionic compounds depends on the ionic radius and the charge on the cation.

The lattice energies of both the oxides and carbonates become less negative as you go down group 2. They do not, however, fall at the same rate – the oxide lattice energy falls faster than the carbonate value. The value of the overall enthalpy change of the decomposition reaction gradually becomes more positive as you go down the group.

$$\Delta H^{\ominus}_{dec} = \Delta H^{\ominus}_{lat}\,[XCO_3(s)] - \Delta H^{\ominus}_{lat}\,[XO(s)]$$

This is a positive number. (Ions are bring pulled apart)

The number you are taking away is falling faster than the first number. That means that the overall answer will get more positive.

The size of the lattice energy is governed by several factors, one of which is the distance between the centres of the positive and negative ions in the lattice. Forces of attraction are greatest if the distances between the ions are small. If the attractions are large, then a lot of energy will be given out when the lattice is formed from the ions and so the lattice energy will have a large negative value.

The lattice energies of both carbonates and oxides fall as you go down groups 1 and 2 because the positive ions are getting bigger (see table 2.5.7). The distances between ions are increasing and so the attractions become weaker.

The lattice energies fall at different rates because of the different sizes of the two negative ions – oxide and carbonate. The oxide ion is relatively small for a negative ion (0.140 nm), whereas the carbonate ion is much larger.

In the oxides, when you go from magnesium oxide to calcium oxide for example, the distance between ions increases from 0.212 nm (0.140 + 0.072 nm) to 0.240 nm (0.140 + 0.100 nm) – an increase of about 13%.

In the carbonates, the distance between ions is dominated by the much larger carbonate ion. Although the carbonate ion increases this distance by the same amount for both magnesium carbonate and calcium carbonate, as a percentage of the total distance the increase will be much less for calcium carbonate. Fig. 2.5.12 shows how marble (a form of calcium carbonate) can be decomposed only by strong heating.

fig. 2.5.12 **Calcium carbonate has high thermal stability.**

When heated to the temperature of an ordinary Bunsen burner flame, the group 1 metal carbonates are stable, with the exception of lithium carbonate, $LiCO_3$ which decomposes to give lithium oxide and carbon dioxide:

$$Li_2CO_3(s) \rightarrow Li_2O(s) + CO_2(g)$$

The group 2 metal carbonates decompose to form stable oxides, with the formation of carbon dioxide gas, eg:

$$CaCO_3(s) \rightarrow CaO(s) + CO_2(g)$$

The temperature at which the group 2 carbonates start to decompose increases down the group (see table 2.5.8). Beryllium carbonate is so unstable that it does not exist at room temperature.

Compound	Temperature at which carbonate starts to decompose (°C)
Magnesium carbonate	540
Calcium carbonate	900
Strontium carbonate	1280
Barium carbonate	1360

table 2.5.8 **The group 2 metal carbonates become more stable to thermal decomposition as you move down the group.**

The nitrates of the s-block metals decompose on heating in the temperature of a Bunsen burner flame. Group 1 metal nitrates, with the exception of lithium nitrate, $LiNO_3$ form their corresponding nitrites and these are then stable to heat. The relatively small decrease in size from the NO_3^- ion to the NO_2^- ion is sufficient with the relatively small group 1 cations to achieve thermal stability. For example:

$$2NaNO_3(s) \rightarrow 2NaNO_2(s) + O_2(g)$$

In contrast, lithium nitrate and all the group 2 metal nitrates decompose on heating to form their corresponding oxide, the larger cations needing the much smaller O^{2-} ion to give them stability:

$$4LiNO_3(s) \rightarrow 2Li_2O(s) + 4NO_2(g) + O_2(g)$$

$$2Mg(NO_3)_2(s) \rightarrow 2MgO(s) + 4NO_2(g) + O_2(g)$$

Thermal stability of s-block hydroxides

The hydroxides of the s-block elements follow the same pattern as the carbonates and nitrates. All group 1 metal hydroxides are stable up to quite high temperatures, with lithium hydroxide the first to decompose at around 650 °C. All of the group 2 metal hydroxides decompose to form the corresponding oxide and water:

$$2LiOH(s) \rightarrow Li_2O(s) + H_2O(g)$$

$$Ca(OH)_2(s) \rightarrow CaO(s) + H_2O(g)$$

Questions

1 Describe the trends in ease of thermal decomposition of the nitrates and carbonates down group 2.

2 Write symbol equations for the decomposition of potassium nitrate, calcium nitrate and strontium carbonate.

Group 7: the halogens

The elements of group 7 of the periodic table are better known as the **halogens**. The name dates back to the early years of the nineteenth century when Jöns Jacob Berzelius used it to show that chlorine, bromine and iodine all occur in the sea as salts. The term 'halogen' is derived from the Greek and means 'salt producing'. It remains appropriate – the halogens are very reactive and readily form salts.

Physical properties of the halogens

The halogens are a family of non-metallic elements which exist as diatomic molecules. They are all very reactive and are strong oxidising agents. Chlorine and bromine are poisonous. The chemical behaviour of the halogens is governed by the seven electrons in their outer shell – the addition of only one further electron by either ionic or covalent bonding will give the halogen atom a noble gas configuration. The most common oxidation state for the group 7 elements is therefore –1, although other oxidation states do exist.

As the atomic number increases down the group not only do the group 7 elements become less reactive, they also become less volatile and darker in colour. Chlorine is a greenish yellow gas. Bromine (which takes its name from the Greek word *bromos* meaning stench) is a dark red liquid giving off a dense reddish brown vapour, while iodine is a shiny, greyish-black crystalline solid.

The physical properties of the elements of group 7 are summarised in **table 2.5.9**, in which the trends discussed earlier in this section can be clearly seen.

The solubility of the halogens in water decreases down the group. Chlorine reacts in water forming a solution that is sometimes called chlorine water – a mixture of hydrochloric acid and chloric(I) acid.

Element	Chlorine	Bromine	Iodine
Atomic number	17	35	53
Electronic configuration	2,8,7	2,8,18,7	2,8,18,18,7
Electronegativity	3.0	2.8	2.5
Density (g cm^{-3})	1.56*	3.12	4.93
Melting temperature (°C)	–101	–7	114
Boiling temperature (°C)	–35	59	184
Atomic radius (nm)	0.099	0.144	0.133
Electron affinity (kJ mol^{-1})	–349	–325	–295
Enthalpy of atomisation (kJ mol^{-1})	122	112	107
Hydration energy of ion [$X^-(g) \rightarrow X^-(aq)$] (kJ mol^{-1})	–364	–335	–293
Lattice energy of NaX(s) (kJ mol^{-1})	–780	–742	–705

* At 238 K

table 2.5.9 **Physical properties of the group 7 elements.**

$$Cl_2(g) + H_2O(l) \rightarrow HCl(aq) + HClO(aq)$$
$$Cl_2(g) + H_2O(l) \rightarrow 2H^+(aq) + Cl^-(aq) + ClO^-(aq)$$

Chloric(I) acid is the substance that gives a solution of chlorine its bleaching properties.

Bromine dissolves and reacts with water in a similar way, but to a lesser extent.

Iodine is almost insoluble in water but is soluble in potassium iodide solution because of the formation of I^{3-} ions:

$$I_2(s) + I^-(aq) \rightarrow I^{3-}(aq)$$

Halogens are non-polar so they are more soluble in hydrocarbon solvents than in water (see chapter 2.3), as shown in **fig. 2.5.13**. In each tube there is water (lower layer) and cyclohexane (upper layer). Chlorine is added to the first tube, bromine to the second and iodine to the third. You will notice, especially with bromine and iodine, that most of the halogen has dissolved in the cyclohexane layer.

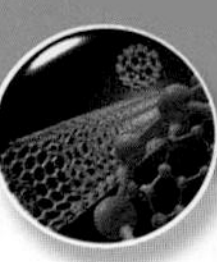

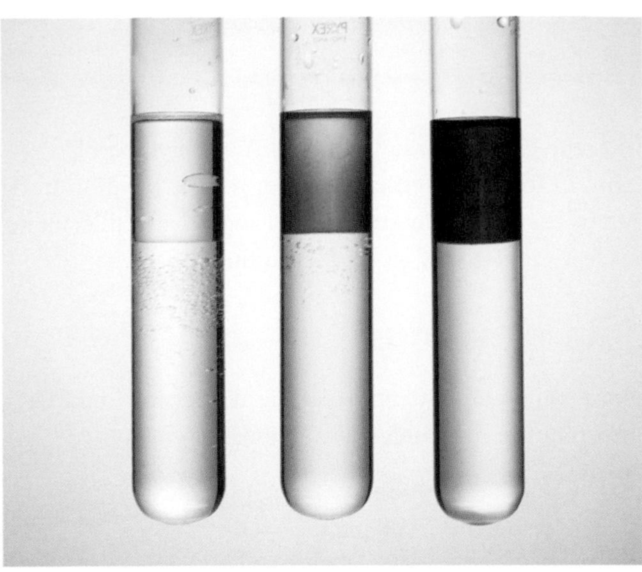

fig. 2.5.13 Halogens are more soluble in organic solvents such as cyclohexane than in water.

HSW Properties of fluorine and astatine

The trend in electron affinity is linked to a gradation of physical and chemical properties from fluorine (at the top of group 7) to astatine (at the bottom).

Fluorine is a pale yellow gas consisting of diatomic molecules. It reacts explosively with hydrogen under all conditions. It reacts with metals and non-metals. It is more reactive than chlorine.

Astatine is intensely radioactive. The most stable isotope has a half-life of only 8.3 hours. It is found only very rarely in certain uranium deposits and has no known significance outside the research laboratory.

Oxidation reactions of the halogens

Halogens are strong oxidising agents, and this is shown in the reactions with metals, with non-metals and with iron(II) ions in solution.

With metals

The halogens react strongly with all of the more electropositive elements. This means that when they react with metals they remove some or all of the outer electrons and become reduced to negative halide ions themselves. This is seen particularly clearly in the reactions of the halogens with the group 1 and 2 metals. The vigour of the reaction between a particular halogen and metal is a function of the reactivity of the particular metal and the particular halogen (chlorine always reacts more vigorously than bromine, for example).

The reactions with sodium (group 1) and magnesium (group 2) are shown here:

$$2Na(s) + Cl_2(g) \rightarrow 2NaCl(s)$$

$$Mg(s) + Cl_2(g) \rightarrow MgCl_2(s)$$

Iron reacts with chlorine and bromine to form iron(III) halides, but with iodine it forms iron(II) iodide.

$$2Fe(s) + 3Cl_2(g) \rightarrow 2FeCl_3(s)$$

$$Fe(s) + I_2(s) \rightarrow FeI_2(s)$$

With non-metals

In reaction with the non-metals, the halogens usually achieve a noble gas configuration through covalent bonding. This may be a single covalent bond such as that formed in the hydrogen halides, H—X. Table 2.5.10 summarises the reactions of hydrogen with halogens, eg:

$$H_2(g) + Cl_2(g) \rightarrow 2HCl(g)$$

Halogen X reacting with hydrogen: $H_2(g) + X_2(g) \rightarrow 2HX(g)$	Conditions
Chlorine	Explodes in direct sunlight, proceeds slowly in the dark
Bromine	300 °C and platinum catalyst
Iodine	300 °C and platinum catalyst – proceeds slowly and only partially

table 2.5.10 The reactivities of the halogens are demonstrated in their reaction with hydrogen. Chlorine reacts readily in sunlight, while bromine and iodine require a catalyst to bring about the reaction.

The halogens react with other non-metals, including phosphorus. Chlorine reacts with phosphorus to form phosphorus(III) chloride:

$$2P(s) + 3Cl_2(g) \rightarrow 2PCl_3(l)$$

With excess chlorine, phosphorus(V) chloride is formed:

$$2P(s) + 5Cl_2(g) \rightarrow 2PCl_5(s)$$

With iron(II) chloride solution

When a halogen is added to pale green iron(II) chloride solution, the halogen oxidises the green iron(II) ions to brown iron(III) ions, eg:

$$2FeCl_2(aq) + Cl_2(g) \rightarrow 2FeCl_3(aq)$$

Iodine/thiosulfate titrations

Earlier in this chapter you studied acid–base titrations. Another type of titration is a redox titration, in which the reaction is not a neutralisation but a redox reaction. A common example is the reaction of iodine and thiosulfate ions, which is carried out in a similar way to acid–base titrations.

Iodine/thiosulfate titrations

In chapter 2.4 you saw how ionic half-equations can be written for oxidising and reducing agents. One common oxidising agent is iodine:

$$I_2(aq) + 2e^- \rightarrow 2I^-(aq)$$

The reducing agents you met included thiosulfate:

$$2S_2O_3^{2-}(aq) \rightarrow S_4O_6^{2-}(aq) + 2e^-$$

Combining these two, the overall ionic equation for the reaction of iodine with thiosulfate is:

$$I_2(aq) + 2S_2O_3^{2-}(aq) \rightarrow S_4O_6^{2-}(aq) + 2I^-(aq)$$

This reaction is the basis of the iodine/thiosulfate titration.

Practical

You can investigate the amount of iodine liberated by an oxidising agent (eg hydrogen peroxide solution) using titration with sodium thiosulfate solution.

Finding the purity of potassium iodate(V)

Iodate(V), IO_3^-, reacts as an oxidising agent:

$$IO_3^-(aq) + 5I^-(aq) + 6H^+(aq) \rightarrow 3I_2(aq) + 3H_2O(l)$$

If iodate(V) ions are added to excess potassium iodide solution, iodine is liberated. The mixture can be titrated with standard sodium thiosulfate solution using starch as indicator.

A weighed sample of impure potassium iodate(V) solid (0.80 g) is dissolved in water and made up to 250 cm^3 of solution in a volumetric flask.

25.0 cm^3 of this solution are added to excess potassium iodide solution, to liberate iodine.

In the titration 20.00 cm^3 of 0.1 mol dm^{-3} sodium thiosulfate were needed to react with the liberated iodine.

$$\frac{20.00 \times 0.1 \text{ mol sodium thiosulfate}}{1000} = 0.002 \text{ mol of thiosulfate}$$

From the equation for the reaction of iodine with thiosulfate above, you can see that 0.002 mol of thiosulfate will react with 0.001 mol of I_2. So 0.001 mol is the quantity of iodine that was liberated by 25.0 cm^3 of impure potassium iodate(V) solution.

From the equation for the reaction of iodate(V) with iodide above, you can see that the number of moles of iodate(V) required to liberate 0.001 mol of iodine is given by:

$$\text{number of moles } IO_3^- = \frac{0.001 \text{ mol}}{3} = 0.000\,333 \text{ mol}$$

$M(KIO_3) = 214$, so:

Mass of potassium iodate(V) in 0.000 333 mol

$= 0.000\,333 \times 214 \quad = 0.0713$ g

The mass of potassium iodate(V) in the sample weighed out is 10 times this, since 25 cm^3 of the 250 cm^3 solution was used in the reaction:

mass KIO_3 = 0.0713 g × 10 = 0.713 g

Percentage purity of potassium iodate(V)

$$= \frac{\text{mass calculated}}{\text{mass weighed out}} \times 100\%$$

$$= \frac{0.713 \times 100}{0.80} = \mathbf{89.1\%}$$

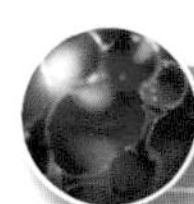

HSW Uncertainty in the results

There must be some doubt in the certainty of the result above. For example, if the number of moles of potassium iodate(V) were adjusted from 0.000 333 to 0.000 35 the final percentage would be 93.6%. A calculation like this involving several steps can lead to different results depending upon the use of significant figures.

Finding the percentage of copper in brass

An application of this redox titration is in the analysis of brass. Brass is an alloy of copper and zinc (see fig. 2.5.14).

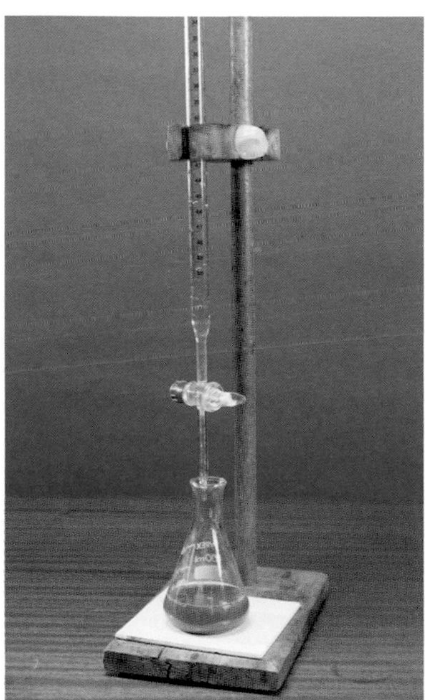

fig. 2.5.14 Titrating to find the percentage of copper in brass.

A weighed sample of brass is reacted with concentrated nitric acid to produce a mixture of copper(II) nitrate and zinc(II) nitrate. The solution is then neutralised. and excess potassium iodide solution is added to liberate iodine:

$$2Cu(NO_3)_2(aq) + 4KI(aq) \rightarrow 2CuI(s) + 4KNO_3(aq) + I_2(aq)$$

The liberated iodine is titrated with standard sodium thiosulfate solution. This gives a measure of how much copper was in the original sample.

Questions

1 Write an ionic equation for the reaction of copper(II) nitrate with potassium iodide solution. Explain the changes that have happened in terms of oxidation and reduction.

2 Suggest why it is necessary to neutralise the solution in the brass experiment before titrating.

3 What assumption about zinc nitrate is made in the titration for the brass experiment?

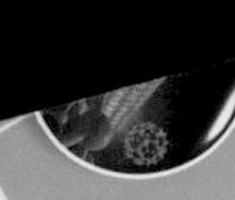

Reactions of the halogens and halides

Disproportionation

With chlorine, bromine and iodine, the products of the reaction with sodium hydroxide solution depend on the temperature at which it takes place. In cold (15 °C) dilute alkali, a mixture of halide, X^- and halate(I), XO^- ions is formed:

$$X_2 + 2OH^-(aq) \rightarrow X^-(aq) + XO^-(aq) + H_2O(l)$$

$$Cl_2(g) + 2NaOH(aq) \rightarrow NaCl(aq) + NaClO(aq) + H_2O(l)$$

These halate(I) ions may then decompose to give more halide and halate(V), XO_3^- ions. The decomposition takes place at different rates depending on both the halogen involved and the temperature:

$$3XO^-(aq) \rightarrow 2X^-(aq) + XO_3^-(aq)$$

$$3NaClO(aq) \rightarrow 2NaCl(aq) + NaClO_3(aq)$$

Sodium chlorate(I) → sodium chloride + sodium chlorate(V)

$$3NaIO(aq) \rightarrow 2NaI(aq) + NaIO_3(aq)$$

Sodium iodate(I) → sodium iodide + sodium iodate(V)

In the case of chlorine, the decomposition occurs only very slowly at 15 °C, but at 70 °C it is very rapid. This provides a way of obtaining the two different products, simply by varying the temperature at which the reactions take place. The same is true for the reactions of bromine with alkalis, although in this case both reactions occur rapidly at 15 °C and the temperature needs to be lowered to 0 °C to prevent decomposition of the BrO^- ion. Both reactions occur rapidly even at 0 °C in the case of iodine, and so separating the different products is considerably less easy.

All these reactions of chlorine, bromine and iodine with alkali are examples of **disproportionation** (see chapter 2.4). When chlorine reacts with alkali, Cl^- ions and ClO^- ions are formed. This involves a change in oxidation number from 0 (in Cl_2) to –1 (in Cl^-) and +1 (in ClO^-). Similarly, the decomposition of ClO^- also results in a simultaneous increase and decrease in the oxidation number of chlorine:

$$\underset{+1}{3ClO^-(aq)} \rightarrow \underset{-1}{2Cl^-(aq)} + \underset{+5}{ClO^{3-}(aq)}$$

Reactions of the hydrogen halides

The hydrogen halides result from the reaction of the halogens with hydrogen, as you saw earlier in the chapter. The hydrogen halides react with ammonia gas to form ammonium halides, eg hydrogen chloride and ammonia gases react to form ammonium chloride fumes (**fig. 2.5.15**):

$$NH_3(g) + HCl(g) \rightarrow NH_4Cl(s)$$

Hydrogen chloride dissolves in water to form hydrochloric acid.

fig. 2.5.15 **Formation of ammonium chloride.**

Redox reactions of the halogens with potassium halides

Potassium halides will react with another halogen in a **displacement (redox) reaction**. For example, chlorine reacts with potassium iodide solution to form potassium chloride and iodine. The equation for the reaction is:

$$2KI(aq) + Cl_2(g) \rightarrow 2KCl(aq) + I_2(aq)$$

Figure 2.5.18 shows the reaction when chlorine is bubbled through potassium iodide solution. This occurs because chlorine is a stronger oxidising agent than iodine.

fig. 2.5.16 **The reaction of chlorine with potassium iodide solution is a redox reaction.**

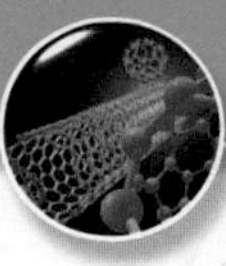

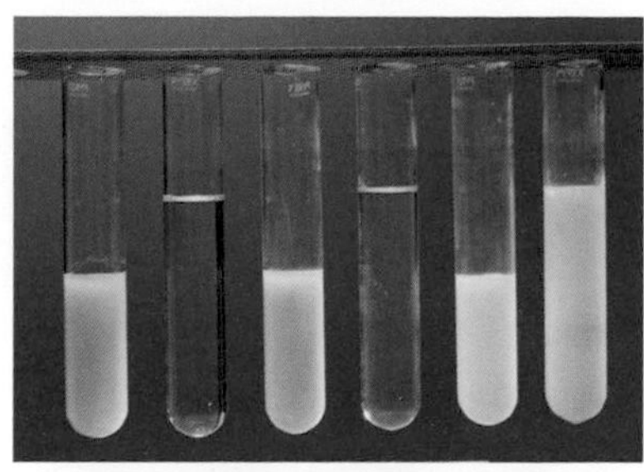

fig. 2.5.17 The first test tube contains a precipitate of silver chloride, which dissolves in dilute ammonia solution. The third test tube contains silver bromide precipitate, which dissolves in concentrated ammonia solution (test tube 4). The fifth test tube contains silver iodide, which does not dissolve in concentrated ammonia solution (test tube 6).

Reactions with concentrated sulfuric acid

Concentrated sulfuric acid reacts with a halide to form a hydrogen halide, eg:

$$NaCl(s) + H_2SO_4(l) \rightarrow NaHSO_4(s) + HCl(g)$$

With sodium bromide and sodium iodide, however, the concentrated acid acts as an oxidising agent, oxidising the product to bromine or iodine respectively:

$$2HBr(g) \rightarrow Br_2(g) + 2H^+(aq) + 2e^-$$

$$2H^+(aq) + H_2SO_4(aq) + 2e^- \rightarrow 2H_2O(l) + SO_2(g)$$

overall: $2HBr(g) + H_2SO_4(aq) \rightarrow Br_2(g) + 2H_2O(l) + SO_2(g)$

Hydrogen iodide is oxidised even more thoroughly by concentrated sulfuric acid. The sulfuric acid is reduced in this case, partly to hydrogen sulfide, because hydrogen iodide is an even stronger reducing agent than hydrogen bromide:

$$H_2SO_4(l) + 8H^+(aq) + 8e^- \rightarrow H_2S(g) + 4H_2O(l)$$

$$8I^-(aq) \rightarrow 4I_2(aq) + 8e^-$$

overall: $H_2SO_4(l) + 8H^+(aq) + 8I^-(aq) \rightarrow H_2S(g) + 4H_2O(l) + 4I_2(aq)$

Testing for halides: the silver halides

If you have an unknown halide salt and want to see what sort of halide it is, you can use silver nitrate solution to do a standard test. This analysis of the halide ions is based on the different solubilities of the silver halides (see fig. 2.5.17). Dilute nitric acid is added to the unknown halide solution to prevent the precipitation of any other silver salts. Silver nitrate solution is then added.

If the unknown halide is a chloride, Cl^- a white precipitate of silver chloride results which in turn is soluble in dilute ammonia solution:

$$Ag^+(aq) + Cl^-(aq) \rightarrow \underset{\text{white}}{AgCl(s)}$$

$$AgCl(s) + 2NH_3(aq) \rightarrow [Ag(NH_3)_2]^+(aq) + Cl^-(aq)$$

If left in sunlight, partial decomposition of the silver chloride into silver and chlorine makes the precipitate turn greyish in colour.

A bromide, Br^- gives a cream precipitate of silver bromide which in turn dissolves in concentrated ammonia solution in the same way as the chloride:

$$Ag^+(aq) + Br^-(aq) \rightarrow \underset{\text{cream}}{AgBr(s)}$$

$$AgBr(s) + 2NH_3(aq) \rightarrow [Ag(NH_3)_2]^+(aq) + Br^-(aq)$$

Iodide, I^- gives a yellow precipitate of silver iodide which is insoluble in concentrated ammonia solution.

$$Ag^+(aq) + I^-(aq) \rightarrow \underset{\text{yellow}}{AgI(s)}$$

Questions

1 A student testing for halides acidifies with dilute hydrochloric acid rather than dilute nitric acid. Why does this cause a problem?

2 How could sodium bromide and sodium chloride be distinguished with concentrated sulfuric acid?

HSW Halogens and the water supply

The halogens are a group of very reactive elements, which form very stable compounds. Yet the halogens are a group of great importance, both in biological systems and in industrial applications. Scientists have found a variety of ways of using these poisonous elements to improve the human condition – although some ethical issues have raised their heads as a result.

Chlorine

As you have seen earlier in this chapter, chlorine is a very effective oxidising agent. This property is vitally important when it comes to providing us with clean drinking water. In countries like the UK, all of our drinking water is chlorinated. Water from reservoirs and rivers is filtered to remove solid particles and then treated with chlorine to kill any bacteria in it. Small amounts of chlorine remain in the treated water and this prevents possible recontamination by bacteria. When chlorine dissolves in water, an equilibrium reaction is set up in which hydrochloric acid and chloric(I) acids are produced. It is the latter which is responsible for the oxidising power of aqueous chlorine. Most people don't even think about the importance of chlorine in the water – except perhaps in swimming pools when its use is more obvious.

However, in the developing world around 1.5 million people, mainly children, die every year as a result of drinking dirty, contaminated water. The United Nations has set a target to halve the number of people without access to clean drinking water by 2015. Chlorine and chlorine compounds will play a big part in this process. In some places big water tanks can be built to supply a whole village. Calcium chlorate(I) – involving both group 2 and group 7 elements! – is used to sterilise the water which can then be supplied to the villagers through pipes. But in some situations it isn't even possible to provide tanks. Millions of people worldwide simply collect their drinking water from rivers and lakes – often the same rivers and lakes which are used as toilets and for washing both people and clothes. Even well water is often contaminated by dirty water from nearby. So an alternative strategy is to supply people with calcium or sodium chlorate(I) which they can add to their water to make it safe to drink. Chlorine already saves millions of lives around the world through clean drinking water – in future it may well save many more.

fig. 2.5.18 **When natural disasters strike anywhere in the world, chlorine-based chemicals can sterilise any drinking water available and prevent the spread of diseases such as cholera.**

Fluorine

Chlorine was first added to drinking water in 1897. No-one questions it, and it is internationally agreed to be the best way to provide clean drinking water. However, when it comes to adding another type of halogen salt to the drinking water, opinions are more divided.

Fluoride ions have been shown to help in the development of healthy tooth enamel. Enamel is the hard, shiny white covering of the teeth that protects them from attack by acids formed in the mouth by the action of bacteria on food. It is the hardest substance formed by the body. The normal structure of the tooth enamel involves an open lattice of a compound known as **hydroxyapatite**. This is made up of calcium along with phosphorus, oxygen and hydrogen (see fig 2.5.19). However scientific studies have shown that trace concentrations of fluoride ions present at the time of tooth formation in the gums enable the body to produce even harder enamel. When fluoride ions are present, fluoroapatite is formed. The fluoride ions effectively block up the holes in the enamel structure, as illustrated by fig 2.5.19, making the teeth much less vulnerable to decay. Only traces of fluoride are needed – a good thing as fluoride ions are toxic at higher concentrations.

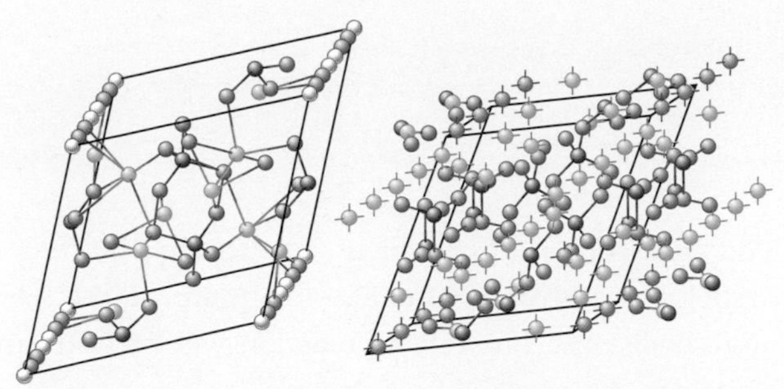

Unit cell of hydroxyapatite.
calcium
phosphorus
oxygen
hydrogen

Structure of fluorapatite.
calcium
phosphorus
oxygen
fluorine

fig. 2.5.19 **Extra fluoride ions lead to a denser enamel, which is less easily attacked by acid from the breakdown of sugary food by bacteria.**

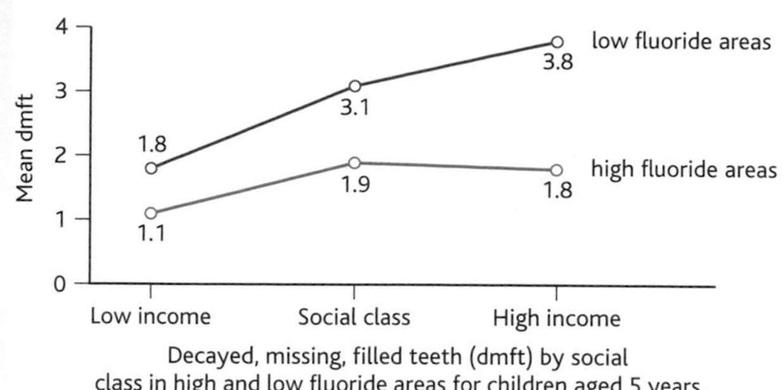

fig. 2.5.20 **Based on evidence such as this, the case for adding fluoride ions to the drinking water supply seems very clear.**

The apparent benefit of fluoride ions to the teeth came from observations that children who grew up with naturally high levels of fluoride ions in their drinking water appeared to have less tooth decay and fewer fillings and extractions than children who lived in areas with less natural fluoride. There is also evidence that older people are less likely to lose their teeth if they live in areas with naturally high fluoride levels. A team at York University looked at evidence from a number of studies and reached a number of conclusions about the value of adding fluoride ions to the water supply in areas where the levels are naturally low. Children from poorer families, who are perhaps most likely to eat cheap processed foods high in hidden sugar, and who may not be able to afford, or be aware of, the benefits of fluoride toothpaste, seemed to benefit most.

Fluoridation of the water supply could even out some of these inequalities. There is a large body of evidence which suggests that fluoridation of drinking water, so that everyone gets about 1ppm, would have many benefits throughout the community and could save a great deal of money in the long term on dental care for both the young and the old.

However, not all of the evidence for fluoridation is so positive. Australia, New Zealand and the US are some of the countries which already have fluoride ions added to their water supply. Some scientists there feel that the possible toxic effects of fluoride, from discolouring the teeth to possible effects on the skeleton, causing arthritis-like symptoms, have not been investigated sufficiently and that any negative evidence is ignored in the face of vested interests. Some people feel that adding fluoride to water removes an element of personal choice and responsibility – although interestingly that argument is not put forward for adding chlorine to water. The debate continues!

Questions

1 Why do you think no one queries the addition of chlorine to the water supply, when fluoridation can cause heated debate?

2 Investigate some of the evidence for and against the fluoridation of water. Evaluate the research on both sides and draw your own personal conclusions, based on the evidence you have found.

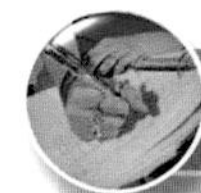

Examzone

You are now ready to try the first Examzone test for Unit 2 (Examzone Unit 2 Test 1) on page 252. This will test you on what you have learnt in Unit 2 chapters 1 to 5.

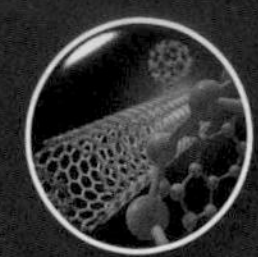
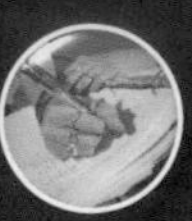

6 Kinetics

Rate of reaction

Reaction rates are of fundamental importance to the research chemist, in the chemical industry and in the living world. The **rate of reaction** describes the speed with which reactants disappear and products form for a particular reaction. Information about what affects reaction rates gives chemists the opportunity to speed up or slow down a particular reaction in the lab. In industry it allows operating conditions to be chosen that make the process take place with maximum efficiency and economy. Studying the rates of reactions also gives us information about how reactions actually take place.

What factors affect the rate of a chemical reaction?

A variety of factors can influence the rates of chemical reactions, speeding them up or slowing them down. There are five main factors that affect how fast a chemical ıeaction takes place:

- concentration
- temperature
- catalysts
- pressure
- surface area

We shall look at each in turn.

Concentration and pressure

Figure 2.6.1 shows two containers of gaseous mixtures of hydrogen and bromine. You can see that there is a higher concentration of particles in (b) than in (a).

(a)

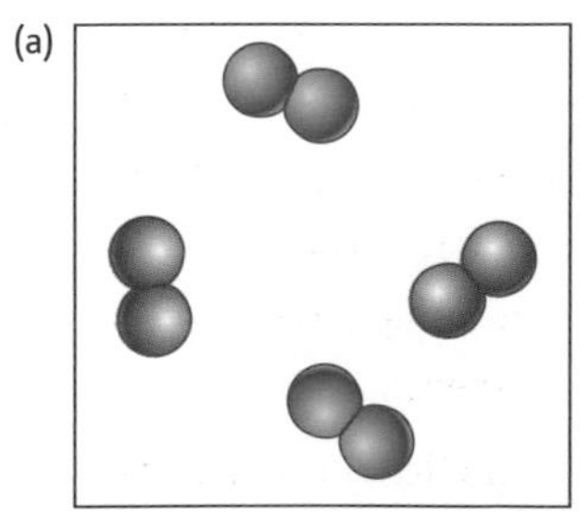

(b)

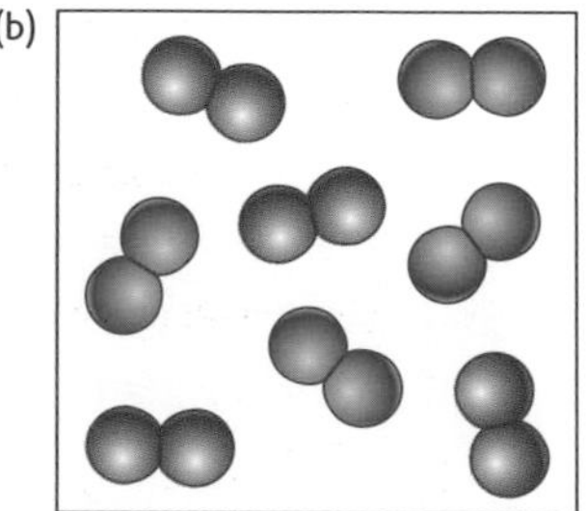

fig. 2.6.1 Two containers with mixtures of hydrogen (red) and bromine (brown). The gases have a higher concentration in (b) than in (a).

In both containers the particles are moving randomly in all directions. If a particle of hydrogen and a particle of bromine collide with sufficient energy, a reaction will take place to form hydrogen bromide.

You can see that in container (b) there are more particles in a given volume, and so there will be more collisions. The rate of reaction will be faster in container (b).

In a mixture of gases, such as a mixture of hydrogen and bromine, increasing the pressure has the same effect as increasing the concentration in a solution. It increases the number of particles per unit volume, leading to an increased reaction rate.

HSW Collision theory

Chemists find it useful to use a **collision theory** model to explain what is happening during a chemical reaction at the molecular level. Looking at a static diagram like **fig. 2.6.1** it is easy to forget that the particles shown are moving randomly in all directions and in three dimensions. The number of collisions that involve sufficient energy to lead to a reaction is related to the number of collisions overall. Increasing the number of collisions by getting the particles to be closer together, or making them move faster, will not only result in more collisions but also, in consequence, more successful collisions. But this is not the whole story. When particles collide they must have more than just sufficient energy. They must collide with the correct orientation. That means one part of a molecule must collide with a certain part of another molecule with sufficient energy before the reaction will take place. You will see the importance of this later when you study reaction mechanisms.

Temperature

Collision theory says that particles have to collide with a certain energy for a reaction to take place. If the temperature is raised, the number of particles is unchanged but the average energy of each particle is greater – they are moving faster on average. This means that more collisions will have sufficient energy to result in a reaction, and so the reaction will be faster at a higher temperature.

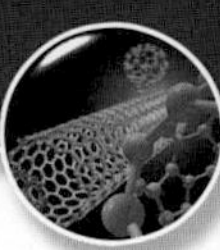

Surface area

For any reaction to happen, the particles of the reactants must be brought into contact with each other. If the reactants are in the same phase (either liquids or gases) their reacting particles can intermingle freely, giving them the maximum opportunity to react together. This is why the majority of chemical reactions are carried out in aqueous or organic solution, or in the gaseous state. A reaction with reactants in the same phase is known as a **homogeneous reaction**.

However, if one of the reactants is a solid then the reacting particles can only meet at the surface of the solid. If the surface area of the solid can be increased, then the rate at which the reaction can take place will be increased too (see **fig. 2.6.2**). Reactions between two substances in different phases are known as **heterogeneous reactions** (see **fig. 2.6.3**). Heterogeneous reactions can be pretty explosive – there have been a number of occasions where the tiny size of particles of grain dust have allowed the combustion reaction between the solid carbohydrate and the gaseous oxygen present in a flour or grain silo to progress so rapidly that it has been completely destroyed!

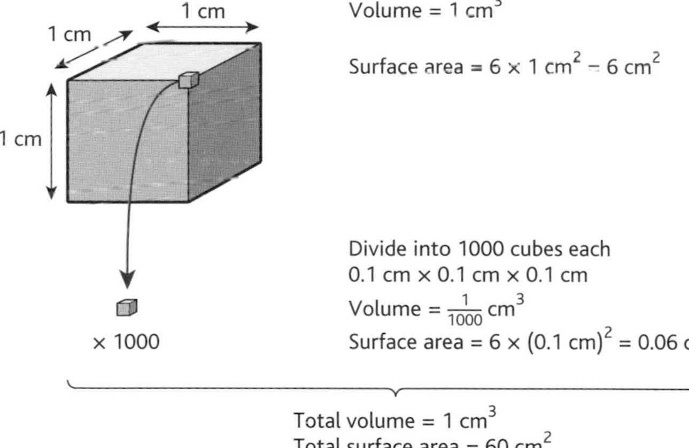

fig. 2.6.2 If you divide a 1 cm cube into 0.1 cm cubes you increase the surface area from 6 cm^2 to 60 cm^2.

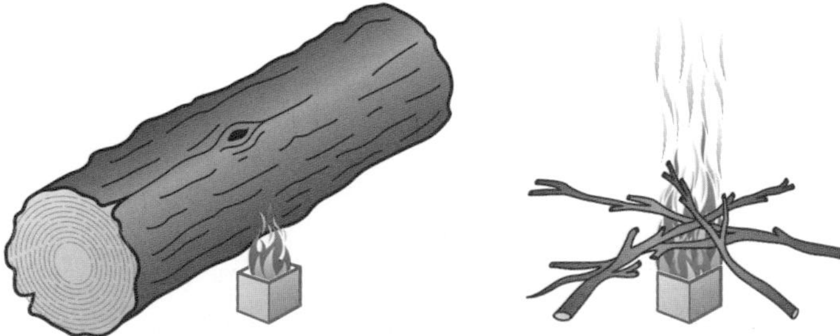

fig. 2.6.3 In the heterogeneous reaction between wood in the solid state and oxygen in the gaseous state the surface area of the wood exposed to the air makes all the difference to the rate of the combustion reaction.

Catalysts

A **catalyst** is a substance that changes the rate of a chemical reaction (usually speeding it up) without being used up or undergoing any permanent chemical change. Catalysts can enable reactions to occur that would simply be too slow to be recognised without them. Almost every major industrial chemical process depends on catalysts to enable a profitable reaction rate to be maintained without resorting to excessive and expensive conditions of temperature and pressure. The enzymes that control the biochemistry of every living cell are also specialised catalysts, made of protein. You will return to how catalysts work later in this chapter.

fig. 2.6.4 Enzymes – biological catalysts – are widely used in the industrial preparation of baby foods.

Questions

1. A student is studying the reaction between calcium carbonate and hydrochloric acid. He wants to find out whether increasing the concentration of the acid has more effect on the rate of reaction than raising the temperature. What advice would you give him when planning his experiment?
2. Food decays by a series of chemical reactions. Why is it that food can be stored much longer in a refrigerator than at room temperature?
3. Flour dust in a flour mill can explode in air. Suggest why this may happen, when a bag of flour in the kitchen will not explode.

Measuring reaction rates

You have seen that five main factors affect the rate of a reaction. How can you measure their effect? To follow the progress of a chemical reaction quantitatively, chemists take measurements and calculate the reaction rate. There are various methods but they all depend on measuring how quickly the reactants are used up, or how quickly the products are formed. This amounts to measuring the change in concentration of either one of the reactants or one of the products with time.

The rate of a chemical reaction can be expressed as:

$$\text{rate of reaction} = \frac{\text{change in concentration}}{\text{change in time}}$$

So, for example, if the concentration of a product of a reaction increases by 0.25 mol dm^{-3} each second, then we can say that the rate of formation of this product is 0.25 mol dm^{-3} s^{-1}. If the measured concentration is increasing (ie you are monitoring a product), the rate is expressed as a positive number. If the measured concentration is decreasing (ie you are monitoring a reactant), the rate is expressed as a negative number. So, for example, if 0.5 mol dm^{-3} of a reactant is used up every second then the rate of reaction is –0.5 mol dm^{-3} s^{-1}.

These values are average rates of reaction for the period of time over which the measurements were taken.

Measuring the volume of a gas

In some reactions it is easy to see how the rate changes; eg in a reaction that produces bubbles of gas:

$$Mg(s) + 2HCl(aq) \rightarrow MgCl_2(aq) + H_2(g)$$

you can vary the reaction conditions – eg the concentration of the acid, the temperature or the surface area of the magnesium – and see what effect this has on the rate by simply counting the number of bubbles at regular time intervals, eg every 15 s. For more accurate results you can measure the volume of gas given off at regular time intervals throughout the reaction, using the apparatus shown in **fig. 2.6.5(a)**. You can then plot a graph of the total volume of gas produced against time (see **fig. 2.6.5(b)**).

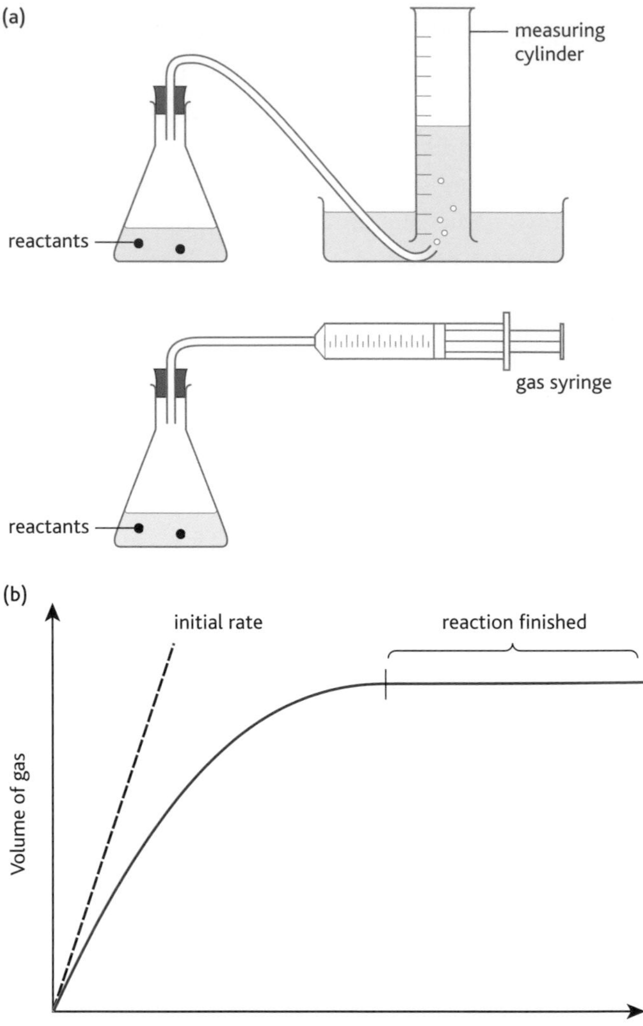

fig. 2.6.5 (a) Two practical methods of measuring the volume of gas produced every 15 seconds. (b) Graph of volume of gas collected against time.

In **fig. 2.6.5(b)** you can see that the graph is steepest at the start of the reaction and the steepness (rate) reduces throughout the reaction. If you draw a tangent to the curve at the start of the reaction (as shown by the dotted line) you can work out the gradient of the line. This is the **initial rate of reaction**. The reaction rate changes throughout the reaction, as does the concentration of reactants and products, so it is most accurate for comparing the initial rates of different reactions.

You could study the decomposition of hydrogen peroxide using the same apparatus.

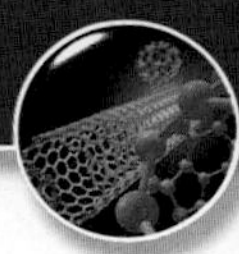

Measuring the change in mass of the reaction mixture

Another way of monitoring the change in concentration of a gaseous product is to measure the mass of the total reaction mixture as the reaction progresses, and see how quickly it decreases. For example, the rate of the reaction between calcium carbonate and dilute hydrochloric acid:

$$CaCO_3(s) + 2HCl(aq) \rightarrow CaCl_2(aq) + H_2O(l) + CO_2(g)$$

can be monitored by measuring the mass of the apparatus every 15 seconds. The mass decreases as carbon dioxide escapes from the solution. This method can be used to investigate the effect of changing the temperature, the concentration of the acid or the particle size of the calcium carbonate.

Monitoring a colour change

Colour changes provide another way of measuring the rate of a reaction. In the 'iodine clock' reaction, you can compare the effect of altering the concentration of one of the reactants, hydrogen peroxide.

A solution of hydrogen peroxide and sulfuric acid is added to a solution of iodide ions, thiosulfate ions and starch. A slow reaction forms iodine, which reacts rapidly with thiosulfate until the thiosulfate is used up. The excess iodine then reacts with the starch to form a blue-black complex. Here are the equations for the reactions:

$$2H^+(aq) + H_2O_2(aq) + 2I^-(aq) \rightarrow I_2(s) + 2H_2O(l)$$

$$I_2(aq) + 2S_2O_3^{2-}(aq) \rightarrow S_4O_6^{2-}(aq) + 2I^-(aq)$$

The time is measured from the mixing of the solutions until the solution turns blue-black.

Other methods of finding the rate of reaction

Some other commonly used methods of monitoring the rate of a reaction while investigating the effects of different factors are summarised below.

- **Titrimetric analysis** involves removing small portions (**aliquots**) of the reaction mixture at regular intervals. These aliquots are usually added to another reagent which immediately stops or **quenches** the reaction so that there are no further changes to the concentrations in the reaction mixture until further analysis can be carried out. The quenched aliquots are then titrated to find the concentrations of known compounds in them.
- **Colorimetric analysis** is particularly valuable where one of the reactants or products of the reaction is coloured. The colour changes throughout the reaction can be detected using apparatus called a photoelectric **colorimeter**. The colour changes measured in this way can then be used to calculate changes in concentration.
- **Conductimetric analysis** involves measuring the conductivity changes in a reaction mixture over time. These reflect the changes in the ions present in the solution and so can be used to measure the changes in concentration of the various components of the mixture.

Questions

1 a Which method in **fig. 2.6.5(a)** do you think is likely to give the more accurate results? Suggest why.

b If the gas produced were soluble, which method would be better and why?

2 Why would the method involving measuring the mass of the reaction not work well in the reaction of magnesium and dilute hydrochloric acid? Think about the density of hydrogen.

Activation energy

Collisions between reacting particles

You have seen how collision theory says that for a reaction between two particles to occur, those particles must collide in the correct orientation and with sufficient energy. The reaction rate is a measure of how frequently effective collisions occur – an effective collision being one that results in the formation of product particles. The factors you have met that increase the rate of collisions will also increase the rate of a reaction.

Activation energy

Few collisions between particles actually lead to a chemical change. Why is this? As you have seen, in order to react they must collide with a certain minimum kinetic energy. This is known as the **activation energy, E_A**. If two slow-moving particles collide, even if they are in the right orientation, they will simply bounce apart as a result of the repulsion of their negative electron clouds. An increase in temperature will increase the kinetic energy of the particles and so increase the rate of reaction, because a greater proportion of particles will have sufficient energy to overcome the activation energy and react (see **fig. 2.6.6**).

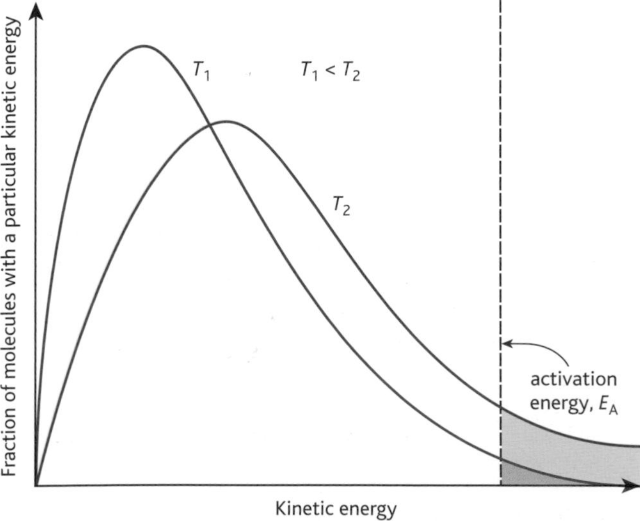

fig. 2.6.6 These curves show the kinetic energy distributions of the particles for a reaction mixture at two different temperatures. The shaded areas are proportional to the total fraction of particles that have the minimum activation energy. You can see that at the higher temperature, T_2 more particles have enough energy to react.

What happens during a reaction?

When particles collide with the correct orientation, they slow down, stop and then fly apart again. In an unsuccessful collision the particles separate unchanged, whereas in a successful collision the activation energy barrier is crossed and the particles that separate are chemically different from the original ones.

As the particles slow down and stop they possess less and less kinetic energy and their potential energy increases. The relationship between the activation energy and the total potential energy of the colliding reactants is shown in **fig. 2.6.7**. This is a **reaction profile**, which follows the path of a reaction as reactant particles come together, collide and produce product particles which then separate.

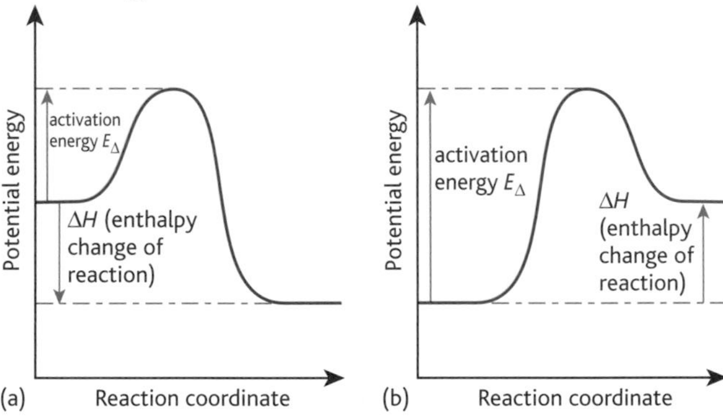

fig. 2.6.7 Reaction profile diagrams for (a) an exothermic reaction and (b) an endothermic reaction. The progress of the reaction is shown on the horizontal axis, the **reaction coordinate**.

As two particles collide, they slow down and their kinetic energy is converted to potential energy – they begin to climb the activation energy hill. If their combined initial kinetic energies are less than the activation energy, E_A then they cannot reach the top of the hill so they fall back as reactants, gaining their original kinetic energy again. On the other hand if the combined energies are equal to or greater than E_A and the particles are correctly orientated, then they can overcome the activation energy barrier and form product particles.

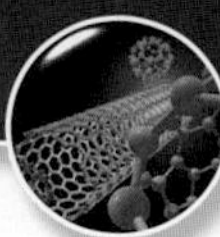

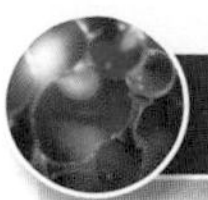

HSW How do catalysts work?

The activation energy model can be used to explain how a catalyst increases the rate of a reaction. One simple explanation is that a catalyst lowers the activation energy, so more collisions have enough energy to result in a reaction. A more sophisticated model suggests that if the particles collide with sufficient energy and in the correct orientation, they form an activated complex. The **activated complex** is not a chemical substance that can be isolated, but consists of an association of the reacting particles in which bonds are in the process of being broken and formed. **Figure 2.6.8** shows the activated complex in the decomposition reaction of hydrogen iodide into hydrogen and iodine.

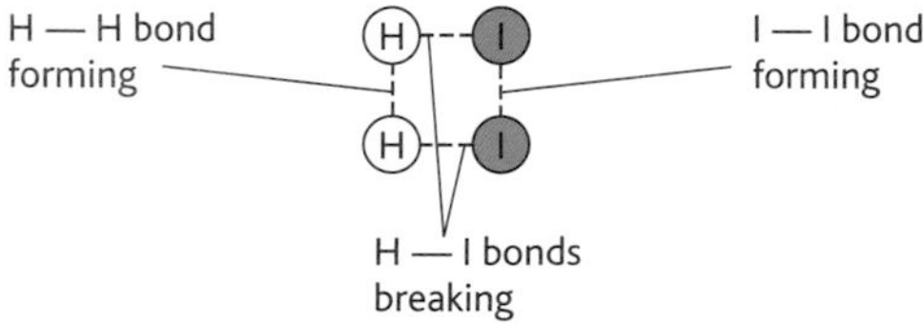

fig. 2.6.8 The activated complex in the decomposition of hydrogen peroxide.

Fig. 2.6.9 shows a more sophisticated model for the reaction profile of an exothermic reaction involving a catalyst.

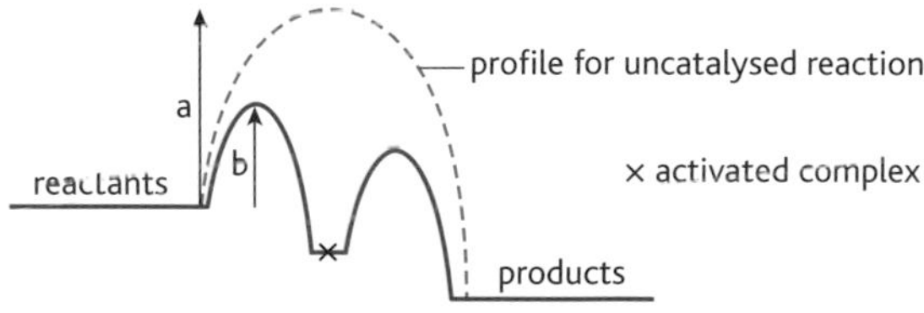

fig. 2.6.9 A catalyst lowers the activation energy of a reaction by forming an activated complex, a low-energy intermediate stage.

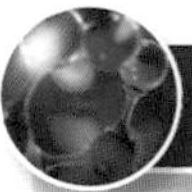

HSW The Maxwell-Boltzmann model

The particles in a reaction mixture do not all have the same kinetic energy. There is a spread, with some particles moving very slowly, some very quickly and the majority being somewhere in the middle. **Figure 2.6.10** shows the **Maxwell–Boltzmann model**, a mathematical picture of this distribution of velocities. This model was created independently in the mid-nineteenth century by Ludwig Boltzmann (in Vienna) and James Clerk Maxwell (in Cambridge).

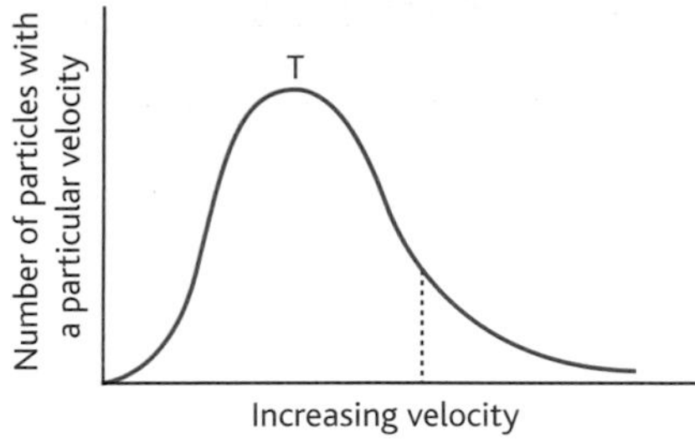

fig. 2.6.10 The Maxwell–Boltzmann distribution shows the range of velocities of the particles in a reaction mixture.

The vertical dotted line shows the velocity the particles must have if a collision is to bring about a reaction – it represents the activation energy. This model can help explain why an increase in temperature speeds up a reaction rate.

Figure 2.6.11 shows the Maxwell–Boltzmann distribution for a reaction mixture at three different temperatures. You can see that as the temperature rises, the graph becomes flatter and a much greater proportion of particles are moving fast enough to overcome the activation energy (dotted line). This leads to more collisions and a faster reaction.

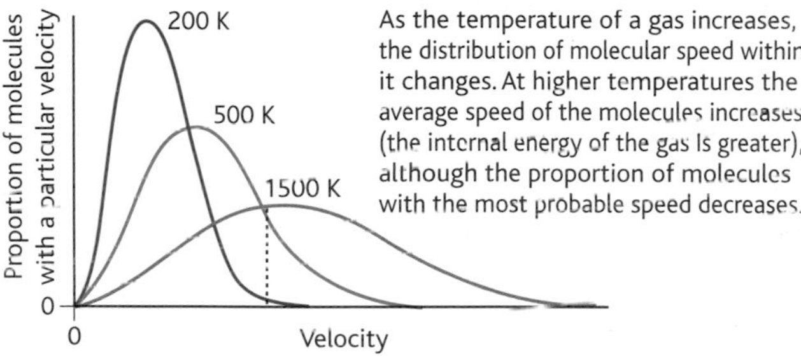

fig. 2.6.11 Seventy years after Boltzmann published his theoretical model, it was shown practically that the distribution of velocities in a gas agreed with his predictions.

So the Maxwell–Boltzmann distribution is an example of a mathematical model that allows you to understand and predict the effect of changing conditions on the rate of reaction.

Questions

1 In a particular exothermic reaction, the activation energy is high. What does this suggest about the likely rate of reaction?

2 What two factors determine whether two colliding particles will react?

3 Suggest problems that Boltzmann and Maxwell might have had in collaborating.

7 Chemical equilibria

Dynamic equilibrium

Equilibrium – a balance

In chapter 2.6 you thought about the rates of chemical reactions, and factors that affect how quickly reactions go. Here you will be looking at not how fast, but how far – the extent to which a reaction goes – or how much of the reactants is converted to products. Many reactions are **equilibria** – their reactants and products exist together, rather than going to **completion** with all the reactants turning readily to products. You will also be looking at the factors that affect the **position of equilibrium** – how far a reaction goes towards completion.

In understanding a chemical equilibrium, you can picture a liquid put in an empty box which is then sealed. The particles of the liquid evaporate and collect in the empty space above it. These particles fly around and collide with each other, with the walls of the container and with the surface of the liquid. When a gaseous particle collides with the liquid surface it tends to rejoin the liquid. Its kinetic energy is transferred to the particles in the liquid so that it no longer possesses enough energy to escape from the liquid again.

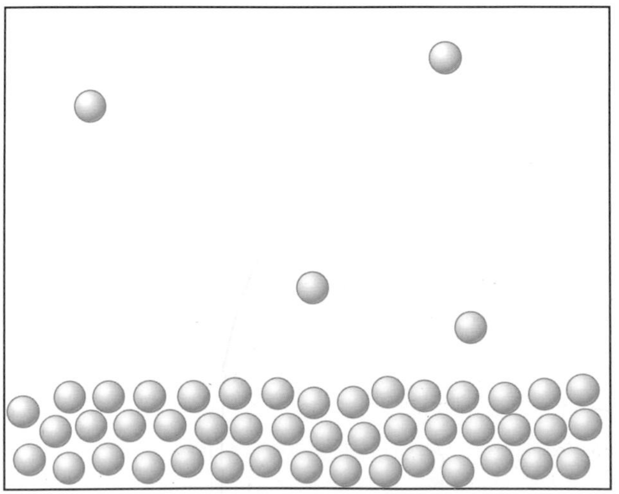

fig. 2.7.1 Particles can leave the liquid if they have enough energy, and particles can leave the vapour and rejoin the liquid, giving up some of their energy to the other particles in the liquid.

At first, far more particles leave the liquid than enter it, because there are very few particles in the vapour phase. As the number of particles in the vapour phase increases, the rate at which particles rejoin the liquid increases, because there are more particles in the vapour phase colliding with the liquid surface. Eventually a point is reached where the rate at which particles are leaving the liquid is the same as the rate at which particles are joining it.

When this happens we have reached a point where two opposing effects, evaporation and condensation, are occurring at equal rates so that they cancel out. Clearly we have an equilibrium – but more importantly, this is an example of a **dynamic equilibrium**. A dynamic equilibrium involves two opposing processes that occur at equal rates. An equilibrium of this type is often established in chemical processes too.

Another way of describing a dynamic equilibrium is to say that it has constant **macroscopic** properties (those properties that an external observer can see and measure) while at the same time **microscopic** processes (processes on a molecular scale) continue to occur. For our example of the liquid in the box, the total amounts of liquid and vapour are constant (the macroscopic observation), while at a microscopic level individual particles are continually moving between the phases.

Nitrogen dioxide and dinitrogen tetroxide – a steady state

In heavily polluted air, a brown haze may sometimes be seen. The gas responsible for this is nitrogen dioxide, NO_2 formed by oxidation of the nitrogen monoxide, NO emitted by car engines.

Nitrogen dioxide is in equilibrium with dinitrogen tetroxide, N_2O_4 which is colourless:

$$2NO_2(g) \rightleftharpoons N_2O_4(g)$$

In an investigation, 0.0800 mol of nitrogen dioxide are sealed in a container and kept at a constant 25 °C. Eventually the amount of nitrogen dioxide drops to a steady 0.0132 mol while the amount of dinitrogen tetroxide rises to a steady 0.0334 mol. The left-hand side of **fig. 2.7.2** shows this.

Suppose that now we take a container of the same size and put 0.0400 mol of dinitrogen tetroxide in it. (This is exactly the amount of dinitrogen tetroxide that 0.0800 mol of nitrogen dioxide would produce if it reacted completely to form dinitrogen tetroxide.) If we keep this container at 25 °C, we find that once again we end up with 0.0132 mol of NO_2 and 0.0334 mol of N_2O_4 when a steady state has been reached – the right-hand side of **fig. 2.7.2** represents this situation.

You can see that the same final situation comes about whether we start with pure nitrogen dioxide or pure dinitrogen tetroxide. In both cases the amount of reactant (NO_2 or N_2O_4) falls and the amount of product (N_2O_4 or NO_2) rises until equilibrium is attained. At this point the amount of nitrogen dioxide reacting to form dinitrogen tetroxide is exactly equalled by the amount of dinitrogen tetroxide reacting to form nitrogen dioxide.

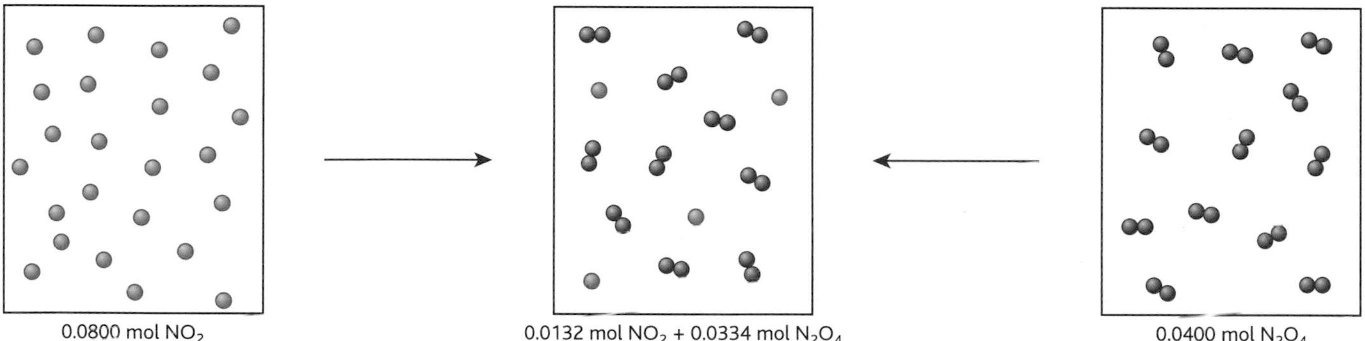

fig. 2.7.2 The final equilibrium state is the same no matter which gas we start with provided the temperature and pressure are constant.

This example shows how a dynamic equilibrium is established in a chemical process. As the reaction proceeds, a plot of the concentrations of the reactants and products looks like the one shown in **fig. 2.7.3**. These concentrations change fairly rapidly at first, and then reach steady values. At this point the rate of conversion of reactants into products exactly equals the rate of conversion of products into reactants, and we have a **steady state**. Another way of saying this is to say that at equilibrium the rate of the forward reaction and the rate of the backward reaction are equal.

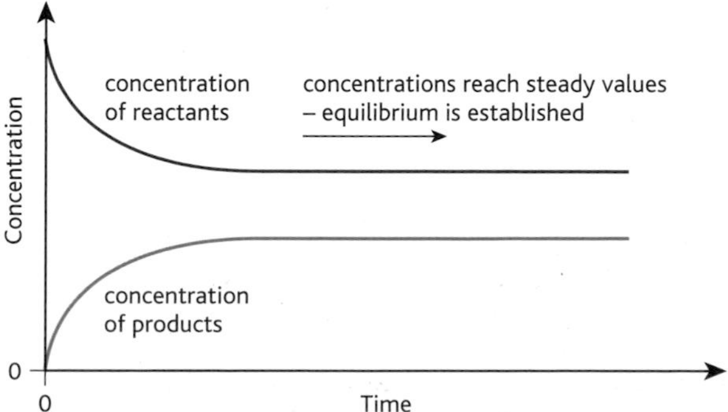

fig. 2.7.3 As equilibrium is reached the concentrations of reactants and products in a chemical process reach steady values.

The reaction of nitrogen and hydrogen to produce ammonia

You may have studied the Haber process at GCSE. It is a good example of an industrial process where it is important to control an equilibrium reaction in order to get a good yield of product.

$$N_2(g) + 3H_2(g) \rightleftharpoons 2NH_3(g)$$

The forward reaction is exothermic.

fig. 2.7.4 Ammonia-based fertilisers have been used for over 100 years to increase crop yields and help feed the ever-expanding world population.

What affects a dynamic equilibrium?

The position of equilibrium is not fixed for a reaction, but changes as you change the reaction conditions.

When a system in dynamic equilibrium is upset, it responds in such a way as to return it to equilibrium again. The tendency of systems to behave in this way was noted by the French chemist Henri Le Chatelier, who in 1888 proposed that:

- Whenever a system which is in dynamic equilibrium is disturbed, it tends to respond in such a way as to oppose the disturbance and so restore equilibrium.

This is known as **Le Chatelier's principle**.

Three main factors affect the position of equilibrium, or the ratio of reactants:products in the equilibrium mixture:

- temperature
- pressure
- concentration.

You will look at each in turn.

Temperature

In the Haber process example, the forward process is exothermic. If you increase the temperature, the equilibrium responds to oppose the change – so to oppose the rise in temperature, the reaction moves to the left (the endothermic direction), resulting in less product and more reactants in the equilibrium mixture.

For an endothermic reaction, the opposite is true. Increasing the temperature drives the reaction further to the right, resulting in more products and less reactants in the mixture.

Pressure

For a gaseous reaction, the pressure affects the concentration of particles (the number of particles per unit volume). The effect varies depending on whether the number of moles of product is the same as the number of moles of reactant or not. In the ammonia reaction, there are more moles of reactants (4 mol) than product (2 mol). One mole of any gas occupies the same volume, so at constant pressure the reactants take up twice the volume of the product. Increasing the pressure has more effect on the reactants than on the products, effectively increasing their concentration (see below), so an increased pressure drives the reaction to the right, forming more product.

Concentration

For a reaction in aqueous solution, increasing the concentration of the reactants will drive the reaction to the right – once again the system is responding to oppose the change. Conversely, increasing the concentration of the products drives the reaction to the left.

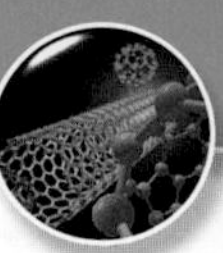

Iodine(I) chloride and iodine(III) chloride

If chlorine gas is passed through a U-tube containing solid iodine, a brown liquid, iodine(I) chloride, ICl is formed with a brown vapour above it. The bottom of the U-tube gets hot – the reaction is exothermic. If more chlorine gas is passed though the U-tube a yellow solid, iodine(III) chloride, ICl_3 is formed:

$$I_2(s) + Cl_2(g) \rightarrow 2ICl(l)$$

$$ICl(l) + Cl_2(g) \rightleftharpoons ICl_3(s)$$

If the chlorine supply is removed, and the U-tube is tipped horizontal, the yellow crystals disappear and a brown liquid is seen. The equilibrium is moving back to the left to reform the iodine(I) chloride and chlorine. These were separated from the reaction mixture when the U-tube was tipped. If the chlorine supply is attached again the yellow solid forms as the equilibrium moves to the right, opposing the increase in chlorine concentration and producing iodine(III) chloride again.

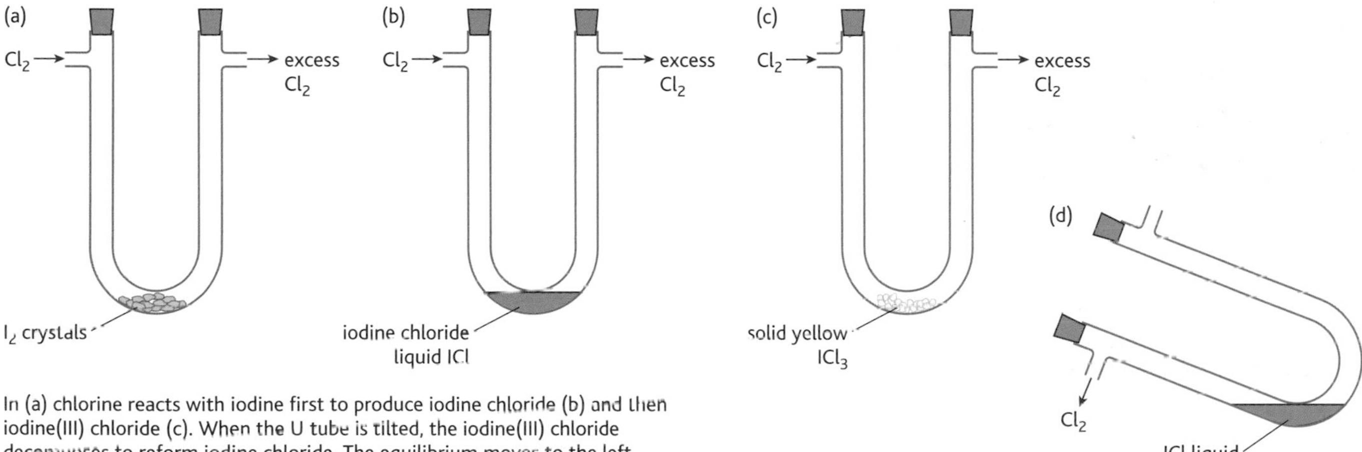

fig. 2.7.5 Investigating the equilibrium between iodine(I) chloride, iodine(III) chloride and chlorine.

Questions

1 The equation shows the reversible reaction between nitrogen dioxide and dinitrogen tetroxide:

$$\underset{\text{yellow}}{N_2O_4(g)} \rightleftharpoons \underset{\text{brown}}{2NO_2(g)}$$

If a sealed bulb containing an equilibrium mixture of N_2O_4 and NO_2 is heated, the mixture goes darker in colour.

- **a** Is the equilibrium moving left or right as the temperature rises?
- **b** Suggest a reason why the equilibrium might move in this direction.
- **c** If the pressure acting on the equilibrium mixture increased, how would you expect the mixture to change colour? Explain your answer.

2 Explain the results of the iodine(I) chloride experiment in terms of Le Chatelier's principle.

Predicting changes in the position of equilibrium

HSW Making the most of equilibria in industry

The Haber process – a compromise

Controlling equilibrium reactions in industrial processes is vital if a chemical manufacturer is to maximise profits. Returning to the Haber process, to get the maximum yield of ammonia, a high pressure would push the equilibrium to the right to make more ammonia. Because the forward reaction is exothermic, running the reaction at a low temperature would tend to produce more ammonia (as the equilibrium responds by trying to oppose the low temperature). However, if the temperature is too low the reaction is slow to come to equilibrium. A compromise is struck at 450 °C – traditionally an iron catalyst is used to speed up the reaction. The catalyst does not produce a higher yield of ammonia by moving the position of equilibrium, but it makes the reaction reach equilibrium faster. Because the ammonia is removed as it is produced, the low concentration of ammonia also pushes the reaction in the forward direction.

The role of a catalyst

As you have seen, in the Haber process the usual catalyst is iron. However, research has shown that ruthenium catalyst has a 40% better efficiency. A more expensive catalyst is justified because it is not used up in the process. However, ruthenium does not enable the process to take place at a lower temperature and chemists are working constantly to find a better catalyst that will work at a lower temperature.

Although theory suggests that a greater yield is obtained at higher pressures, increasing the pressure is an expensive process so a compromise pressure is used. The product is continually removed to push the equilibrium to the right. Also, in the Haber process all of the nitrogen and hydrogen are not combined when the mixture of gases goes through the catalyst chamber the first time, but they are recycled through to increase the yield.

The atom economy for this process is 100%, because all of the reactants are turned into products and there are no waste products.

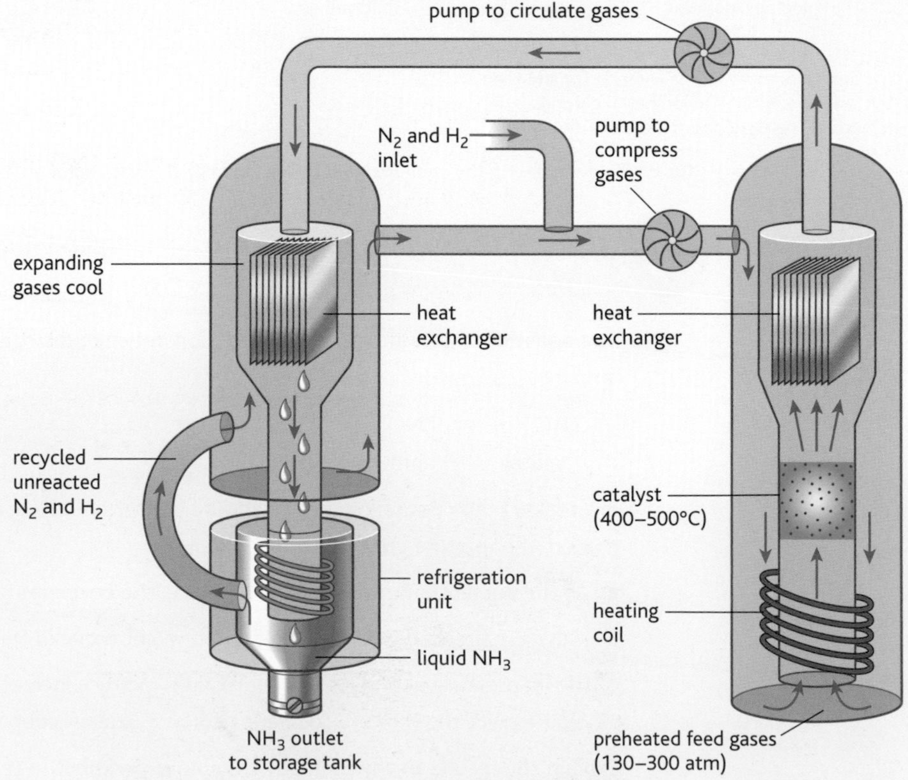

fig. 2.7.6 The Haber process is used to make ammonia by controlling an equilibrium reaction.

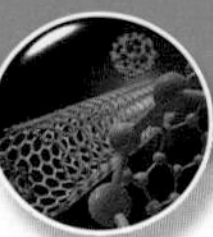

Stored methane in equilibrium

There are huge amounts of methane trapped as solid methane hydrate in ice structures deep in the oceans (see **fig. 2.7.7(a)**). Methane hydrate is in equilibrium with gaseous methane and water:

methane hydrate(s) ⇌ methane(g) + water(l)

The forward reaction is endothermic. According to Le Chatelier's principle, if the Earth's temperature rises, the equilibrium will move to the right to oppose this change, releasing more methane into the atmosphere.

If there is an increase in pressure, since methane is a gas, the equilibrium will move to the left. Moving to the left turns the methane gas back into solid methane hydrate, which does not exert a pressure, so this change in the equilibrium reduces the pressure.

It has been suggested that the sudden release of large amounts of methane gas from methane hydrate deposits could be a cause of past and future climate change. As greenhouse gases such as methane cause a rise in the temperature at the surface of the Earth, the sea temperature rises as well. If these ice structures melt as a result, then huge reservoirs of methane could be released into the atmosphere – to act as greenhouse gases and make the situation even worse.

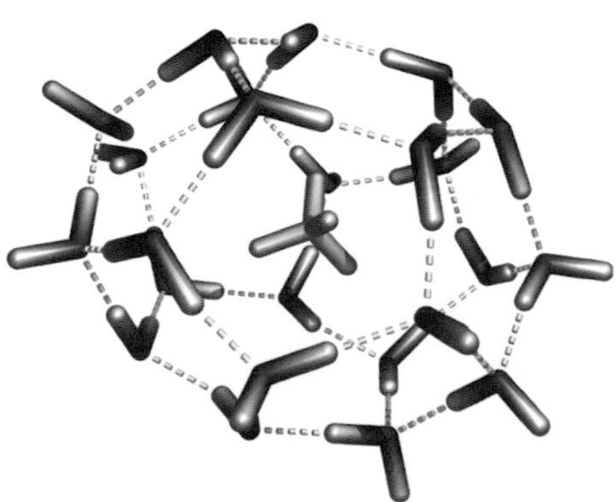

fig. 2.7.7 (a) Model of methane hydrate.

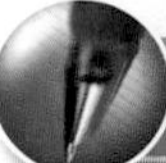

Predicting colour changes

This equation summarises the conversion of chromate(VI) ions to dichromate(VI) ions:

$$2CrO_4^{2-}(aq) + 2H^+(aq) \rightleftharpoons Cr_2O_7^{2-}(aq) + H_2O(l)$$

chromate(VI) ions	dichromate(VI) ions
yellow	orange

A solution containing an equilibrium mixture was treated with:

a excess acid

b excess alkali.

Predict the colour changes you would expect in each case.

a The solution turns orange.
According to Le Chatelier's principle, the equilibrium moves to the right to oppose this change (an increase in concentration of a reactant) and use up the excess acid.

b The solution turns yellow.
The alkali will neutralise the acid, so reducing the concentration of H^+ on the left-hand side of the equation. According to Le Chatelier's principle, the equilibrium will move to the left to oppose this change and produce more acid.

Questions

1 The equation for the reaction of ethanoic acid and ethanol is:

$$CH_3COOH(l) + C_2H_5OH(l) \rightleftharpoons CH_3COOC_2H_5(l) + H_2O(l)$$

a Predict whether the equilibrium position will move if the pressure is increased. Suggest a reason for your prediction.

b Predict the effect on the position of the equilibrium of adding:

(i) excess $CH_3COOH(l)$
(ii) excess $C_2H_5OH(l)$
(iii) additional water
(iv) a catalyst.

fig. 2.7.7 Heating methane hydrate releases CH_4, which burns while the ice melts.

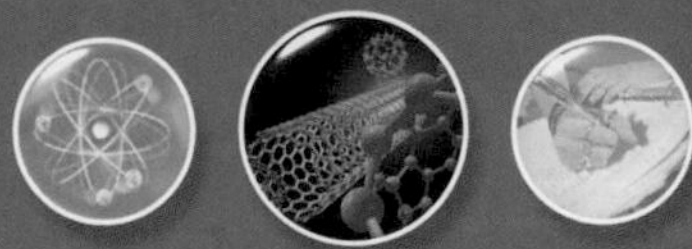

8 Organic chemistry – alcohols and halogenoalkanes

The alcohols – a homologous series

In chapter 1.5 you met the main groups of organic compounds, and learned how organic compounds are named. The final two chapters in unit 1 introduced two organic families, the alkanes and alkenes. Here you will study two more homologous series – the alcohols and halogenoalkanes.

Alcohols are extremely important organic compounds in industry. This is because they are good solvents and also make suitable raw materials for converting to other compounds.

The alcohol functional group

As you saw in chapter 1.5, **alcohols** have an alkyl group, which can be represented by R, along with an –OH or **hydroxyl** functional group:

R—O—H

The naming of alcohols is relatively simple – the name of the alkyl group containing the largest number of carbon atoms is used with the suffix -ol. A number may be used in front of the -ol to indicate which carbon atom it is attached to. Side chains are named in the usual way. **Table 2.8.1** and **fig. 2.8.1** show some examples of alcohols.

fig. 2.8.1 Ball and stick models of methanol, ethanol, propan-2-ol and propan-1-ol (clockwise from the top).

Alcohol	Displayed formula	Skeletal formula
Methanol	H | H—C—OH | H	OH
Ethanol	H H | | H—C—C—OH | | H H	OH
Propan-1-ol	H H H | | | H—C—C—C—OH | | | H H H	OH
Butan-1-ol	H H H H | | | | H—C—C—C—C—OH | | | | H H H H	OH
Butan-2-ol	H H H H | | | | H—C—C—C—C—H | | | | H H OH H	OH
2-methylbutan-2-ol	H | H—C—H H H | H | | | | H—C—C—C—C—H | | | | H H OH H	OH

table 2.8.1 Some examples from the homologous series of alcohols.

Types of alcohols

Alcohols have one of three types of structure – they may be primary, secondary or tertiary. The definition of these three types of alcohol depends on the number of alkyl groups present on the carbon atom attached to the functional group (see **fig. 2.8.2**). These differences in structure affect the way in which the alcohols react.

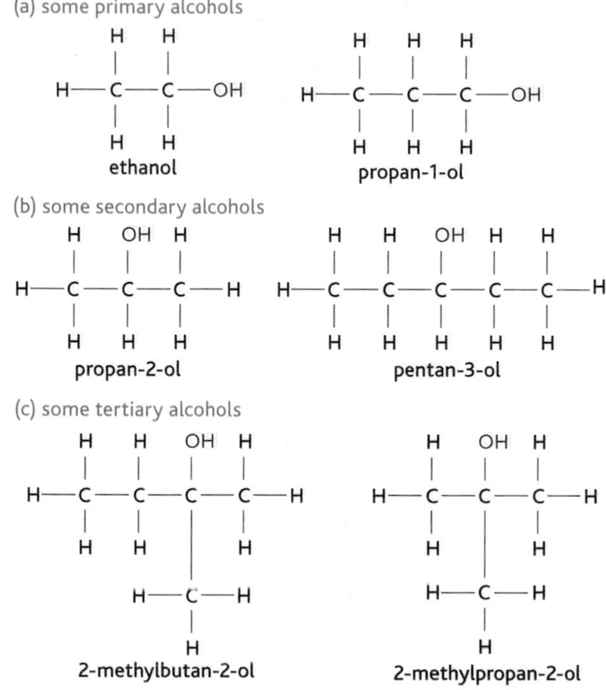

fig. 2.8.2 Primary, secondary and tertiary alcohols show distinct chemical properties.

- **Primary alcohols** have the general structure RCH_2OH. This means there are two hydrogen atoms attached to the carbon joined to the –OH group.
- **Secondary alcohols** have two alkyl groups and one hydrogen atom attached to the carbon that is joined to the –OH group, giving them the general structure RR_1CHOH.
- **Tertiary alcohols** have three alkyl groups and no hydrogen atoms attached to the carbon that is joined to the –OH group, so they have the general structure RR_1R_2COH.

Combustion of alcohols

Alcohols burn in a plentiful supply of air or oxygen to produce carbon dioxide and water vapour and release energy, eg:

$$CH_3CH_2OH(l) + 3O_2(g) \rightarrow 2CO_2(g) + 3H_2O(l)$$

Bioethanol is being produced by fermentation of plant material in large quantities to be used as a motor fuel, usually mixed with petrol. This reduces reliance on crude oil products and cuts air pollution. However, producing more and more bioethanol reduces the land available for growing food crops.

Reaction with sodium

When a small piece of sodium is put in water it rushes about on the surface, because sodium is marginally less dense than water (water has a density of 1.0 g cm^{-3} compared with 0.97 g cm^{-3} for sodium). The vigorous reaction results in the formation of hydrogen gas and a solution of sodium hydroxide with the evolution of heat. The water is acting as an acid and losing a proton, H^+. When a similar piece of sodium is put in ethanol it sinks – sodium is more dense than ethanol (density 0.79 g cm^{-3}). A steady stream of hydrogen is then produced. A solution of sodium ethoxide is formed with the evolution of heat.

These two reactions are obviously very similar, with ethanol, like water, acting as an acid. Because the reaction between sodium and ethanol is less vigorous than that between sodium and water it is reasonable to assume that ethanol is a weaker acid than water.

$$2Na(s) + 2CH_3CH_2OH(l) \rightarrow \underset{\text{sodium ethoxide (dissolved in ethanol)}}{2CH_3CH_2O^-Na^+(alc)} + H_2(g)$$

Figure 2.8.3 shows this reaction of sodium with ethanol. This reaction is often used in the laboratory to remove unwanted pieces of sodium from reaction mixtures.

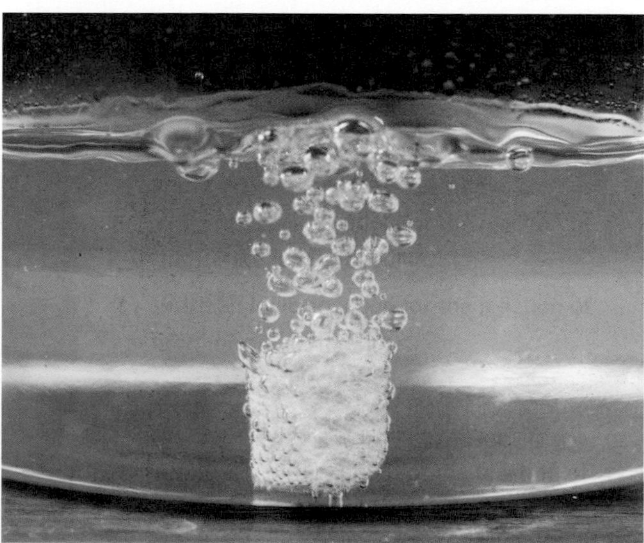

fig. 2.8.3 The reaction of sodium with ethanol.

Testing for alcohols

You may carry out reactions of alcohols, eg methanol, ethanol, propan-1-ol and propan-2-ol, using microscale methods. This could include reactions with phosphorus(V) chloride to show the presence of the –OH group.

Substitution reactions to form halogenoalkanes

Alcohols react with phosphorus(V) chloride, PCl_5 to produce a chloroalkane and hydrogen chloride gas, eg:

$$CH_3CH_2OH(l) + PCl_5(s) \rightarrow \underset{\text{chloroethane}}{CH_3CH_2Cl(l)} + POCl_3(l) + HCl(g)$$

Corresponding reactions occur with other phosphorus halides.

This reaction is used as a qualitative test for the presence of the –OH group – if you add phosphorus(V) chloride to an unknown liquid, the evolution of hydrogen chloride gas is evidence for the presence of the –OH group.

Oxidation of alcohols

Oxidation reactions of alcohols involve the interaction of both the carbon skeleton and the —OH functional group. These are very useful for finding out whether an alcohol is primary, secondary or tertiary because these all give different reactions with common oxidising agents such as acidified potassium dichromate(VI).

- Primary alcohols are easily oxidised to form aldehydes, and these **aldehydes** are themselves rapidly oxidised to **carboxylic acids** (see **fig. 2.8.4**). This means that the most common product of the oxidation of a primary alcohol is a carboxylic acid unless the aldehyde is separated during the reaction.

butan-1-ol → butanal + H_2O → butanoic acid

fig. 2.8.4 The oxidation of a primary alcohol gives an aldehyde followed by a carboxylic acid.

- Secondary alcohols are also readily oxidised to form **ketones**, but these ketones do not in general undergo any further oxidation and so they are retrieved as the final product of oxidation (see **fig. 2.8.5**).

butan-2-ol → butanone + H_2O

fig. 2.8.5 The oxidation of a secondary alcohol gives a ketone.

Testing for primary, secondary and tertiary alcohols

You may distinguish primary, secondary and tertiary alcohols using potassium dichromate(VI) using microscale techniques.

- Tertiary alcohols are not readily oxidised by any of the common oxidising agents.

Figure 2.8.6 shows the results of oxidising primary, secondary and tertiary alcohols with acidified potassium dichromate(VI). With the primary and secondary alcohols, the orange dichromate(VI) is reduced to dark green chromium(III).

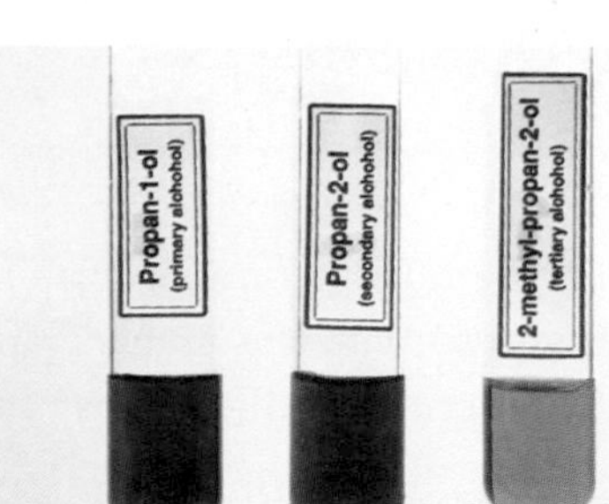

fig. 2.8.6 Testing primary, secondary and tertiary alcohols with potassium dichromate(VI).

Extracting the products of oxidation of alcohols

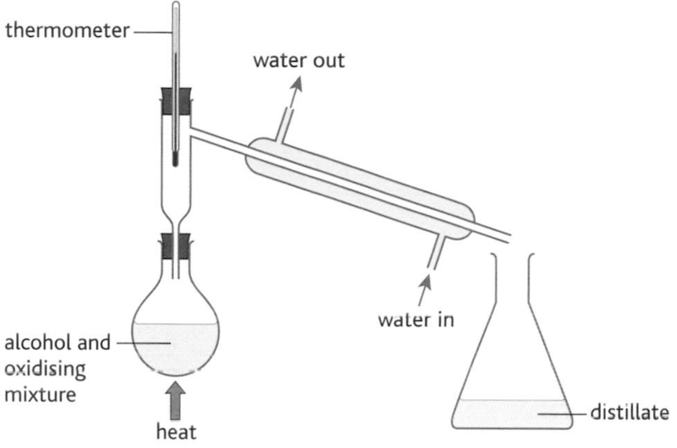

fig. 2.8.7 This apparatus is used for preparing a ketone from a secondary alcohol.

The apparatus in **fig. 2.8.7** can be used to prepare a ketone from a secondary alcohol. If butan-2-ol is heated with acidified potassium dichromate(VI) solution, butanone is produced and this distils off and is collected as the distillate in the receiver.

The boiling temperatures of butan-2-ol and butanone are 99.5 °C and 80 °C respectively. Providing the temperature on the thermometer is kept below about 95 °C, only butanone will distil off.

The same apparatus can be used for the oxidation of a primary alcohol, eg ethanol. **Table 2.8.2** shows the boiling points of ethanol and the products of oxidation.

Compound	Boiling temperature (°C)
Ethanol	78
Ethanal	21
Ethanoic acid	118

table 2.8.2 The boiling temperatures of ethanol and its oxidation products.

Using this apparatus in **fig. 2.8.7**, the ethanol is oxidised to ethanal, and providing the temperature on the thermometer is above 21 °C and below about 70 °C the distillate is ethanal. It is removed from the mixture before it has the chance to be oxidised further.

In order to convert ethanol into ethanoic acid, the apparatus in **fig. 2.8.8** can be used. The flask contains ethanol and excess acidified potassium dichromate(VI). The vertical **reflux condenser** prevents ethanol, ethanal and ethanoic acid vapours from leaving the flask. After the flask has been heated for some time, all of the ethanol will be converted into ethanoic acid. The apparatus in **fig. 2.8.7** can then be used to distil off the ethanoic acid.

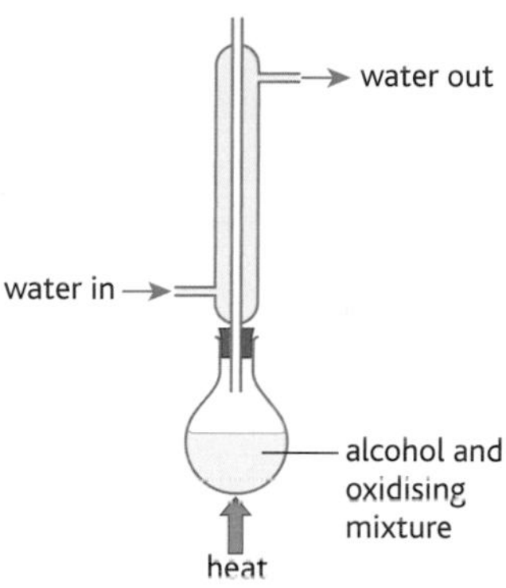

fig. 2.8.8 Reflux apparatus for preparing a carboxylic acid from a primary alcohol.

Questions

1 Draw the displayed formulae of four alcohols with a formula C_4H_9OH and write the name of each alcohol. Classify each alcohol as primary, secondary or tertiary.

2 Write an equation for the reaction of methanol with oxygen. Suggest what might be formed if there is a limited supply of oxygen.

3 Iodomethane can be produced when methanol is heated with phosphorus and iodine. Phosphorus and iodine react to form phosphorus(III) iodide. Write equations for these reactions.

4 Write equations for the two stages in the oxidation of ethanol.

5 A liquid produces hydrogen chloride fumes with phosphorus(V) chloride but is not oxidised by acidified potassium dichromate(VI). What does this suggest about the identity of the liquid?

The halogenoalkanes – another homologous series

Halogenoalkanes

Halogenoalkanes are a family of chemicals that are found relatively rarely in the natural world. Most of them are synthesised in the chemical laboratory or on a larger scale in chemical plants. Halogenoalkanes play a wide variety of roles in the world, many of them to the benefit of the human race. For example, they are important in medicine, agriculture and the production of plastics. However, they are also implicated in much environmental damage, both on the surface of the planet and in the atmosphere around it.

A halogenoalkane has a hydrocarbon skeleton with a halogen functional group. It is formed when a halogen atom replaces a hydrogen atom in a hydrocarbon molecule. The halogen may be fluorine, chlorine, bromine or iodine. The properties of an organic halogen compound are therefore affected by three things:

- the type of hydrocarbon skeleton
- the halogen or halogens attached
- the position of the halogen in the molecule.

It is also possible to have halogenoalkanes containing more than one halogen atom, eg 1,2-dibromoethane, CH_2BrCH_2Br and even for there to be different halogen atoms within the same molecule. **Figure 2.8.9** shows some examples of halogenoalkanes.

(a)

bromomethane | dichloromethane | 1,3-dichloropropane

fig. 2.8.9 (a) Displayed formula of bromomethane, dichloromethane and 1,3-dichloropropane. (b) Space-filling model of bromoethane.

Structural isomers are a very common feature of halogenoalkanes, and so both the drawing and the naming of the compounds need particular care. Changing the position of a halogen atom within a halogenoalkane makes a great difference to the properties of the molecule.

Like the alcohols, the halogenoalkanes may be classified as primary, secondary or tertiary compounds.

- **Primary halogenoalkanes**, eg 1-chloropentane have two hydrogen atoms bonded to the carbon atom carrying the halogen (see **fig. 2.8.10(a)**).

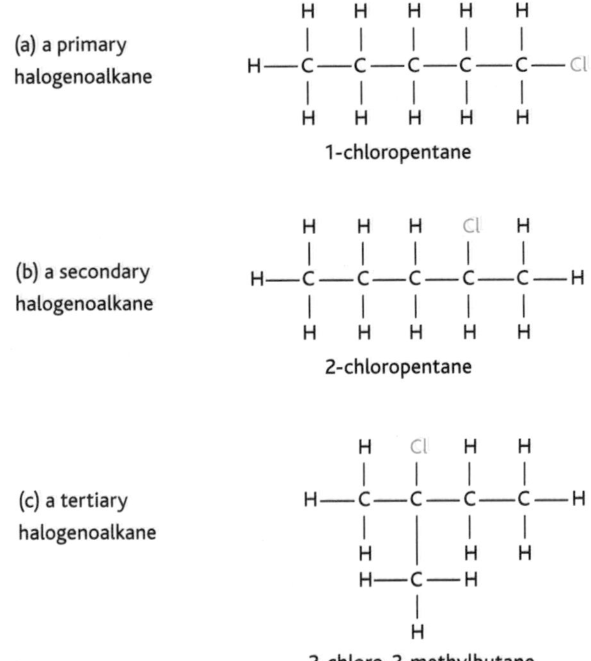

fig. 2.8.10 (a) Primary, (b) secondary and (c) tertiary halogenoalkanes.

- **Secondary halogenoalkanes**, eg 2-chloropentane have one hydrogen atom bonded to the carbon atom carrying the halogen (see **fig. 2.8.10(b)**). The halogen-carrying carbon is bonded to two other carbon atoms.
- **Tertiary halogenoalkanes**, eg 2-chloro-2-methylbutane, have no hydrogen atoms bonded to the carbon carrying the halogen. This carbon atom is bonded to three other carbon atoms (see **fig. 2.8.10(c)**).

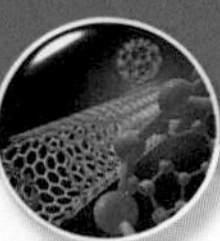

HSW Halogenoalkanes and anaesthesia

People have been developing surgery – attempting to remove diseased parts of the body or mend damaged internal regions – for centuries. However, developing fine surgical skills is almost impossible when working on a conscious patient anaesthetised only by alcohol and held down by straps and other helpers! It was only with the development of anaesthetics that surgery could move from the glorified butchery of the barber-surgeons to the sophisticated techniques increasingly used today. These range from operations carried out through minute openings made in the body wall (keyhole surgery) to massive transplant operations involving many hours in the operating theatre.

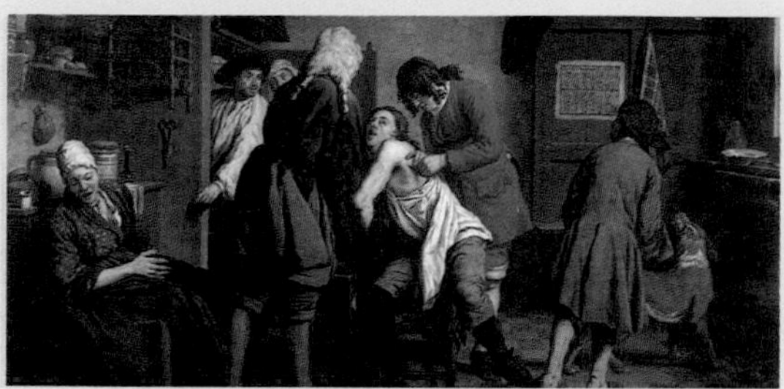

fig. 2.8.11 Surgery without anaesthetic was fast, brutal and often resulted in death.

The history of modern anaesthesia can be said to have begun in 1841 when the American Charles Jackson discovered that ethoxyethane ('ether') acts as an anaesthetic. Crawford Long used it to perform surgery in 1842, but because he did not publish his work immediately the credit goes to William Moreton, who used ethoxyethane for dental surgery in 1846. Ethoxyethane is an effective anaesthetic but it is highly flammable. Meanwhile in 1844 Horace Wells used nitrous oxide (dinitrogen oxide or 'laughing gas') as a dental anaesthetic for the first time. Nitrous oxide is not flammable or toxic but it produces only a very light anaesthesia. The final discovery of this extraordinary decade in the history of anaesthesia was when the Scot Sir James Simpson showed that the halogenoalkane trichloromethane ('chloroform') was a superior anaesthetic to both ethoxyethane and nitrous oxide. Trichloromethane received the royal stamp of approval when used by Queen Victoria herself during the deliveries of several of her children.

Trichloromethane produces a deep level of anaesthesia, but it can also cause liver damage so it is no longer used. The risks outweigh the benefits and better compounds have been developed. Interestingly, although trichloromethane is no longer used for pain relief during childbirth as it is too toxic, nitrous oxide is still very much present as part of the 'gas and air' mixture used by many women during labour.

It was not until the middle of the twentieth century that further strides were made in the development of anaesthetics. Demand grew for compounds that could be inhaled easily and that produced deep sleep yet were non-toxic and non-flammable. By this time organic chemists recognised two important features of organic compounds which helped them in their search:

- The substitution of a chlorine atom into a molecule of the alkane family results in a compound with anaesthetic properties – trichloromethane was a clear example. Increasing the number of chlorine atoms in the compound increased the depth of anaesthesia given, but unfortunately it also made the compound more toxic.
- Carbon–fluorine bonds are very stable and so their presence in a compound leads to non-flammable, non-toxic and unreactive properties.

Using this information, chemists came up with 'Halothane' (fluothane or 2-bromo-2-chloro-1,1,1-trifluoroethane). With this effective compound giving deep yet safe anaesthesia, along with similar compounds which followed, modern surgery really took off.

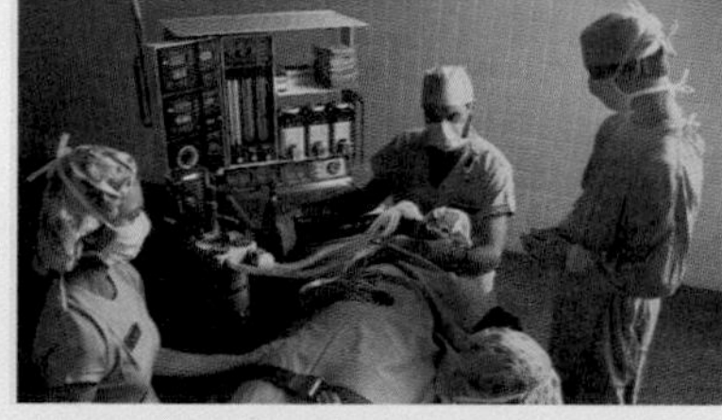

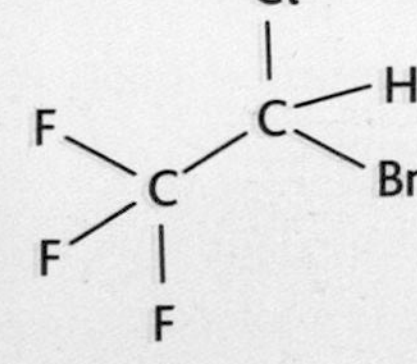

fig. 2.8.12 Halothane (fluothane) is a deceptively simple molecule which has allowed surgery to progress so that operations are very safe for the patient and an enormous range of surgical procedures have become possible.

Questions

1 Draw two displayed structures of $C_2H_4Br_2$. Name these compounds.

2 Perchloroethylene (or tetrachloroethene) is the most widely used dry cleaning fluid. Suggest the molecular formula and displayed formula of this.

Uses of halogenoalkanes

Refrigerants

Refrigerants are liquids which circulate inside a refrigerator. They change phase from a gas to a liquid and back, transferring heat from the inside of the refrigerator to the outside in the process. A good refrigerant should be a liquid that:

- changes state from gas to liquid and back at a temperature below the temperature of the refrigerator
- has a high enthalpy of vaporisation
- has a moderate density in liquid form and a relatively high density in vapour form
- is not toxic, flammable or corrosive.

Early refrigerators used ammonia or sulfur dioxide gas as the refrigerant. But these are toxic and were soon replaced. The compounds developed for this purpose were often halogenoalkanes.

Compounds containing both C—Cl and C—F bonds are called **chlorofluorocarbons** or **CFCs**. Within the CFCs, a set of molecules known by the trade name 'Freon' are widely used. These are based around methane and ethane, in which some or all of the hydrogen atoms are replaced by chlorine or fluorine atoms.

So many halogenoalkane refrigerants have been developed that a naming system has been devised by DuPont. The meaning of the codes is as follows:

- rightmost digit: number of fluorine atoms per molecule
- tens digit: one plus the number of hydrogen atoms per molecule
- hundreds digit: the number of carbon atoms minus one (missed out for halogenomethanes which have only one carbon atom)
- thousands digit: number of double bonds in the molecule (this is omitted when zero)
- any remaining bonds not accounted for are occupied by chlorine atoms
- a suffix of a lower case letter a, b or c indicates unbalanced isomers (see **fig. 2.8.13**).

For example, R134a has 4 fluorine atoms, 2 hydrogen atoms and 2 carbon atoms. The 'a' suffix indicates that the isomer is unbalanced by 1 atom, giving 1,1,1,2-tetrafluoroethane. R134 without the 'a' suffix would be 1,1,2,2-tetrafluoroethane (see **fig. 2.8.13**).

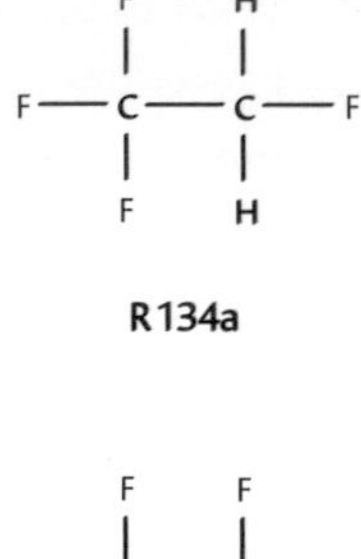

H—C—C—H (F F above, F F below)

R134

fig. 2.8.13 The molecular structures of the halogenoalkanes R134a and R134.

The problem with CFCs

Surrounding the Earth, in the upper atmosphere, is a layer of ozone gas (O_3). This absorbs short wavelength ultraviolet light very strongly, protecting life at the surface from its harmful effects. Without this layer there would be more cases of skin cancer and eye cataracts.

Because of their unreactivity, chlorofluorocarbons released into the atmosphere do not decay, and find their way eventually into the upper atmosphere. When they reach the ozone layer, two reactions happen.

The first of these involves homolytic breakdown of the CFC molecules under the influence of ultraviolet light:

$$CCl_2F_2 \rightarrow \cdot CClF_2 + Cl\cdot$$

This reaction produces the highly reactive chlorine free radical. The atom may simply recombine with its original molecule, or it may attack a molecule of ozone:

$$Cl\cdot + O_3 \rightarrow ClO\cdot + O_2$$

$$ClO\cdot + O_3 \rightarrow Cl\cdot + 2O_2$$

Net reaction: $2O_3 \rightarrow 3O_2$

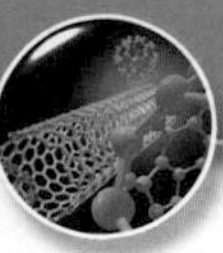

Notice that the chlorine atom is not used up in this reaction, so one CFC molecule can destroy literally thousands of ozone molecules.

Decreases in ozone concentrations in the ozone layer were first detected by Sherwood Rowland and Mario Molina in the late 1970s, and these have since been confirmed. The most massive decreases happen over Antarctica, where ozone concentrations temporarily fall each spring, resulting in an ozone 'hole' (see **fig. 2.8.14**). The size of this hole appears to be increasing, and a similar hole now appears to have developed over the Arctic as well.

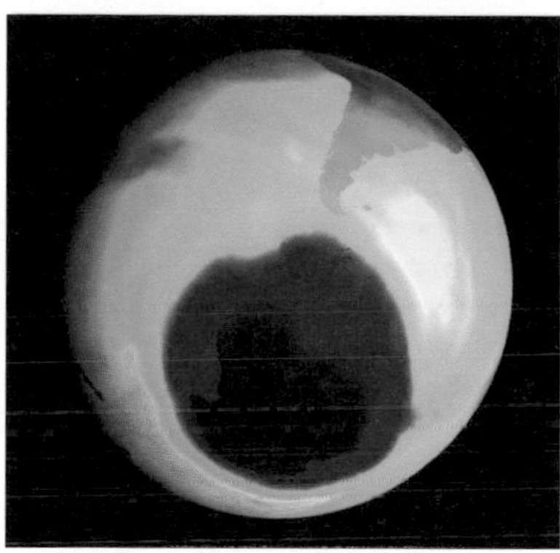

fig. 2.8.14 The ozone hole over the Antarctic in 2001.

In 1987 an international treaty called the Montreal Protocol set out plans for a reduction in the use of CFCs. Today about 190 nations have agreed to phase out their production and use of CFCs. However, high levels of CFCs in the atmosphere are likely to persist for some time. Scientists are not sure how quickly a reduction in the level of ozone depletion will be seen. There is some evidence it may be sooner than at first thought.

It is important that all countries tackle this because it is a global phenomenon. What happens in one country affects everybody. The Montreal Protocol appreciates that meeting the targets of not using harmful CFCs and replacing them with safer alternatives is easier for developed countries. For example, the Protocol recommended stopping the use of CFCs in developed countries by 1995 but in developing countries by 2010. For more about CFCs, see p.238.

HSW Scientists can't see into the future!

Thomas Midgley (born in 1889) was a very good industrial chemist who solved some of the most pressing problems of his day. Unfortunately the solutions he found left more problems for everyone who followed after him, difficulties he could never have foreseen ...

The early refrigerators used unpleasant chemicals which could be fatal when they leaked. Midgely was commissioned to find a better refrigerant. He invented both tetrafluoromethane and dichlorofluoromethane – the first of the CFCs. In a demonstration to fellow scientists, Midgley breathed in the gas and then exhaled it gently over a candle flame, which went out, showing his new compound was both non-toxic and non-flammable. The new, effective and safe refrigerants became widely used for many years.

What Thomas Midgley could not have known was that the CFCs he developed were also capable of damaging the protective layer of ozone around the earth. Twenty-first century scientists realise that CFCs react with and destroy ozone and there is international cooperation to reduce the use of these chemicals. Unfortunately it will take a long time before atmospheric ozone is restored to pre-Midgley levels! Midgley also developed tetraethyllead, the poisonous anti-knocking agent which used to be added to petrol.

Scientist can answer questions – but they also raise new issues. How many of our new scientific developments will turn out to have done more harm than good? In spite of all our modern safeguards, some almost certainly will!

What are the safe alternatives to CFCs?

Hydrochlorofluorocarbons, commonly known as **HCFCs**, are a group of synthetic compounds containing hydrogen, chlorine, fluorine and carbon. They are not found anywhere in nature. HCFC production increased after countries agreed to phase out the use of CFCs. Unlike the CFCs however, most HCFCs are broken down in the lowest part of the atmosphere, and pose a much smaller risk to the ozone layer. Unfortunately HCFCs are also very potent greenhouse gases, despite their very low atmospheric concentrations, measured in parts per trillion (million million).

An example of an HCFC is HFC134a, which is CF_3CH_2F. You will notice this contains no C—Cl bonds but has the even stronger C—F bonds.

Butane is also an alternative to CFCs, but of course it is flammable. Hydrochlorofluorocarbons and butane are also used as propellants in aerosol containers.

HSW Fire retardants

Halogenoalkanes used to be used as **fire retardants** in some types of fire extinguisher. For example, fire extinguishers to be used around electrical equipment should not contain water-based chemicals. The halogenoalkane extinguishers were green in colour and contained liquid bromochlorodifluoromethane (BCF or Halon 1211) under pressure. This liquid vaporises when sprayed onto the flames as it has a boiling temperature of 4 °C, forming a heavy gas which does not burn, so it formed a layer over the flames, excluding air or oxygen.

The two problems associated with this type of fire extinguisher were:

- It contributed to the depletion of the ozone layer and so had to be replaced following the Montreal Protocol.
- Because of the narcotic effects of the gas people sometimes stole the fire extinguishers to inhale the gas. There were 16 deaths caused by this gas in 1990.

These Halon-type fire extinguishers have now been replaced by extinguishers containing liquid carbon dioxide. Carbon dioxide fire extinguishers are heavier and more expensive than the Halon extinguishers.

fig. 2.8.15 Fire retardants are included within the resin system used for circuit boards.

Printed circuit boards, by their nature, can ignite in use. Because of this it has been standard practice to include fire retardants within the resin system used for the circuit board (see **fig. 2.8.15**).

TBBPA is the primary flame retardant used in polymers for printed circuit board production. TBBPA is an organic molecule that includes approximately 59% bromine and therefore is a 'halogenated fire retardant'. There has been a move towards 'halogen free' printed circuit boards. Epoxy polymers always contain a small amount of halogen because they are formed as a by-product of the chemical processes of manufacture.

Synthetic fibres and fabrics are used in a growing variety of applications ranging from clothing to carpets, blankets and many industrial uses. The flammability of these fabrics is important not only in everyday clothing but also for specialised applications such as those encountered in the space industry. Flammability is a measure of the ability of the fabric to burn and propagate a flame or fire. Most fabrics decompose when they are exposed to fire. However, many of them do not actually burn. The aim of using flame retardants is to prevent the propagation of a fire.

There are usually five methods for making a fabric flame retardant:

- finishing the fabrics with flame-retardant chemicals
- adding a flame retardant to the polymer the fabric is made of, before it is made into converted fibres
- copolymerisation – polymerising a flame-retardant polymer with the polymer used to make the fabric
- using a polymer that contains chlorine, bromine and fluorine, which give inherent flame retardancy to the polymer and which are also the basis for flame-retardant additives added to the polymer to make the final fabric flame retardant
- chemically modifying the polymer, eg chlorinating polyester. For example, suppose you have a polyester fibre that is going to go into making a shirt that has to be flame retardant – the polyester is chlorinated by chemical methods and this increases the fire resistance of the fabric.

In general, combustibility increases with increase in hydrogen content. A decrease in hydrogen content will also decrease the height of the flame. As you have seen, the presence of elements like nitrogen, chlorine, bromine and fluorine tend to make the polymer flame retardant.

The use of flame-retardant materials has obvious benefits, eg making children's nightclothes less likely to catch fire where there are naked flames, but these flame retardants can affect the environment.

Other uses of halogenalkanes

As well as being important as anaesthetics, halogenoalkanes are used as insecticides. As you saw in chapter 1.5, DDT (dichlorodiphenyltrichloroethane) was widely used as an insecticide. However, DDT does not break down in the environment but builds up in higher members of a food chain. The eggshells of birds of prey containing large amounts of DDT are fragile.

Questions

1 Draw the displayed formula and write the chemical name for R12, R22 and R152.

2 Give two reasons why chlorinated hydrocarbons are not used today in fire extinguishers.

3 Suggest why chlorine-based insecticides last a long time in the soil.

Reactions of the halogenoalkanes

Substitution reactions

The chemistry of the halogenoalkanes is largely based on two factors:

- which halogen atom is present – organic fluorine compounds are very stable so here you will concentrate on the chloro-, bromo- and iodo- compounds
- the position of the carbon–halogen bond within the molecule.

Organic halogen compounds undergo **substitution reactions**, and also to a lesser extent under different conditions **elimination reactions**.

These substitution reactions are nucleophilic. You will find out more about the mechanisms of these reactions of halogenoalkanes in chapter 2.9.

With the hydroxide group

In these reactions of halogenoalkanes (represented as RX) the –OH group substitutes for the halogen, giving an alcohol and a hydrogen halide – a reaction that is sometimes called **hydrolysis** ('splitting with water'). The reaction with water is slow at room temperature – a more rapid reaction occurs with OH^- ions:

$$RX + H_2O \rightarrow ROH + HX$$
$$RX + OH^- \rightarrow ROH + X^-$$

For example:

$$CH_3Cl + H_2O \rightarrow CH_3OH + HCl$$
chloromethane methanol

$$CH_3CH_2CH_2Br + H_2O \rightarrow CH_3CH_2CH_2OH + HBr$$
1-bromopropane + water propan-1-ol + hydrogen bromide

The C—Cl bond is more polarised than C—Br and C—I bonds as a result of the greater electronegativity of the chlorine atom (see chapter 2.2), so it seems reasonable to expect the chloroalkanes to be hydrolysed faster than the bromo- or iodoalkanes. In fact, the opposite is the case. This is because of the greater strength of the C—Cl bond compared with the C—Br and C—I bonds (see **table 2.8.3**). This makes the chloroalkanes less reactive than bromoalkanes or iodoalkanes. This order of reactivity holds good for most of the other reactions of the group, and shows that it is the bond strength that is the determining factor in the rate of reaction of the halogenoalkanes rather than the bond polarity.

This also shows that fluorocarbons are very unreactive due to the strength of the C—F bond. We will see the significance of this later.

Bond	Average bond energy (kJ mol^{-1})
C—F	467
C—Cl	346
C—Br	290
C—I	228

table 2.8.3 Bond energies for the chlorine–halogen bonds.

With alcoholic potassium hydroxide

Potassium hydroxide dissolves in ethanol as well as in water. With aqueous solutions of potassium hydroxide, halogenoalkanes undergo substitution reactions to form an alcohol. With alcoholic potassium hydroxide, elimination reactions occur, eg:

$$CH_3CHBrCH_3 + OH^- \rightarrow CH_3CH{=}CH_2 + H_2O + Br^-$$
2-bromopropane + hydroxide ions → propene + water + bromide ions

In this reaction a molecule of hydrogen halide is eliminated from the halogenoalkane, forming an alkene as a product. When species such as OH^- or RO^- are present, these may act as a base by accepting a proton rather than as a nucleophile.

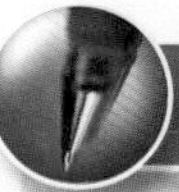

Reactions of the halogenoalkanes

You may carry out reactions of halogenoalkanes in the laboratory. As we have seen, the progress of these reactions is often judged by the formation of a silver halide. The halogen atom bonded to a carbon atom makes a halogenoalkane much more reactive than an alkane. This makes them useful intermediates in organic synthesis.

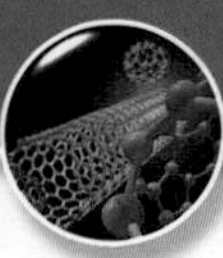

The reaction of potassium hydroxide with halogenoalkanes depends greatly on the reaction conditions (see **table 2.8.4**).

Type of reaction	Product	Conditions
Substitution	Alcohol	Reflux/heat with an aqueous solution of potassium hydroxide
Elimination	Alkene	Reflux/heat with an alcoholic solution of potassium hydroxide

table 2.8.4 Conditions for substitution and elimination reactions of the halogenoalkanes.

With alcoholic ammonia

Ammonia acts as a nucleophile because of the lone pair of electrons on its nitrogen atom. This means that it can easily replace a halogen atom in a halogenoalkane:

$$\underset{\text{iodoethane}}{CH_3CH_2I} + NH_3 \rightarrow \underset{\text{ethylamine}}{CH_3CH_2NH_2} + HI$$

This is not the end of the story, however. Ethylamine possesses a lone pair of electrons, just like ammonia. It reacts further with iodoethane:

$$\underset{\text{ethylamine}}{CH_3CH_2NH_2} + \underset{\text{iodoethane}}{CH_3CH_2I} \rightarrow \underset{\text{diethylamine}}{(CH_3CH_2)_2NH} + HI$$

In turn, there are two further stages:

$$\underset{\text{diethylamine}}{(CH_3CH_2)_2NH} + \underset{\text{iodoethane}}{CH_3CH_2I} \rightarrow \underset{\text{triethylamine}}{(CH_3CH_2)_3N} + HI$$

$$\underset{\text{triethylamine}}{(CH_3CH_2)_3N} + \underset{\text{iodoethane}}{CH_3CH_2I} \rightarrow \underset{\text{tetraethylammonium iodide}}{(CH_3CH_2)_4N^+I^-}$$

To carry out this reaction, iodoethane is heated with a concentrated solution of ammonia in ethanol. The reaction is carried out in a sealed tube rather than under reflux – the ammonia would simply escape up the condenser as a gas because it does not condense.

Questions

1 Describe and explain the reactions that can occur when sodium hydroxide reacts with 1-chloropropane. Describe the conditions and write symbol equations.

2 Write equations for the reactions of iodomethane with an alcoholic ammonia solution.

3 Why is it possible to get a substitution reaction with iodomethane and potassium hydroxide, but not an elimination reaction?

4 Suggest why ethylamine is never prepared by reaction of iodoethane and ammonia. If there was no alternative, what could you do to maximise the yield of ethylamine?

Identifying and preparing halogenoalkanes

How do different halogenoalkanes react?

When halogenoalkanes react with water to form an alcohol, one colourless liquid reacts to form another colourless liquid. To follow the reaction you need to detect either the formation of one of the products or the disappearance of one of the reactants. A useful indicator here is silver nitrate, which lets you see the appearance of the halide ions formed in the reaction, and also gives some idea of the reaction rate.

Silver nitrate is an ionic salt that dissolves in water to form Ag^+ and NO_3^- ions. Neither of these ions reacts with halogenoalkanes. However, the Ag^+ ion does react with halide ions to form insoluble products which appear as precipitates. Each silver halide has its own characteristic colour (see **fig. 2.8.16**).

So the reaction of a halogenoalkane with water can be followed by spotting the formation of a silver halide precipitate – if the time taken for the precipitate to appear is measured, then some idea of the rate of the reaction can be obtained.

fig. 2.8.16 The difference made by the different halogens on the rate of reaction of the halogenoalkanes can be shown clearly using silver nitrate.

In **fig. 2.8.16**, 1 cm^3 of silver nitrate solution was placed in each of test tubes A, B and C, together with 1 cm^3 of ethanol (as a solvent for both the silver nitrate and the halogenoalkane). All three tubes were placed in a water bath at 50 °C. After 10 minutes, 5 drops of 1-chlorobutane, 5 drops of 1-bromobutane and 5 drops of 1-iodobutane were added to tubes A, B and C respectively, and the tubes were replaced in the water bath. The precipitate formed most rapidly in tube C, more slowly in B and least slowly in tube A.

The result of the investigation confirms that the rate of hydrolysis of the halogenobutanes increases in the order:

1-chlorobutane $<$ 1-bromobutane $<$ 1-iodobutane (slowest)

The effect of the carbon skeleton

You can carry out a similar experiment to compare the effects of altering the carbon skeleton of a halogenoalkane. **Figure 2.8.17** shows three halogenoalkanes that have the same number of carbon, hydrogen and bromine atoms.

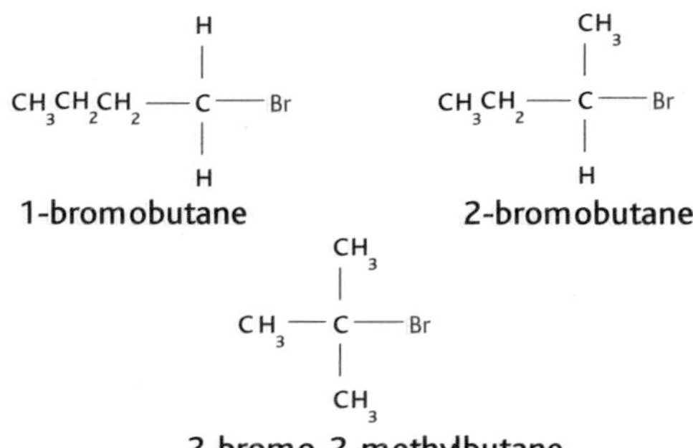

fig. 2.8.17 Isomeric halogenoalkanes.

1-bromobutane is a primary halogenoalkane, 2-bromobutane is a secondary halogenoalkane and 2-bromo-2-methylpropane is a tertiary halogenoalkane.

Three test tubes labelled A, B and C are set up as before containing the three bromobutanes plus 1 cm^3 of ethanol which is the solvent. This time it is not necessary to warm the solutions. Just add 1 cm^3 of silver nitrate solution and shake the tubes. Leave them to stand and time how long it takes for them to go cloudy.

You have seen that when testing for a bromide the formation of the cream precipitate is instantaneous. This is because it is an ionic reaction in solution:

$$Ag^+(aq) + Br^-(aq) \rightarrow AgBr(s)$$

With these reactions, a covalent C—Br has to be broken each time and so this will be slower than the formation of the silver bromide.

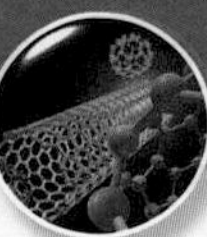

The results of this experiment show that the reaction with 2-bromo-2-methylpropane (tertiary halogenoalkane) is faster than with 2-bromobutane (secondary halogenoalkane) which is in turn faster than with 1-bromobutane (primary halogenoalkane).

In general halogenoalkanes react faster in substitution reactions where the halogen is attached to a branched chain.

Preparing a halogenoalkane from an alcohol

1-bromobutane can be prepared in the laboratory using butan-1-ol with sodium bromide, water and concentrated sulfuric acid. The sodium bromide and concentrated sulfuric acid react to produce hydrogen bromide. Hydrogen bromide reacts with butan-1-ol to form 1-bromobutane:

$$NaBr(s) + H_2SO_4(l) \rightarrow NaHSO_4(s) + HBr(g)$$

$$CH_3CH_2CH_2CH_2OH(l) + HBr(g) \rightarrow CH_3CH_2CH_2CH_2Br(l) + H_2O(l)$$

Sodium bromide, butan-1-ol and water are placed in the flask (see **fig. 2.8.18**).

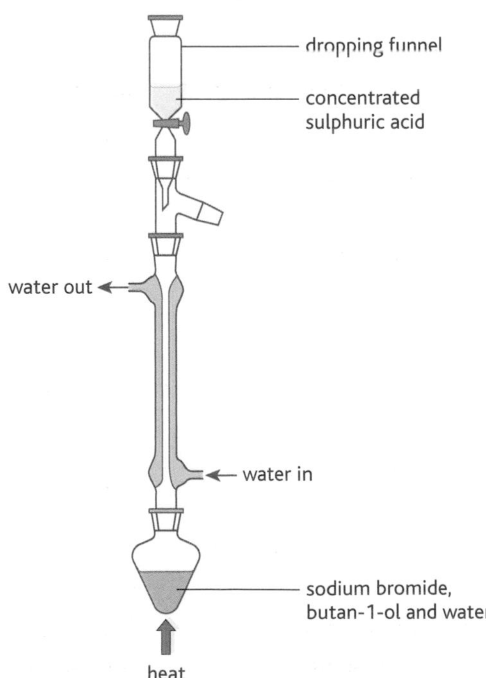

fig. 2.8.18 Apparatus for preparing a halogenoalkane from an alcohol.

The flask is put into a beaker of cold water and concentrated sulfuric acid is added slowly from the flask. The flask is cooled because the reaction at this stage is exothermic.

The dropping funnel is removed and the apparatus refluxed on a water bath for 30 minutes.

The apparatus in **fig. 2.8.19** is set up, the mixture is distilled and the distillate is collected. The distillate will be collected as two layers – a lower organic layer and an upper aqueous layer.

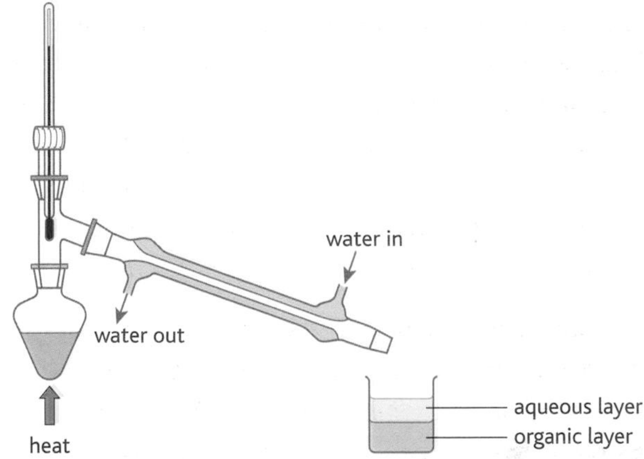

fig. 2.8.19 Apparatus for distilling a halogenoalkane from the reaction mixture.

1-bromobutane distils into the organic layer. **Table 2.8.5** shows the impurities that may be in the distillate, and which layer they are likely to be in.

Organic layer	Aqueous layer
Unreacted butan-1-ol	Unreacted butan-1-ol
Bromine	Oxides of sulfur
But-1-ene	Hydrogen bromide
	Water

table 2.8.5 Impurities that may be in 1-bromobutane prepared from an alcohol.

The aqueous layer is discarded. The organic layer can be purified and finally redistilled to produce pure 1-bromobutane (boiling temperature 102 °C).

Questions

1 In the investigation of the reactions of 1-chlorobutane, 1-bromobutane and 1-iodobutane, what was done to ensure a fair test?

2 What type of reaction is taking place when but-1-ene is formed as an impurity in the preparation of 1-bromobutane?

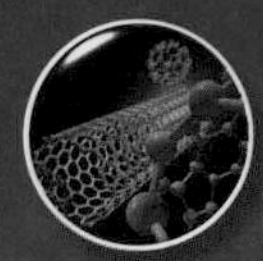
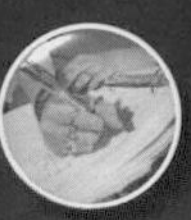

9 Mechanisms

Classifying reactions

Whether they are studying organic or inorganic reactions, in order to find out how a reaction takes place chemists first identify the products of the reaction. They then start thinking about the steps that take place for the reaction to occur – the **reaction mechanism**.

HSW Evidence for mechanisms

It is impossible to study reaction mechanisms directly because the particles are too small to observe. The most common approach is to look for evidence, then propose a mechanism that supports this evidence.

Chemists can generally only disprove a mechanism, rather than proving that a particular mechanism is correct. What sort of evidence do they collect?

In the reaction between methane and chlorine, as you saw in chapter 1.6, some ethane is produced. The mechanism must explain how this molecule with two carbon atoms is formed from a molecule with one carbon atom. The mechanism for the reaction of methane and chlorine involves $CH_3\bullet$, a methyl free radical. Two of these radicals can collide and form ethane, CH_3CH_3. So the existence of ethane in the products supports the free radical mechanism, but cannot prove it.

Chemists also use kinetic (rate of reaction) studies. You will see later that two very similar reactions can have different kinetics. These data help chemists to work out what is happening at a particle level.

Throughout units 1 and 2 you have met a number of organic reactions. To help organise all the different reactions, and also to help predict the outcome of an unknown reaction, chemists classify these reactions into different types, including:

- addition
- elimination
- condensation (addition followed by elimination)
- substitution
- oxidation
- reduction
- hydrolysis
- polymerisation.

Here you can look at each of these reaction types in turn.

Addition reactions

In an **addition reaction** two or more substances react to form a single product only:

reactant 1 + reactant 2 → product

For example, bromine undergoes an addition reaction with ethene:

$CH_2{=}CH_2$ + Br—Br $\quad$ CH_2BrCH_2Br
ethene $\quad$ 1,2-dibromoethane

You can see from this example that the organic reactant has a double bond, which becomes a single bond after the addition. Addition may also occur across a triple bond to produce a double bond or a single bond.

Elimination reactions

An **elimination reaction** is an organic reaction in which a small molecule is removed from a larger molecule leaving a double bond in the larger molecule.

For example, the dehydration of ethanol is an elimination reaction:

$CH_3CH_2OH \rightarrow CH_2{=}CH_2 + H_2O$

Condensation reactions

A **condensation reaction** involves an addition reaction followed by an elimination reaction. As a result, two reactants combine to form a larger molecule with the elimination of a small molecule such as water or hydrogen chloride.

An example is the formation of an ester from a carboxylic acid and an alcohol, which is an equilibrium reaction:

$$CH_3COOH + CH_3CH_2OH \rightleftharpoons CH_3COOCH_2CH_3 + H_2O$$
$$\text{ethanoic acid + ethanol} \rightleftharpoons \text{ethyl ethanoate + water}$$

Substitution reactions

In a **substitution reaction** one atom or group of atoms is replaced by another atom or group of atoms.

reactant 1 + reactant 2 → product 1 + product 2

For example, in chapter 1.6 you studied the reaction of ethane and chlorine, in which a hydrogen atom in ethane is replaced by a chlorine atom:

$$\begin{array}{ccccc} & H & & H & \\ & | & & | & \\ H— & C & — & C & —H \\ & | & & | & \\ & H & & H & \end{array} + Cl—Cl \rightarrow \begin{array}{ccccc} & H & & H & \\ & | & & | & \\ H— & C & — & C & —Cl \\ & | & & | & \\ & H & & H & \end{array} + HCl$$

Another example is the reaction of bromoethane with sodium hydroxide solution, in which a bromine atom is replaced by an –OH group:

$$\begin{array}{ccccc} & H & & H & \\ & | & & | & \\ H— & C & — & C & —Br \\ & | & & | & \\ & H & & H & \end{array} + OH^- \rightarrow \begin{array}{ccccc} & H & & H & \\ & | & & | & \\ H— & C & — & C & \quad OH \\ & | & & | & \\ & H & & H & \end{array} + Br^-$$

Oxidation and reduction reactions

You studied the topic of **oxidation** and **reduction** in chapter 2.4. An example of an organic redox reaction is the conversion of a secondary alcohol to a ketone:

$$\underset{\text{butan-2-ol}}{\begin{array}{ccccccccc} & H & & H & & H & & H & \\ & | & & | & & | & & | & \\ H— & C & — & C & — & C & — & C & —H \\ & | & & | & & | & & | & \\ & H & & H & & OH & & H & \end{array}} \rightarrow \underset{\text{butanone}}{\begin{array}{ccccccccc} & H & & H & & & & H & \\ & | & & | & & & & | & \\ H— & C & — & C & — & C & — & C & —H \\ & | & & | & & \| & & | & \\ & H & & H & & O & & H & \end{array}} + H_2O$$

The alcohol is oxidised by an oxidising agent such as acidified potassium dichromate(VI). The orange dichromate(VI) ion, $Cr_2O_7^{2-}$ is reduced to dark green chromium(III), Cr^{3+}.

Hydrolysis reactions

A **hydrolysis** reaction is the splitting up of a molecule by reaction with water. Organic compounds are often refluxed with water to bring about the reaction. This is normally a slow process but is faster if a dilute acid or alkali solution is used, for example the hydrolysis of bromoethane:

$$CH_3CH_2Br + H_2O \rightarrow CH_3CH_2OH + HBr$$

Sodium hydroxide solution also helps the reaction go more quickly:

$$CH_3CH_2Br + NaOH \rightarrow CH_3CH_2OH + NaBr$$

You saw above that a condensation reaction between a carboxylic acid and an alcohol forms an ester. The reverse reaction, the breaking down of the ester to reform the carboxylic acid and alcohol, is a hydrolysis reaction:

$$CH_3COOCH_2CH_3 + H_2O \rightleftharpoons CH_3COOH + CH_3CH_2OH$$
$$\text{ethyl ethanoate + water} \rightleftharpoons \text{ethanoic acid + ethanol}$$

Polymerisation reactions

A **polymerisation reaction** involves the joining of small molecules (monomers) together into a long chain. Sometimes a single monomer is involved; other polymers are made up of more than one monomer.

If M represents the monomer, the polymer could be represented by:

$$—M—M—M—M—M—M—M— \quad \text{or} \quad –[M]_n–$$

There are two types of polymerisation, depending on the reaction that joins the monomers – **addition polymerisation** and **condensation polymerisation**.

Questions

1 What type of reaction is shown in the following equation?

$CH_3I + NH_3 \rightarrow CH_3NH_2 + HI$

2 Naturally occurring esters are split up by boiling with sodium hydroxide solution. What type of reaction is taking place?

Bond breaking

As you saw in chapter 1.6, chemical reactions involve bond fission followed by the formation of new bonds. In a covalent bond two electrons are shared between two atoms. When that bond is broken during a chemical reaction there are two ways in which the electrons may be shared out leading to **homolytic** and **heterolytic fission**.

Homolytic bond breaking

Figure 2.9.1 shows two atoms X and Y bonded by a single covalent bond.

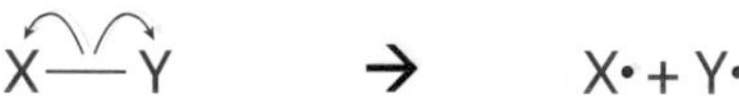

fig. 2.9.1 In homolytic bond fission each new species has one of the electrons from the covalent bond.

In homolytic fission this bond is broken so that both X and Y take a single unpaired electron. The prefix 'homo' means 'the same'.

X• and Y• are referred to as **free radicals** (or just radicals) and a dot represents an unpaired electron. If X and Y contain more than one atom, the dot is placed close to the atom with the unpaired electron, eg:

Cl—Cl → Cl• + •Cl
chlorine

CH_3CH_3 → $•CH_3$ + $•CH_3$
ethane

Free radicals are extremely reactive and as a result are short lived. They tend to react to gain an electron from another species so that they no longer have an unpaired electron, and this produces another free radical from the other species.

Heterolytic fission

Figure 2.9.2 shows the heterolytic fission of a covalently bonded species X—Y. In this case both electrons go to X and none to Y.

X—:—Y X^-: + Y^+

fig. 2.9.2 In heterolytic bond fission ions are formed, with one ion taking both of the electrons from the covalent bond.

The prefix 'hetero' means different. In this case, X has a negative charge. An organic ion like this is called a **carbanion** and a positive organic ion such as Y is called a **carbocation**.

The curly arrow notation shown here is useful when depicting reaction mechanisms. The full-headed arrow shows the movement of a pair of electrons, while half-arrow heads show single electrons moving in homolytic fission. It is important to draw these arrows accurately to show where these electrons start and where they finish during the reaction.

Classifying reagents: electrophiles, nucleophiles and free radicals

When thinking about how organic reactions take place, chemists devise a reaction mechanism. A reaction mechanism often occurs in stages and involves species classified as electrophiles, nucleophiles or free radicals.

An **electrophile** is an atom (or group of atoms) that is attracted to an electron-rich centre, or an atom where it accepts a pair of electrons to form a new covalent bond. Electrophiles either have a positive charge (carbocations) or are electron deficient (see **fig. 2.9.3**)

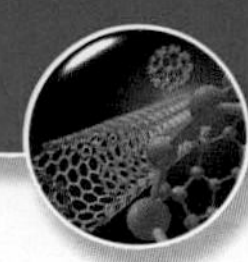

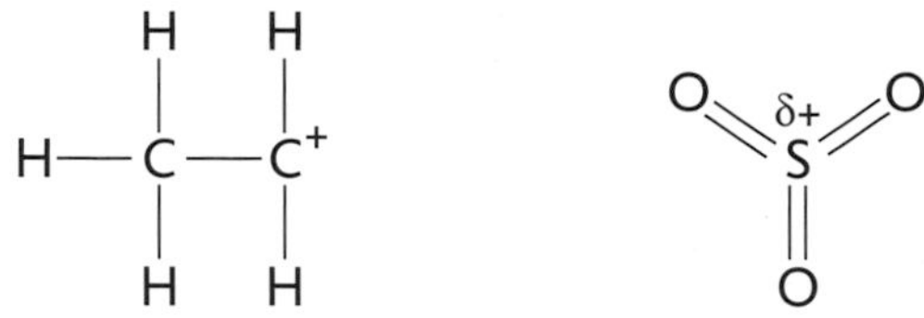

fig. 2.9.3 Examples of electrophiles – these species are attracted to form covalent bonds with electron-rich centres.

A **nucleophile** is an atom (or group of atoms) that is attracted to an electron-deficient centre, or an atom where it donates a pair of electrons to form a new covalent bond. Nucleophiles either have a negative charge (carbanions) or are electron rich, having a lone pair of electrons (see **fig 2.9.4**).

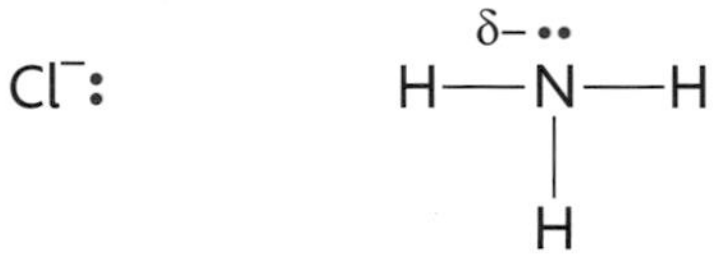

Cl^{-}:   δ– •• H—N—H | H

fig. 2.9.4 Examples of nucleophiles – these species are attracted to form covalent bonds with electron-deficient centres.

Whether a centre in a molecule is likely to be electron rich or electron deficient will depend on the polarity in bonds within the molecule.

Reaction of alkanes with halogens

You will know from chapter 1.6 that the reaction of an alkane with a halogen is a free-radical substitution reaction.

In the presence of ultraviolet light, halogen molecules are split into halogen atoms (free-radicals) and these react with the alkane molecule. This reaction involves homolytic bond fission so does not require an electron-rich or electron-deficient centre.

Reaction of alkenes with bromine and hydrogen bromide

Electrophilic addition

Chapter 1.7 covered these reactions of alkenes, which are **electrophilic addition reactions**. The double bond consists of four electrons (two in a π bond and two in a σ bond) and this provides an electron-rich centre for the attack of an electrophile.

The reaction of propene with hydrogen bromide can form two different bromopropanes (see **fig. 2.9.5**).

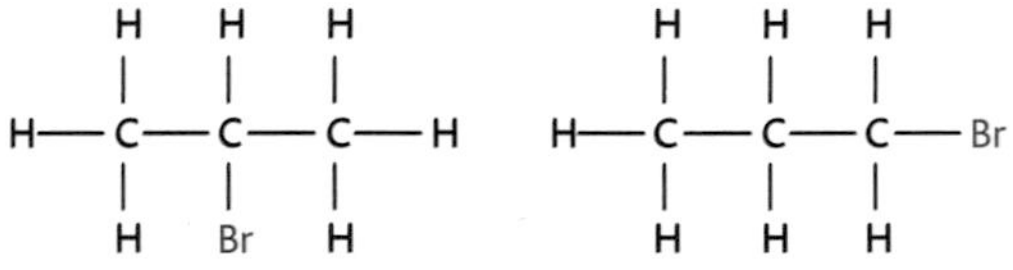

2-bromopropane   1-bromopropane

fig. 2.9.5 These two products are possible from the electrophilic addition of hydrogen bromide to propene.

The carbocation leading to the formation of 2-bromopropane is more stable than its alternative because the two methyl groups donate electron density and stabilise the positive charge. So the major product of the reaction is 2-bromopropane, in line with Markovnikov's rule.

Questions

1. Suggest conditions under which free-radical reactions are likely to take place.
2. What name is given to an atom of group of atoms with a lone pair of electrons?
3. A carbocation is attacked by a molecule. Is the molecule an electrophile or a nucleophile?

Nucleophilic substitution of halogenoalkanes

In chapter 2.8 you saw how halogenoalkanes undergo substitution reactions, eg the reaction of 1-iodopropane with aqueous potassium hydroxide solution:

$$CH_3CH_2CH_2I + OH^- \rightarrow CH_3CH_2CH_2OH + I^-$$

propan-1-ol

HSW How nucleophilic substitution takes place

What is the mechanism of this reaction? **Figure 2.9.6** shows the reaction in more detail.

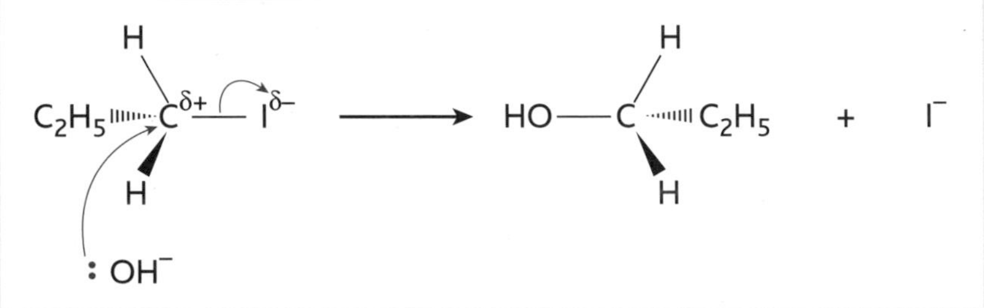

fig. 2.9.6 **The reaction of 1-iodopropane with hydroxide ions.**

The hydroxide ion is a nucleophile, and it attacks the electron-deficient carbon atom attached to the more electronegative iodine. The two electrons from the hydroxide ion form a covalent bond with the central carbon atom. The carbon–iodine bond breaks as the two shared electrons in the C—I bond move towards the iodine atom. This atom leaves as the iodide ion. The breaking of this bond is an example of heterolytic fission.

Comparing the reaction of two bromoalkanes with hydroxide ions

Table 2.9.1 shows data from an experiment to measure the rate of the hydrolysis of bromoalkane A.

Concentration of bromoalkane A ($mol\ dm^{-3}$)	Concentration of hydroxide ions ($mol\ dm^{-3}$)	Initial rate ($mol\ dm^{-3}\ s^{-1}$)
0.1	0.1	1.11×10^{-5}
0.2	0.1	2.22×10^{-5}
0.3	0.1	3.33×10^{-5}
0.1	0.2	1.11×10^{-5}
0.1	0.3	1.11×10^{-5}

table 2.9.1 **Results from an experiment to measure the initial rate of the hydrolysis of bromoalkane A at different concentrations of bromoalkane and hydroxide.**

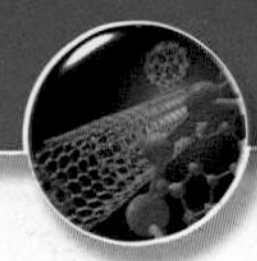

You will notice that the initial rate of reaction depends upon only the concentration of bromoalkane A. The initial rate is independent of the concentration of hydroxide ions.

Table 2.9.2 shows similar data for the hydrolysis of bromoalkane B. Here the initial rate of reaction depends on both the concentration of bromoalkane B and the concentration of OH^- ions.

Concentration of bromoalkane B (mol dm^{-3})	Concentration of hydroxide ions (mol dm^{-3})	Initial rate (mol dm^{-3} s^{-1})
0.1	0.1	1.50×10^{-5}
0.2	0.1	3.00×10^{-5}
0.3	0.1	4.5×10^{-5}
0.1	0.2	3.00×10^{-5}
0.1	0.3	4.5×10^{-5}

table 2.9.2 Results from an experiment to measure the initial rate of the hydrolysis of bromoalkane B at different concentrations of bromoalkane and hydroxide.

So the results of kinetics experiments which show, as you have seen, the collision rates of particles can give information we can use to suggest a mechanism for a reaction. Obviously these two bromoalkanes react by different mechanisms.

Hydrolysis or elimination

There are a number of factors which affect whether a halogenoalkane undergoes addition or elimination in the reaction with OH^-. One important factor is the solvent in which the reaction takes place. Thus, when 2-bromopropane is refluxed with aqueous sodium hydroxide, the product is propan–2–ol.

$$CH_3CHBrCH_3 + OH^- \rightarrow CH_3CHOHCH_3 + Br^-$$

In contrast, refluxing with ethanolic sodium hydroxide produces propene:

$$CH_3CHBrCH_3 + OH^- \rightarrow CH_2{=}CHCH_3 + H_2O + Br^-$$

HSW Mechanisms and the pharmaceutical industry

Knowing the exact way in which a drug works can be of help to chemists synthesising new compounds. Understanding the reaction mechanism of one drug can often suggest the chemical groupings which could be introduced into another molecule to improve the pharmaceutical action and set in motion the development of a new medicine.

Determining the reaction mechanism of a number of drugs has enabled scientists to develop computer programmes which can model new molecules and predict how they will interact with compounds in the human body or in the structure of disease-causing microbes. So understanding reaction mechanisms is helping us to develop better drugs.

Questions

1 In the reaction of bromoalkanes A and B with sodium hydroxide, in one reaction hydroxide ions attack the electron deficient centre in the halogenoalkane molecule. Which bromoalkane reacts in this way? Suggest why.

Chemistry in the ozone layer

The ozone layer is a term widely used in the media – but what is ozone? It is a gas which is naturally found in very small amounts in the stratosphere or upper atmosphere, between 10–50km above the surface of the earth. It is made up of three oxygen atoms. The ozone in the upper atmosphere is produced by the action of UV radiation from the Sun acting on oxygen molecules. This splits oxygen molecules into oxygen free radicals which combine with other oxygen molecules to form ozone.

$$O{=}O + h\nu \rightarrow O\bullet + O\bullet$$
$$O_2 + O\bullet \rightarrow O_3$$

Ozone is also found in the lower atmosphere, but here it is produced as a result of air pollution. However the highest concentration of the ozone in the atmosphere is in the upper atmosphere and it is this that is known as the ozone layer – although as you can see in **fig. 2.9.7** it only contains very low levels of ozone!

As you have seen on p215, the ozone layer helps to protect the surface of the Earth from harmful UV radiation. There is a natural balance between the formation of new ozone and the breakdown of ozone molecules. Anything which alters this balance affects the ability of the ozone layer to absorb UV light. If the rate of breakdown speeds up, the ability of the ozone layer to protect us from the damage caused by UV light from the Sun will be compromised.

Antarctic winter and the ozone hole

If you look back to p215, you can see that CFCs catalyse the breakdown of ozone into oxygen molecules. However, the destruction of the ozone layer doesn't happen all the year round. Why is it that the levels of ozone drop so dramatically in the spring?

The reactions which bring about ozone depletion need light to take place. During the Antarctic winter the Sun never rises. The whole area is in constant darkness. At the same time, in the dark and cold, a whirling vortex of stratospheric winds allow the air to get so cold that polar stratospheric clouds build up. The cloud particles may be water, ice or nitric acid, depending on the conditions. These cloud particles provide a huge surface area for the reactions which lead to the breakdown of ozone. Then when spring comes, so does the sunlight, providing the final link in the chain. It triggers the homolytic breakdown of the CFCs, producing free radicals which do all the damage to the ozone layer, catalysing the breakdown of the molecules. As a result, the ozone hole is at its largest in October, when the full effects of the spring and early summer reactions can be observed.

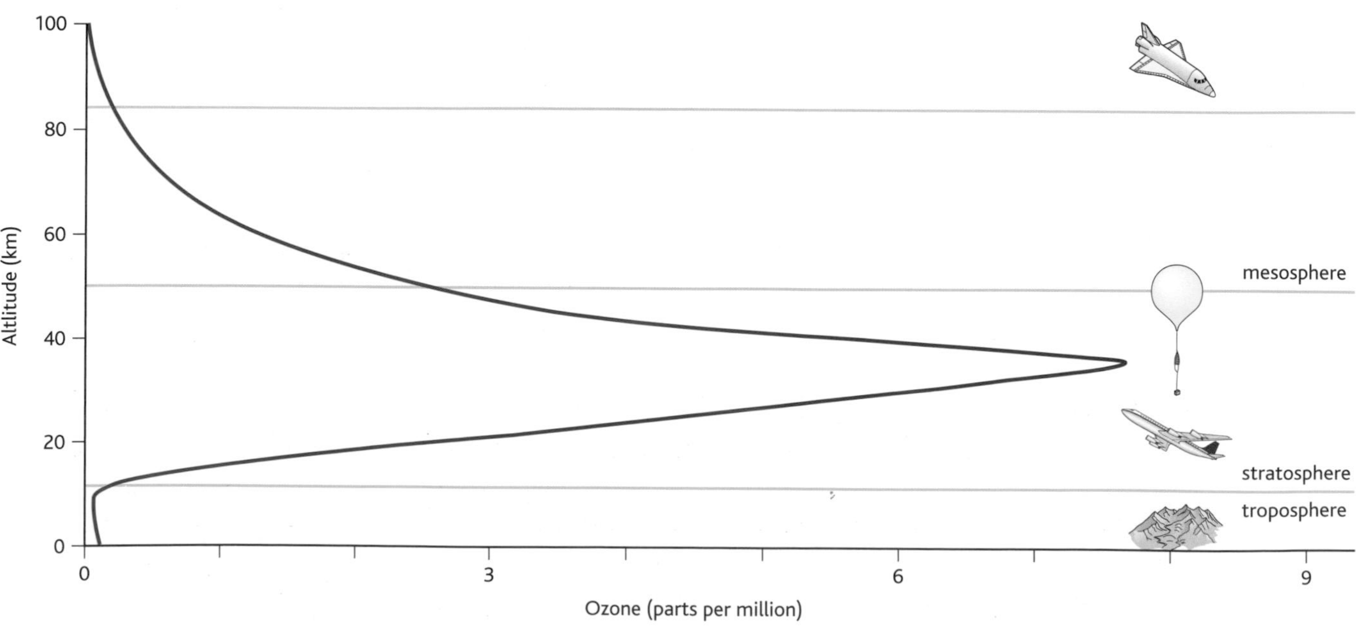

fig. 2.9.7 Levels of ozone peak about 30–35km above the surface of the earth – this is the 'ozone layer'.

HSW Evidence for changes in the ozone hole

The role of the ozone layer is a vital one, yet ozone makes up only 0.000 06% of the atmosphere. The levels of ozone in the atmosphere are measured both using satellite sensors such as the ozone monitoring instrument carried on the Aura satellite, and by ground-based measurements. Ozone may be measured in parts per million, but more often it is measured using Dobson units. A Dobson unit is the number of molecules of ozone that would be needed to create a layer of pure ozone 0.01 mm thick at a temperature of 0 °C and a pressure of 1 atm. The thickness of the ozone layer is around 300 Dobson units. In other words, if all the ozone in the stratosphere was compressed it would form a layer just 3 mm thick around the Earth!

In the area above the Antarctic an 'ozone hole' has developed. The ozone hole isn't actually a hole at all – just an area where the ozone levels are much lower than normal. In these regions the amount of ozone is down to an average of about 100 Dobson units.

Growing or shrinking?

Data on ozone levels has been collected since the 1950s. It is known that no Dobson scores of lower than 220 units were measured over the Antarctic before 1979. Since then the area of the ozone layer which has fewer than 220 Dobson units is regarded as the area of the ozone hole. The ozone hole currently has an area of about 27 million km^2, and Dobson scores of well below 100 have been recorded over recent years.

Since people became aware of the role of CFCs in ozone depletion, worldwide efforts to control the use of these chemicals have had some success, to the extent that many scientists now predict that the ozone hole should stabilise and then gradually start to diminish over the next 20–30 years. At the moment it appears as if the peak was reached in around 2003. However it isn't always easy to collect and interpret data on contentious science like this. **Figure 2.9.8** gives a variety of evidence for you to consider – what conclusions would you draw?

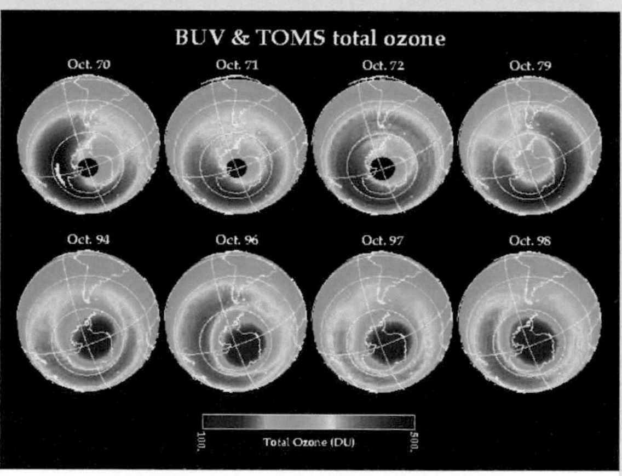

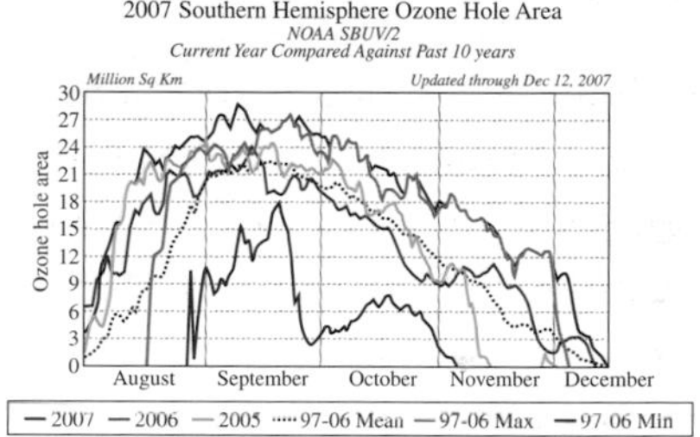

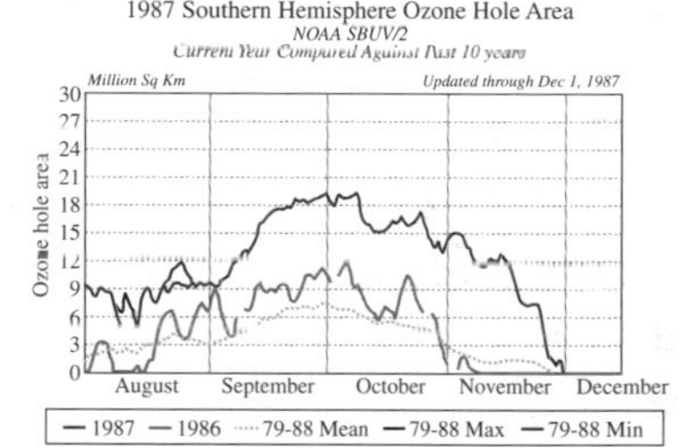

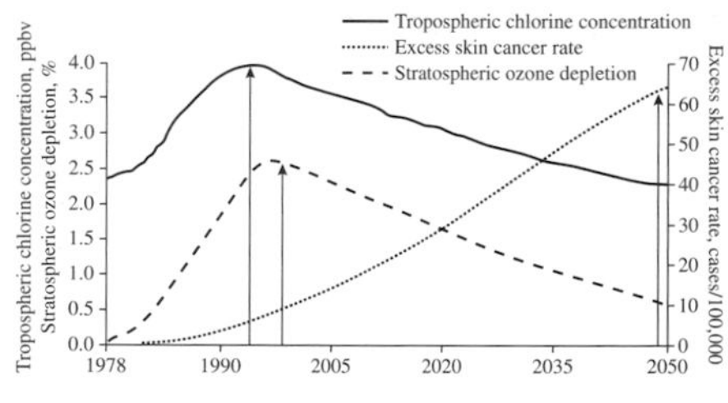

fig. 2.9.8 Data on the ozone whole. You can zoom in on these on the Active Book.

Questions

1 How is the change in size of the ozone hole over the year linked to the mechanism of the reaction which causes ozone depletion in the atmosphere?

2 Look at the data provided in **fig. 2.9.8**. Summarise what it tells you about the ozone hole in the stratosphere.

3 How valid and reliable is this evidence?

4 What further data would you want to see to decide what is happening to the ozone levels in the stratosphere?

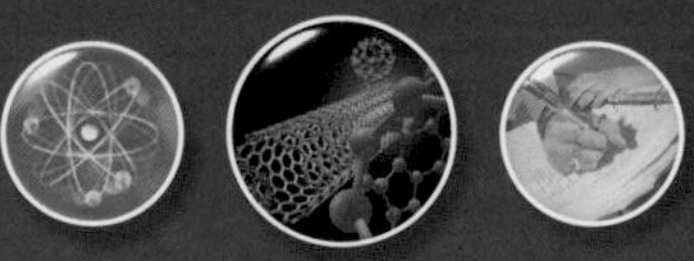

10 Mass spectra and IR absorption

Using mass spectra with organic compounds

The mass spectrometer

You studied the use of a mass spectrometer in chapter 1.3. You will remember that inside the mass spectrometer the sample is first vaporised, then ionised, accelerated and finally deflected by a varying magnetic field. A detector then detects the ions and produces a display which identifies certain peaks. These peaks occur at different mass:charge ratios (m/z).

The mass spectrometer can be used to identify different isotopes in an element, and to find the relative atomic mass of an element and the relative molecular mass of a compound. The mass spectrometer has many applications in chemistry, from pharmaceutical research to space research and from radioactive dating to catching drug cheats in sport.

Using a mass spectrometer to analyse organic compounds

There is a huge number of organic compounds. Identifying an unknown compound by qualitative and quantitative processes is slow and laborious. Using a mass spectrometer, along with infrared spectroscopy, can make this process much easier.

We can represent an organic compound as R—H. When this is put into a mass spectrometer and bombarded with electrons, one product will be R—H^+, which corresponds to the molecule losing a single electron. This peak will be shown on the mass spectrum as the peak with highest m/z value. This gives the relative molecular mass of the compound, and this peak is called the **molecular ion peak** (sometimes known as the parent ion).

Other peaks on the trace are produced by **fragmentation** of the organic molecule (see **fig. 2.10.1**). Only cation fragments are detected by the instrument – anions and radicals are not detected.

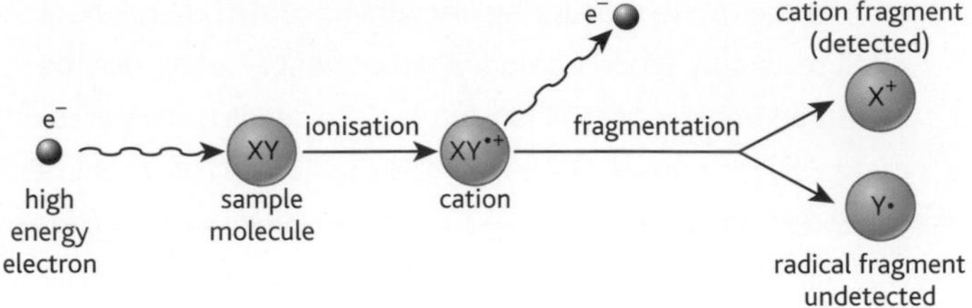

fig. 2.10.1 Fragmentation of an organic molecule in a mass spectrometer. The cation fragment may undergo further fragmentation.

The mass spectrum of butane

Figure 2.10.2 shows the mass spectrum for butane, C_4H_{10}.

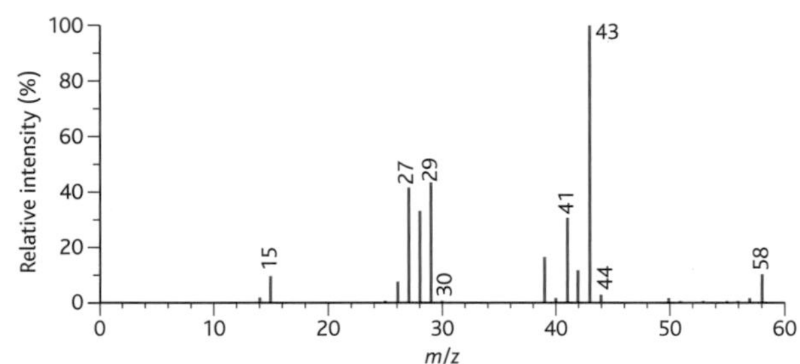

fig. 2.10.2 The mass spectrum of butane. The peak with the largest value of m/z (58) is the molecular ion peak.

The most abundant peak, in this case the one at 43, is called the **base peak**. **Table 2.10.1** shows some of the main peaks and the cations that correspond to them. **Figure 2.10.3** shows how these fragments of butane come about.

m/z	Cation	Fragment lost
58	$CH_3CH_2CH_2CH_3^+$	
43 (base peak)	$CH_3CH_2CH_2^+$	CH_3^-
29	$CH_3CH_2^+$	$CH_3CH_2^-$
15	CH_3^+	$CH_3CH_2CH_2^-$

table 2.10.1 Peaks on the mass spectrum for butane.

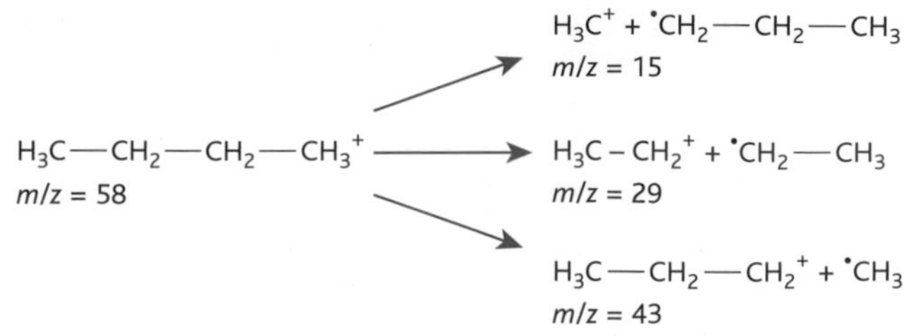

fig. 2.10.3 The fragments formed by butane in the mass spectrometer.

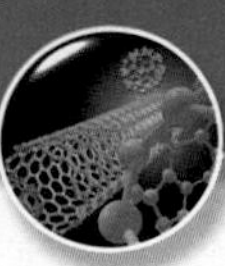

Some peaks on this spectrum are more difficult to interpret. The peak at 28, for example, corresponds to the $CH_3CH_2CH_2CH^{2+}$ fragment (mass 56:charge 2+)

The mass spectrum of ethanol

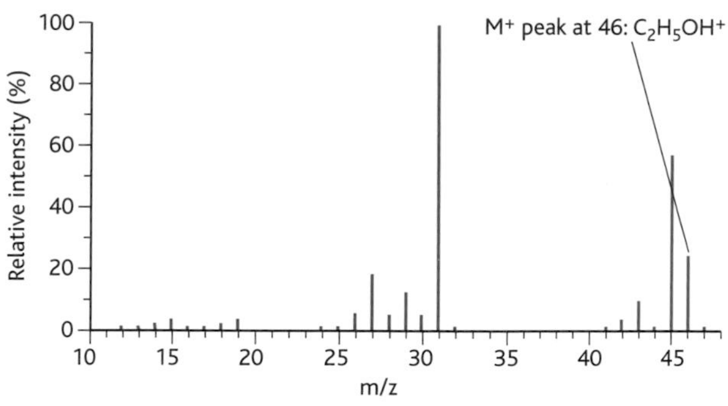

fig. 2.10.4 The mass spectrum of ethanol. The molecular ion peak is at 46.

Figure 2.10.4 shows the mass spectrum of ethanol. In the mass spectrometer the ethanol molecule forms the $CH_3CH_2OH^+$ ion which can then fragment. The relative molecular mass of ethanol is obtained from the molecular ion peak: M_r = 46. The small peak at 47 can be ignored – this is called the $M + 1$ peak and is due to fragments containing a carbon-13 atom. As you saw in chapter 1.3, this isotope makes up about 1% of all carbon atoms.

The peak at 45 corresponds to the loss of a hydrogen atom, forming the fragment $CH_3CH_2O^+$. The base peak at 31 corresponds to the loss of 15 units, namely CH_3. The cation fragment is then CH_2OH^+.

Even with relatively simple molecules like ethanol it is impossible to identify all of the peaks because complicated rearrangements of cation structures can occur.

The mass spectra of propanal and propanone

Figure 2.10.5 reminds you of the displayed formulae of the aldehyde propanal and the isomeric ketone propanone. **Figure 2.10.6** shows the mass spectra of propanal and propanone.

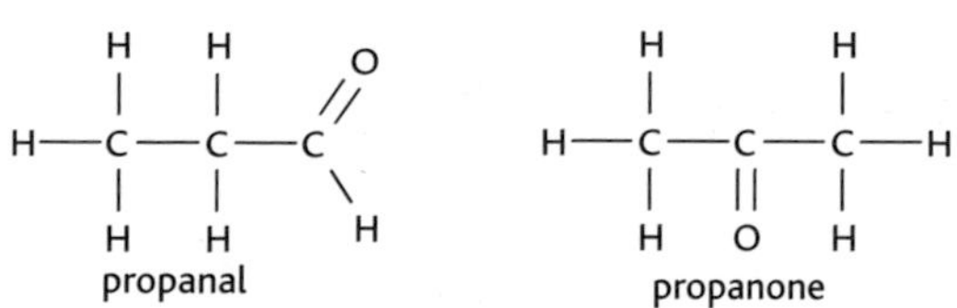

fig. 2.10.5 Propanal and propanone – an aldehyde and a ketone. Although they have the same relative molecular mass, their mass spectra show characteristic differences.

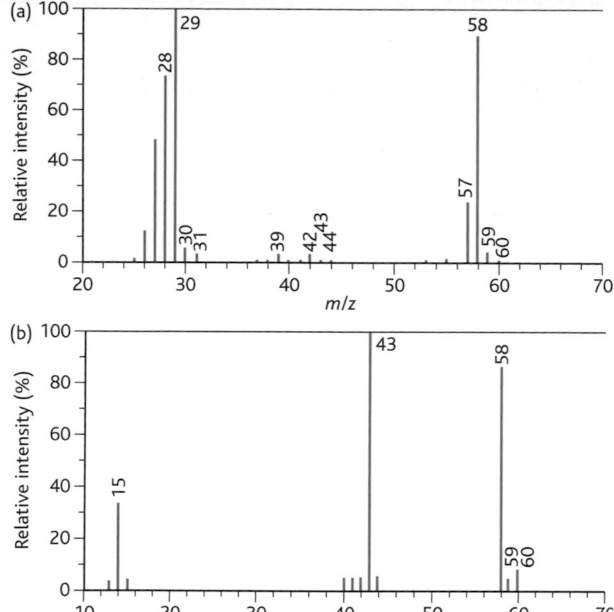

fig. 2.10.6 The mass spectra of (a) propanal and (b) propanone.

In **fig. 2.10.6(a)**, the molecular ion peak is at 58 corresponding to the molecular ion $CH_3CH_2CHO^+$ (M_r = 58). The peak at 57 corresponds to the loss of H–, forming the fragment $CH_3CH_2CO^+$.

The peak at 29 (the base peak) corresponds to both the loss of $CH_3CH_2^-$ (29 units) showing the existence of CHO^+ and CHO^- (29 units) showing the existence of $CH_3CH_2^+$. Both involve the breaking of the bond between the alkyl group and the carbonyl group.

In **fig. 2.10.6(b)**, the mass spectrum of propanone shows major peaks at 58 (the molecular ion peak), 43 (the base peak) and 15. Fragmentation of ketones often occurs at the C—C bond adjacent to the carbonyl group. Because propanone is symmetrical it will form the CH_3CO^+ cation at 43. The base peak at 15 corresponds to CH_3^+.

Questions

1 The mass spectrum of butanone, $CH_3CH_2COCH_3$ contains major peaks at 15, 29, 43, 57 and 72.

a What will be the value of the $M + 1$ peak and why does it occur?

b What is the relative molecular mass of butanone?

c Identify the cation that causes each peak.

2 Write an equation to show the fragmentation of butanone to produce the cation with m/z = 29.

Infrared spectroscopy

fig. 2.10.7 IR spectrometry has many different uses, from testing the quality of tea leaves to solving crimes. For example, hit-and-run drivers can be difficult to track down. But traces of paint from the vehicle will almost always be left on the body of the victim and IR analysis allows police to identify the make, model and year of the car involved. If the car has had a respray or has unusual trim added then it can even be possible to identify the particular car itself.

Absorbing infrared

If a substance is irradiated with infrared (IR) radiation, its molecules will absorb some of this radiation. This absorbed energy causes the bonds in the molecule to vibrate by either stretching or bending.

Different bonds require different amounts of energy to make them vibrate. The frequency of the energy required depends on:

- the bond strength
- the bond length
- the mass of each atom involved in the bond.

An **infrared spectrum** is produced by passing a range of IR frequencies through a sample in an **infrared spectrometer** (see **fig. 2.10.8**). The resulting spectrum is recorded on a chart recorder. The spectrum consists of a series of troughs at frequencies where the sample absorbs the radiation. Confusingly, these troughs are called **peaks**. The frequency of infrared absorption is measured in wavenumbers, cm^{-1} and this is shown on the *x*-axis of an IR spectrum. The **percentage transmission**, which is the percentage of radiation that passes through the sample, is shown on the *y*-axis. In the double-beam spectrometer, one beam of radiation passes through a solution of the sample dissolved in a solvent, and the other passes through the pure solvent in a reference cell. The instrument subtracts the effect of the solvent so the spectrum reflects the infrared absorption of the substance in the sample alone.

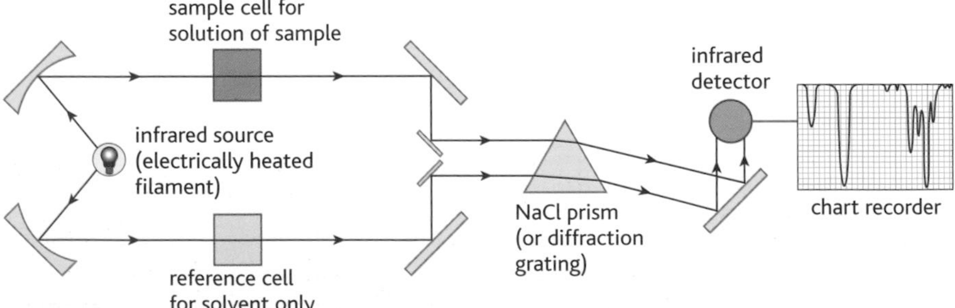

fig. 2.10. 8 A double-beam IR spectrometer.

Bending and stretching

For stronger bonds, a higher frequency of infrared energy is needed to make the bonds vibrate. **Table 2.10.2** shows the bond enthalpies and infrared absorption for the hydrogen halides. You can see that the infrared absorption increases as the bond enthalpy increases.

Compound	Bond enthalpy ($kJ\ mol^{-1}$)	Infrared absorption (cm^{-1})
H—Cl	+432	2886
H—Br	+366	2559
H—I	+298	2230

table 2.10.2 Peaks on the mass spectrum for butane.

Figure 2.10.9(a) shows the vibrations in the H—X bond. The atoms can simply move further apart and closer together. However, molecules that have more than two atoms can vibrate in different ways, eg the three different vibrations for sulfur dioxide, SO_2 are shown in **fig. 2.10.9(b)**. The infrared spectrum of sulfur dioxide (see **Fig. 2.10.10**) shows three peaks, each one corresponding to one of these three types of vibration.

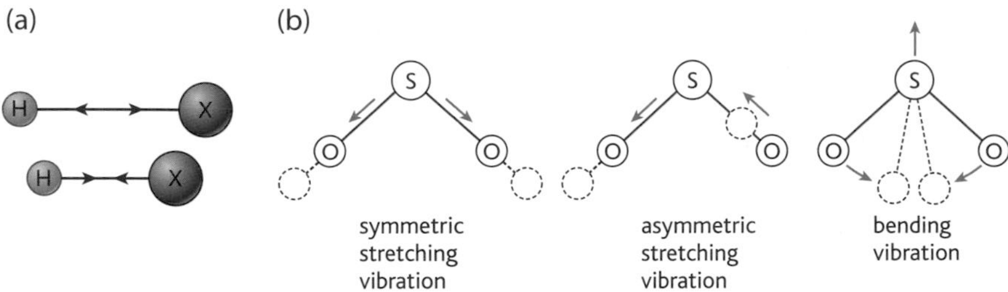

fig. 2.10. 9 (a) The IR-induced vibrations of the H—X bond. (b) In a molecule with more than two atoms, IR absorption brings about more complicated vibrations.

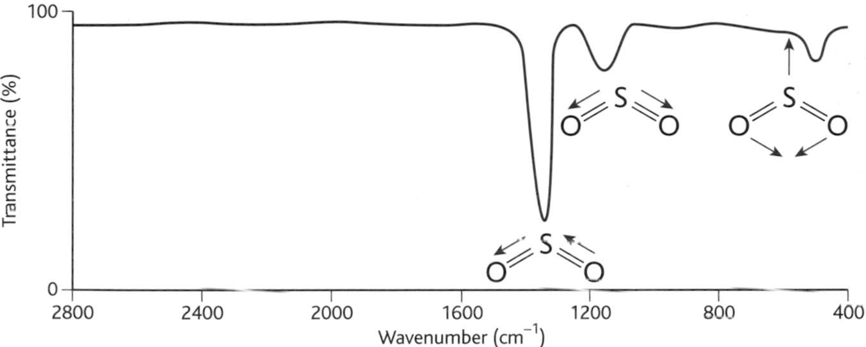

fig. 2.10.10 The IR spectrum for sulfur dioxide has three peaks corresponding to the three types of vibration of the molecule.

Which molecules absorb infrared?

Not all molecules absorb infrared and give an IR spectrum. Only molecules that change their polarity as they vibrate will absorb infrared radiation. So movement of partial charges within the molecule is essential for IR spectroscopy to identify the compound. For example, hydrogen, H_2 and chlorine, Cl_2 molecules do not absorb infrared radiation, but hydrogen chloride, HCl does.

Characteristic infrared absorptions by different functional groups

You can learn how to recognise different functional groups in an IR spectrum by reading the values of the peaks off the spectrum. **Table 2.10.3** gives the common peaks you will see. You do not need to remember these as they will be given when they are needed.

Bond	Intensity (strong/ medium/ weak/ variable)	Functional group	Wavenumber (cm^{-1})
C—H stretching	medium–strong	Alkanes	2850–3000
C—H stretching	medium	Alkenes	3095–3010
C—H stretching	weak	Aldehydes	2820–2900
C—O stretching	variable	Alcohols, esters, acids	1000–1300
O—H stretching	variable	Alcohols (these peaks are broad due to extensive hydrogen bonding)	3200–3750
O—H stretching	weak	Carboxylic acids (these peaks are broad due to extensive hydrogen bonding)	2500–3300
N—H stretching	medium	Amines	3300–3500
C=O stretching	strong	Aldehydes, ketones, carboxylic acids	1680–1740
C—X stretching	strong	Halogenoalkanes	500–1400

table 2.10.3 The IR absorption peaks for bonds and functional groups of organic compounds.

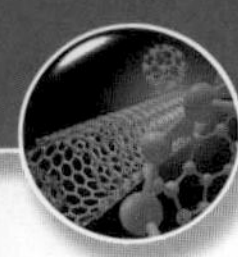

Infrared spectrum of ethanol

Figure 2.10.11 shows the infrared spectrum of ethanol.

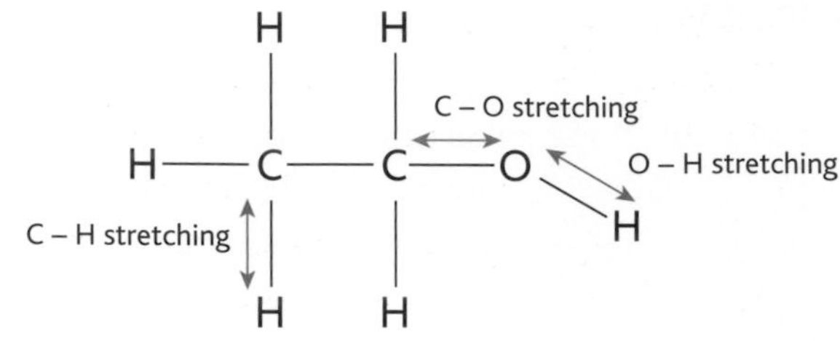

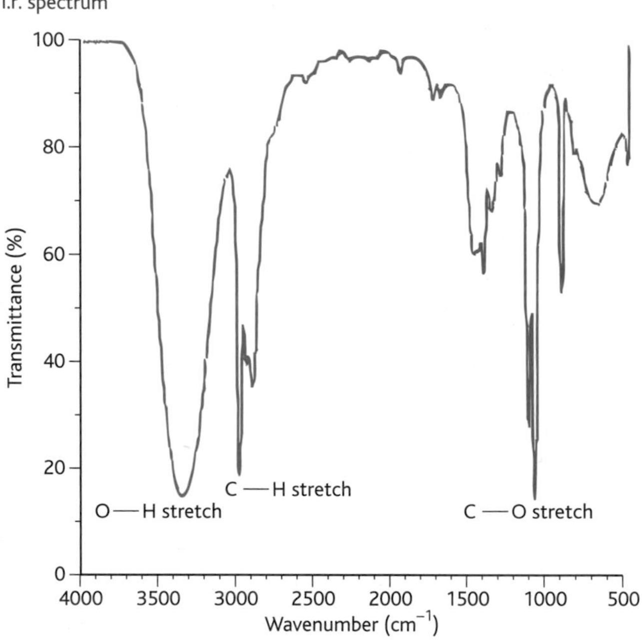

fig. 2.10.11 The IR spectrum of ethanol.

Absorption value (cm^{-1})	Bond present
3350	O—H
2950	C—H
1050	C—O

table 2.10.4 Information about three major peaks in the ethanol spectrum.

The region between 1000 and 1550 cm^{-1} is sometimes called the **fingerprint region**. Here many spectra have complex patterns of absorption. Often a compound can be identified by comparing its spectrum in this region with the spectra of other compounds.

Using an infrared spectrum to identify an unknown liquid

A liquid with molecular formula $C_3H_6O_2$ produces the infrared spectrum shown in **fig. 2.10.12**.

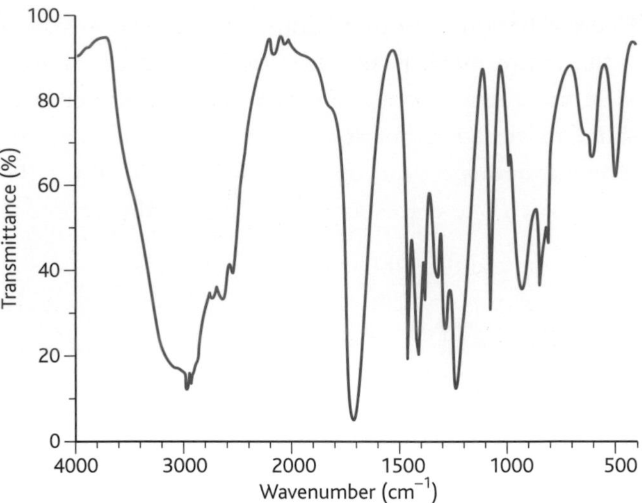

fig. 2.10.12 The IR spectrum of an unknown liquid.

You should notice that two features of this spectrum stand out:

1. The broad peak at 2500–3300 cm^{-1}. Using **table 2.10.3** this suggests the presence of hydrogen bonding in an O—H group.
2. The strong sharp peak at 1680–1720 cm^{-1} suggests a C=O group in a carboxylic acid.

Using this information the compound can be identified as propanoic acid, CH_3CH_2COOH.

Oxidation of ethanol

You will know from chapter 2.8 that ethanol is a primary alcohol which can be oxidised with acidified potassium dichromate(VI), first to an aldehyde (ethanal) and then an acid (ethanoic acid). The oxidation is carried out in such a way that the product distils as it is formed, yielding ethanal (see **fig. 2.10.13**).

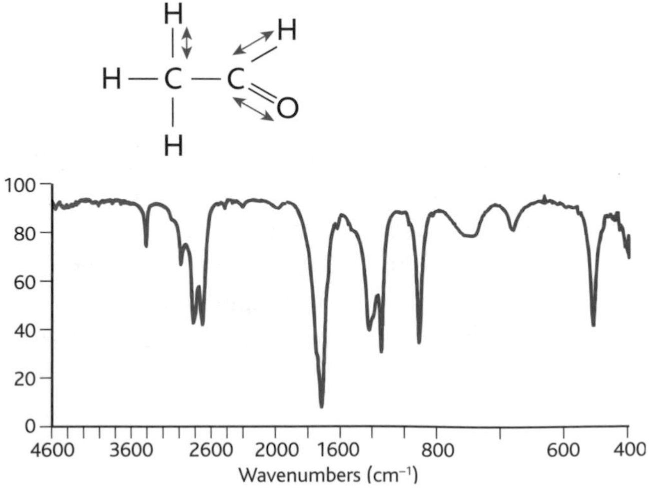

fig. 2.10.13 Ethanal is prepared by the oxidation of ethanol, a primary alcohol.

The infrared spectrum of ethanol was shown in **fig. 2.10.11** with major peaks at 3350, 2950 and 1050 cm^{-1}. Remember that the peak at 3350 cm^{-1} is a broad one due to hydrogen bonding.

The infrared spectrum of ethanal has no broad peak at around 3350 cm^{-1} because the aldehyde has no O—H bond and no hydrogen bonding. Instead there is a major peak at 1680–1750 cm^{-1}. This corresponds to the C=O bond in an aldehyde. You will also see peaks in the area 2800–3000 cm^{-1}. These correspond to stretching of C—H bonds in the CH_3 group and the C—H bond.

As you saw in chapter 2.8, oxidation of ethanol with excess acidified potassium dichromate(VI) under reflux will produce ethanoic acid. **Fig. 2.10.14** shows the displayed formula and infrared spectrum of ethanoic acid.

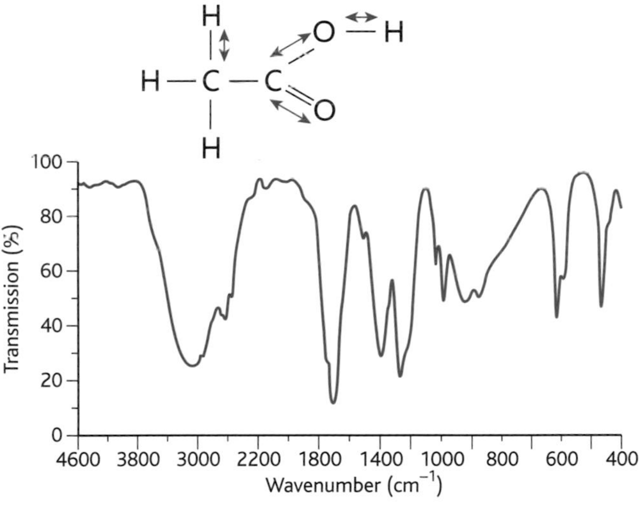

fig. 2.10.14 Displayed formula and IR spectrum of ethanoic acid.

As for **fig. 2.10.12**, notice the broad peak at approximately 3200 cm^{-1} corresponds to O—H stretching in a carboxylic acid. This broad peak was missing from the aldehyde IR spectrum in **fig. 2.10.12**. The peak at about 1750 cm^{-1} in **fig. 2.10.13** corresponds to the stretching of the C=O bond. The C—H bond produces a peak at about 2850–3000 cm^{-1}. This is, however, partly hidden by the broad peak. The peak at approximately 1000–1300 cm^{-1} corresponds to the stretching of the C—O bond.

HSW IR absorption by gases in the atmosphere

In chapter 1.6 you saw the effect of greenhouse gases in the atmosphere, which absorb IR radiation (to cause vibration of the bonds) and to re-radiate it back to the Earth's surface, making the temperature of the Earth higher than it would otherwise be. Greenhouse gases include water vapour, carbon dioxide and methane, but not oxygen or nitrogen. What determines if a gas acts as a greenhouse gas?

Oxygen and nitrogen are diatomic molecules with no polarity (see chapter 2.2). This is why they do not absorb infrared radiation.

Gases such as water vapour, carbon dioxide and methane have molecules composed of different atoms so they all have polarity and can absorb infrared radiation, and so act as greenhouse gases.

Questions

1 Two organic compounds P and Q both have a molecular formula of $C_3H_6O_2$. Compound P absorbs infrared radiation at 1720 cm^{-1} and 1030 cm^{-1}. Compound Q absorbs infrared radiation at 1700 cm^{-1} and shows a broad absorption band between 2600 and 3300 cm^{-1}. Suggest possible structures for P and Q.

2 An organic compound has an IR absorption peak at 1700 cm^{-1}. There is no other peak above 3000 cm^{-1}. Which functional group is present?

3 State the key identifying IR peak for butan-2-ol.

11 Green chemistry

Global warming and climate change

Global warming and the increased greenhouse effect

From chapters 1.6 and 2.9, and from constant reports in the media, you will know about the problems of climate change caused by burning fossil fuels and producing excessive carbon dioxide. Carbon dioxide is the greenhouse gas most often mentioned, but it is not the only culprit. Other greenhouse gases have even greater effects than carbon dioxide. Table 2.11.1 compares the **relative greenhouse factors** of different gases. The higher the greenhouse factor, the greater the effect on global warming. The relative greenhouse factor compares the effect that different gases have in absorbing infrared radiation (see chapter 2.10) compared with carbon dioxide. (Note that water vapour is also a greenhouse gas, but the amount in the atmosphere varies widely).

Greenhouse gas	Formula	Relative greenhouse factor
Carbon dioxide	CO_2	1
Methane	CH_4	30
Nitrogen oxides	NO_x	160
Dinitrogen oxide	N_2O	200
Trichlorofluoromethane	CCl_3F	21000

table 2.11.1 **Comparing greenhouse gases: relative greenhouse factor.**

Global warming potential (GWP) is a measure of how much a given mass of a greenhouse gas is estimated to contribute to global warming. It is determined by:

- their ability to absorb infrared radiation (their relative greenhouse factor)
- their half-life in the atmosphere – a measure of how long they last in the atmosphere before reacting and being broken down.

Table 2.11.2 compares the GWP values for different greenhouse gases.

Greenhouse gas	GWP
Carbon dioxide	1
Methane	20
Dinitrogen oxide	300
CFC12 (CCl_2F_2)	6500

table 2.11.2 **Comparing greenhouse gases: global warming potential.**

The International Panel on Climate Change (IPCC) gives a possible error of plus or minus 35% in these figures. This illustrates how unreliable predictions of long-term climate change are.

Anthropogenic and natural climate change

Anthropogenic climate change is climate change that is due to the activities of human beings (eg burning fossil fuels, deforestation, etc.) while **natural climate change** is due to natural processes occurring on the Earth (such as dissolving of carbon dioxide in sea water and formation of carbonate rocks).

The most significant of the greenhouse gases are carbon dioxide, CO_2 methane, CH_4 and dinitrogen oxide, N_2O. Fig. 2.11.1 shows the levels of these gases in ice cores over 650 000 years. Typical ice cores are removed by drilling deep samples from an ice sheet, most commonly from the polar icecaps of Antarctica or Greenland. The ice has formed from the build-up of layers of snow year after year, so the lower layers are older than the upper layers, and an ice core contains ice formed over a range of years. The properties of the ice or inclusions within the ice can then be used to reconstruct a climatic record over the age range of the core.

When the temperature falls, water molecules containing heavier isotopes will condense faster than normal water molecules. The relative concentrations of the heavier isotopes in the ice cores can be used in a model to reconstruct a picture of the local temperature changes.

In addition, the air bubbles trapped in the ice cores allow scientists to measure the atmospheric concentrations of trace gases, including carbon dioxide, methane and dinitrogen oxide.

You will notice that the levels of these gases fluctuate all the time. The shaded areas show peaks that occurred in the past. In the last couple of hundred years the levels of all three gases have risen alarmingly. These are a result of anthropogenic climate change, and the gases are at their highest levels for over 650 000 years.

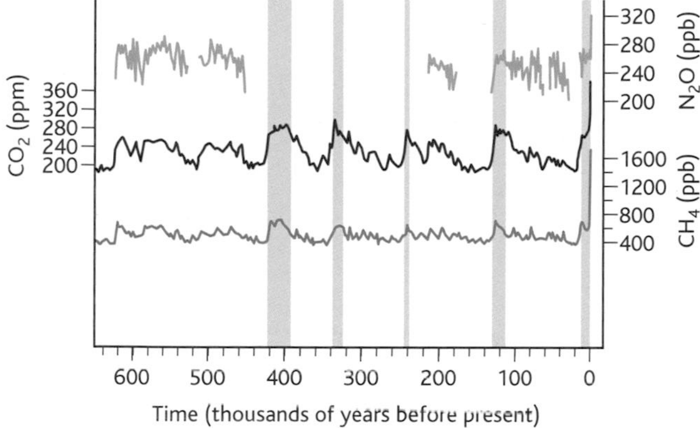

fig. 2.11.1 Measurements of ice cores show fluctuating levels of carbon dioxide, methane and dinitrogen oxide in ice cores, but recent sharp rises are obvious.

A recent report confirms that over the past 8000 years, until just before industrialisation in 1750, the average carbon dioxide concentration in the atmosphere increased by a mere 20 parts per million (ppm). In comparison, the concentration of atmospheric carbon dioxide in 1750 was 280 ppm, which had increased to 379 ppm in 2005. That is a huge increase of 100 ppm in 250 years. At the end of the most recent ice age there was approximately an 80 ppm rise in carbon dioxide concentration – this rise took place over 5000 years, and higher values than at present have only occurred many millions of years ago.

Since 1750, it is estimated that about two-thirds of anthropogenic climate change carbon dioxide emissions have come from burning fossil fuels (coal and petroleum) and about one-third from land-use change (mainly deforestation and agriculture). About 45% of this carbon dioxide has remained in the atmosphere, while about 30% has been taken up by the oceans and the remainder has been taken up by trees and other plants.

Of the carbon dioxide that goes into the atmosphere, about half is removed over a timescale of 30 years; a further 30% is removed within a few centuries; and the remaining 20% will typically stay in the atmosphere for many thousands of years. You saw in chapter 1.6 (fig. 1.6.18) how carbon dioxide levels have increased in the past 40 years, and that as well as the overall rise the levels show seasonal variation.

Air traffic and global warming

A major cause of anthropogenic climate change is air traffic. High-flying jet aircraft produce frozen water vapour trails called **contrails** (condensation trails; see fig. 2.11.2). These are not officially classified as air pollution but as 'artificial clouds'; however, they can contribute to long-term changes in the Earth's climate. Jet engines may also produce nitrogen oxides. As well as being greenhouse gases, these also destroy ozone in the upper atmosphere (see Chapter 2.9).

fig. 2.11.2 Condensation trails from jet engines play their part in the increased greenhouse effect because water vapour is a greenhouse gas. The jet engines also burn huge amounts of fossil fuels and emit correspondingly large amounts of carbon dioxide.

Questions

1 Suggest processes, apart from respiration, that release carbon dioxide into the atmosphere. Do these contribute to natural climate change or anthropogenic climate change?

2 Carbon dioxide has a lower relative greenhouse factor than other gases. Why is it so significant when considering global warming and climate change?

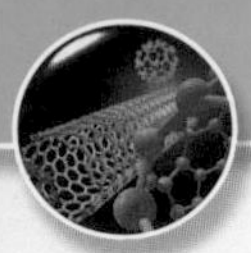

Carbon neutrality and carbon footprint

Carbon neutrality

A fuel is **carbon neutral** if the amount of carbon dioxide absorbed when the raw material is grown, or the fuel is formed, equals the amount of carbon dioxide produced when it is manufactured and then burned. For example, when a tree grows it absorbs carbon dioxide from the atmosphere. When the tree is burned it releases the same amount of carbon dioxide as it absorbed. So this process is carbon neutral. They do act as carbon reservoirs and if all the carbon stored in fossil fuels is released through burning it will have a major impact on the environment.

Chapter 1.6 outlined how petroleum fuels were formed millions of years ago by microorganisms which captured carbon dioxide from the atmosphere. These fuels produce carbon dioxide when they burn. You might argue therefore that petroleum fuels are carbon neutral. However, the carbon dioxide absorbed cannot be set against the carbon dioxide produced within a reasonable time span, say 50 years, so they are not considered to be carbon neutral.

Biofuels, such as bioethanol produced from maize, are also *not* carbon neutral. Although the maize absorbs carbon dioxide when it grows, the stages in the production of the bioethanol require energy. Unless this energy is generated without releasing carbon dioxide, producing biofuels gives out more carbon dioxide than the crop absorbs.

At first sight you might think that burning hydrogen as a fuel should be carbon neutral, since it burns to form only water and no carbon dioxide. However, the hydrogen has to be manufactured because it does not occur naturally. It can be made chemically from natural gas or the electrolysis of water. Either way, there is some carbon dioxide produced in the different manufacturing, construction and distribution steps.

Even a wind farm can be considered to emit carbon dioxide – in generating the energy used to build it and to manufacture the electricity transmission lines, as well as the fuel burned by standby power stations when the wind is too low to produce enough electricity.

Carbon footprint

fig. 2.11.3 A footprint is what we leave behind after we have gone.

A **carbon footprint** is a measure of the impact that human activities have on the environment in terms of the amount of greenhouse gases produced, measured in units of mass of carbon dioxide. The carbon footprint can be seen as the total amount of carbon dioxide and other greenhouse gases emitted over the full life cycle of a product or service. A carbon footprint is usually expressed as a CO_2 equivalent (usually in kilograms or tonnes), which measures the global warming effects of different greenhouse gases over the same timespan. Carbon footprints can be calculated using a **life cycle assessment** (**LCA**) method (see chapter 1.7) or can include just emissions resulting from energy use of fossil fuels.

An alternative definition of a carbon footprint is the total amount of carbon dioxide that results from the actions of an individual (mainly through their energy use) over a period of one year. The average person in the UK has a carbon footprint of 9.4 tonnes of CO_2. This varies according to age, with 50–65 year olds having a higher average carbon footprint of 13.2 tonnes.The carbon footprints of people in different countries are shown in fig. 2.11.4. Notice that in India and China, where increasing amounts of carbon dioxide are being produced as industries are developed, the carbon footprint per head is still shown as low because it is calculated per head of population, and these countries have large populations.

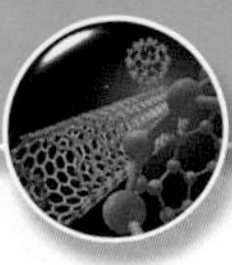

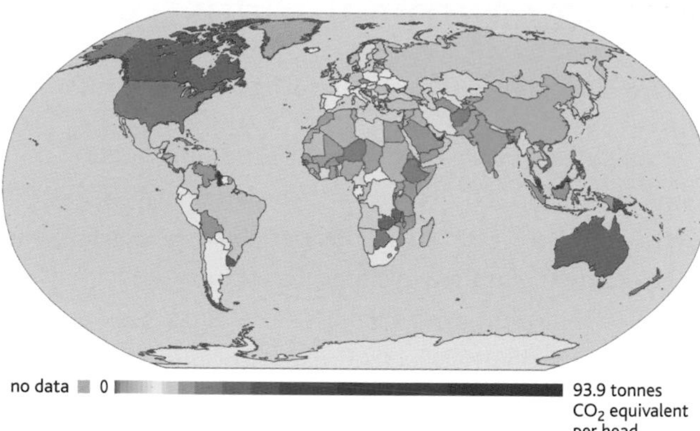

fig. 2.11.4 Greenhouse emissions per head of population by country in 2000.

Carbon offsetting

Companies and individuals may plan to reduce the effect of their carbon footprint by **carbon offsetting**. Traditionally, this often involved paying the cost of planting new trees. However, there are other actions that could have a positive effect, eg paying to build a small dam in a river in Nepal to produce hydroelectricity for a village so the people there do not have to use diesel generators that produce carbon dioxide.

A return flight from London to Sydney, Australia, produces 3.61 tonnes of carbon dioxide per passenger. The cost of offsetting this was calculated to be £40 in early 2008.

CFCs and HCFCs

You know from chapter 2.8 that CFCs and HCFCs do not occur naturally. CFCs persist in the upper atmosphere and destroy the protective ozone, contributing to holes in the ozone layer. This increases the levels of harmful ultraviolet radiation that reach the Earth's surface.

Note that depletion of the ozone layer is not part of the greenhouse effect and global warming – it is absorption of infrared radiation, not ultraviolet, that makes a gas a greenhouse gas. However, as well as reacting with ozone, CFCs and HCFCs are also powerful greenhouse gases, and so they do contribute to global warming.

HSW Science changing society

It was the work of British, Mexican and American scientists – James Lovelock, Mario Molina and Frank Rowland – that clearly established a link between CFCs and the depletion of the ozone layer. Their work eventually led to a change in attitude and the passing of the Montreal Protocol (see page 215).

After the success of Montreal, there have been attempts to set up other agreements. However, when it comes to controlling the emissions of substances such as carbon dioxide, it is not so easy to persuade world leaders to sign up. Some of the measures needed for success – for example reducing dependence on cars and changing the way electricity is generated and used – would be unpopular with voters. So while politicians like to say the right things, they are not always so keen to do them. The Kyoto agreement, an attempt to reduce global carbon dioxide emissions, was doomed to failure when the US, which produces more carbon dioxide per head than anywhere else in the world, refused to sign up. Some US scientists, often funded directly or indirectly by the large oil companies, produced evidence suggesting that global warming was not happening, or that it was not linked to human activities. This is the evidence the US president at the time – himself from Texas and with far-reaching links to the oil industry – chose to believe.

He was not alone in refusing to ratify the treaty. However, as evidence builds from more and more scientists in different countries and in different fields from biology and chemistry to meteorology, politicians are finally beginning to realise that action needs to be taken. Science is changing society again. The question – are we doing too little, too late?

Questions

1 A nuclear power station produces no carbon dioxide, but it is not classified as carbon neutral. Suggest reasons why this might be.

2 'Carbon offsetting is the prerogative of the rich.' Explain this statement and discuss its impact on efforts to reduce global warming.

3 Many people confuse global warming and depletion of ozone layer. Summarise the important features of each.

Reducing hazards and pollution in the chemical industry

The modern chemical industry

Figure 2.11.5 shows a chemical factory. Traditionally factories used up vast quantities of valuable raw materials and energy while producing solid waste, waste water and air pollution. Today factories are designed and built to be 'greener' – to use fewer finite resources and less energy and to produce less pollution.

fig. 2.11.5 Industrial pollution from a chemical factory. Today factories are designed and built to be 'greener'.

Disposing of solid waste

Disposing of unwanted material as waste is an important aspect of modern life. In chapter 1.7 you read about the impact of reusing and recycling rather than simply disposing of waste plastics. Figure 2.11.6 shows shows a factory which processes and recycles old tyres.

fig. 2.11.6 The recycling of tyres to produce rubber crumbs is just one way in which solid waste is increasingly being recycled on an industrial scale. The rubber crumbs produced have many uses including carpet underlay, safe playground surfaces and material to prevent noise pollution.

Much of the domestic material we throw away and waste materials from industry can be recycled, eg glass, metals, paper and cardboard. Plastics are more difficult to recycle as you saw in chapter 1.7. Recycling saves the Earth's natural resources and reduces energy requirements. It uses much less energy, and so it is much cheaper to recycle glass than to make new glass. Recycling aluminium reduces the amount of aluminium that has to be extracted from aluminium oxide by electrolysis. Not only does this reduce the amount of electricity used, it also means that less PFC (perfluorocarbon) gases are emitted. These are produced when aluminium is extracted and are 900 times more potent as greenhouse gases than carbon dioxide.

Material that cannot be recycled has to be disposed of either by incineration (combustion) or in landfill sites. Incineration produces carbon dioxide, but also gives out energy which can be used in energy-recovery systems such as for local heating. Incineration can produce harmful materials such as dioxins if it is not carried out properly. Dioxins may be formed if the temperature in the incinerator is too low.

Suitable landfill sites are becoming less available so disposal of solid waste in this way is set to become more problematic in the future. In a landfill site, bacteria break down waste materials, producing methane gas, so such sites are monitored to prevent methane escaping – not only is it a greenhouse gas but it can also cause explosions. Figure 2.11.7 shows methane produced at a landfill site being burned to prevent it escaping into the atmosphere or causing explosions. Methane can be recovered from a landfill supply and used as fuel. However, it is difficult to get the steady supply that is required.

fig. 2.11.7 Burning methane at a landfill site not only prevents pollution but also generates valuable energy for power supplies.

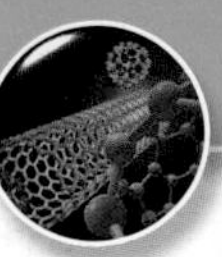

Disposing of waste water

Water containing pollutants cannot legally be discharged into rivers and streams. The quality of water is monitored by the Environment Agency and the Department for Environment, Food and Rural Affairs (Defra). Factories have to pay for every litre of water they use. They therefore develop processes that reuse water whenever possible to save money and resources. Discharged water is usually stored in a lake and then treated to remove harmful materials before being released into a river. Often harmful substances are precipitated and the precipitate is removed by settling or filtration.

Disposing of waste gases

Industrial waste gases include sulfur dioxide, which can cause acid rain and is also a greenhouse gas. Acidic gases may be removed from flue gases using limestone before being discharged into the atmosphere:

$$2CaCO_3(s) + 2SO_2(g) + O_2(g) \rightarrow 2CaSO_4(s) + 2CO_2(g)$$

Although carbon dioxide does not cause acid rain, it is a greenhouse gas. One way of reducing the amount of carbon dioxide we produce is by **carbon capture**.

Carbon capture

One suggestion to solve the problem of rising levels of carbon dioxide in the atmosphere is to liquefy the gas and pump it into old gas wells in the North Sea. The gas will then be stored where the natural gas used to be. It could also be stored under pressure on the sea bed.

Figure 2.11.8 shows a plan to extract natural gas from the North Sea and convert it into hydrogen and carbon dioxide. The hydrogen can be used to produce electricity without pollution while the carbon dioxide can be buried again in porous rocks 1 km below the sea bed.

Some people believe we should start injection of carbon dioxide immediately, and that we should continue until at least 2030. Hopefully by this time we will have developed lower-carbon technology and have reduced carbon dioxide emissions to levels that are not causing environmental damage.

There is a good reason why we should start carbon dioxide storage sooner rather than later. At present almost all the UK offshore oil and gas fields still have their oil drilling platforms in place. These platforms could be modified for carbon dioxide storage at a fraction of the cost of building and installing new facilities. By 2020, many of these platforms will have been removed as the oil and gas supplies run dry. This would leave only a fraction of the storage potential for carbon dioxide that we have today.

Questions

1 How can hydrogen be used to produce pollution-free electricity?

2 Suggest reasons, apart from cost, why carbon capture may not go ahead.

3 Why is removing sulfur dioxide from flue gases with limestone not a total solution to the problem?

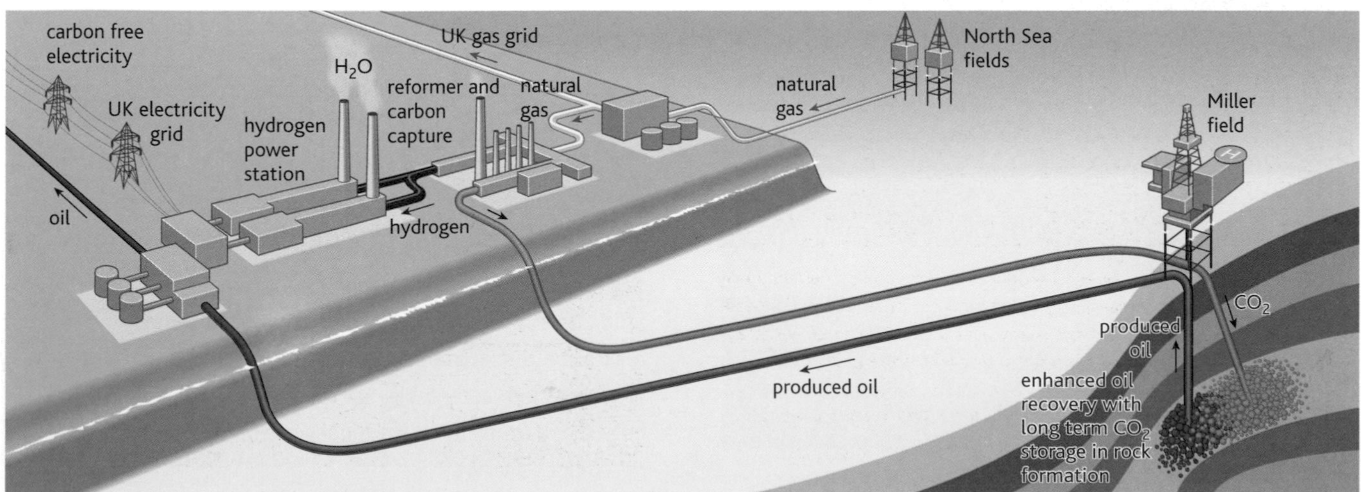

fig. 2.11.8 Carbon capture is the storage of carbon dioxide under the sea bed, rather than emitting it into the atmosphere.

HSW Renewable resources

Traditionally the chemical industry used vast quantities of raw materials such as metal ores and also fossil fuels such as coal, oil and natural gas. These have become much more expensive, encouraging companies to reduce the amounts they use and to look for other raw materials, especially materials that are renewable. For some products, eg ammonia, nitric acid, benzene, sulfuric acid and ethanoic acid, there is no alternative to using fossil fuels in their production.

Although reserves of fossil fuels will last for a long time, the emphasis of research and development is to ensure that these materials are used efficiently and that end-products are recycled. Use of fossil fuels can be minimised, for example by recycling products where possible.

Bio oil from wheat

Material such as wheat straw, which is renewable as a fresh supply is grown each year, can be converted into a valuable oil for fuel by **pyrolysis**. This involves heating the straw in a limited oxygen supply so it does not burn. Sometimes this heating is done with steam. In fast pyrolysis, the straw is heated rapidly for only a few seconds, and this breaks it down into a large number of relatively small molecules. The resulting mixture is then cooled rapidly to prevent further reaction. It produces a dark, oily liquid called **bio oil** or **pyrolysis oil**. This can be used as an alternative to fuel oil made from crude oil. Other products of this process include char (mostly carbon solids) and a mixture of gases. Both of these can be used as fuels on the factory site.

fig. 2.11.9 Crops such as wheat can be used to make biofuels.

Ethanol from organic waste

The valuable chemical feedstock ethanol can be produced by fermentation of plant material. However, fermentation with yeast, the traditional method, does not convert all the sugars to ethanol. KO11 is a genetically modified bacterium developed by the microbiologist Lonnie Ingram in 1987. It converts all sugars into ethanol. The bacterium KO11 would normally produce acids, but the modification produces ethanol instead.

The advantage of using these bacteria to produce ethanol rather than using yeast is that a wider range of sugars can be processed. This enables biomass waste, which would otherwise be thrown away, such as wood waste, corn stalks and rice hulls, and other organic waste to be used for ethanol production.

There is an increasing tendency for farmers to grow crops such as maize for use in making biofuels and other biochemicals. It is important that, in doing this, the amount of food grown for the increasing world population is not forgotten.

Starch

Two other materials produced by plants and which can be used as raw materials in the chemical industry are starch and cellulose.

Starch is the plant's main energy store produced as a product of photosynthesis from carbon dioxide and water. It has a complex structure based on linked glucose rings. It can be extracted from plants as industrial starch.

Until recently starch products have found it hard to compete with similar products made from crude oil products. The increasing price of crude oil and the realisation that starch is renewable is making starch-based products more acceptable. For example, supermarket plastic bags made from starch are biodegradable whereas the oil-based ones are not.

Starch is also an important source of glucose, from which many other materials can be synthesised. Figure 2.11.10 summarises the uses of starch.

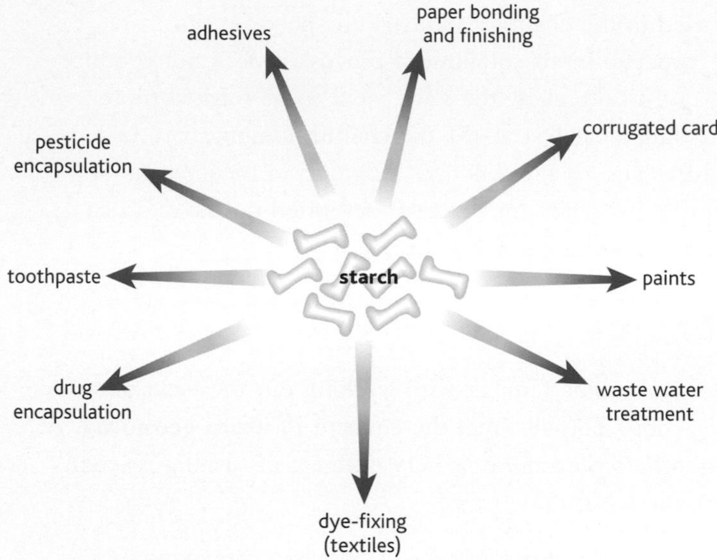

fig. 2.11.10 **Starch is a renewable resource with many applications.**

Figure 2.11.11 shows Zorbix, a starch-based polymer which absorbs water readily and is useful for packaging.

fig. 2.11.11 **Zorbix absorbs up to 50 times its own weight of water. The water is soaked up by a super-absorbent starch-based polymer contained inside the packaging. Zorbix is designed to be as thin as possible, so it can be placed between the pages of wet books and dry them without swelling and damaging them.**

Wood pulp has been used as a source of **cellulose** for a long time. Cellulose has been used to produce polymers such as cellulose ethanoate (acetate) (used for fibres in fabrics and photographic film) and rayon (used as a substitute for silk). Cellulose is also used to make 4-oxopentanoic acid (levulinic acid; see fig. 2.11.12). This is widely used in making pharmaceuticals, plastics and rubbers.

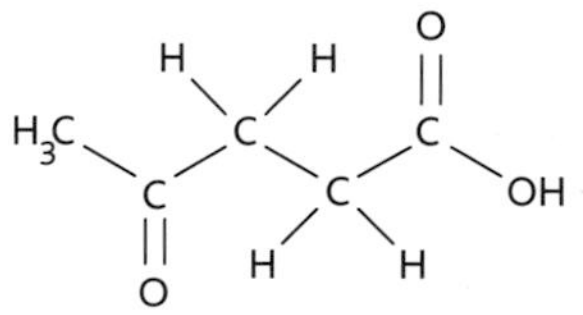

fig. 2.11.12 **The structure of 4-oxopentanoic acid (levulinic acid), a valuable intermediate made from cellulose.**

In 1991 a new fibre was developed from cellulose. Its generic name is lyocell but its trade name is Tencel®. Both rayon and lyocell have fibres made from cellulose, but they have different structures. The advantages of lyocell are:

- It is made using wood pulp from managed forests.
- It uses a non-toxic, biodegradable solvent in its manufacture and the solvent is reused.
- It is biodegradable and can be recycled.

Lyocell is now used in a wide range of clothing and household textiles.

Questions

1 Suggest two reasons for the decline in the use of fossil fuels in the chemical industry.

2 Suggest why lyocell may be preferred to nylon or silk.

Increasing efficiency in the chemical industry

Developing new catalysts

A **catalyst** is used in the chemical industry to speed up the manufacturing process. In an equilibrium process it does not produce a higher yield, but it does allow the same yield to be formed more quickly. Often using a catalyst allows the chemical engineer to use lower temperatures and so save energy. Although in theory catalysts are not used up in a reaction, in practice they often do have to be replaced due to mechanical wear and tear.

Atom economy

In chapter 1.1 you became familiar with working out the yield of a product of a reaction. You also met the concept of **atom economy**. A reaction with good atom economy is very efficient at turning reactants into desired products with little waste:

$$\text{atom economy (\%)} = \frac{\text{mass of atoms in desired product} \times 100\%}{\text{mass of atoms in reactants}}$$

Atom economy

Figure 2.11.13 shows the equation for the production of propan-1-ol from 1-bromopropane. Calculate the atom economy of this process.

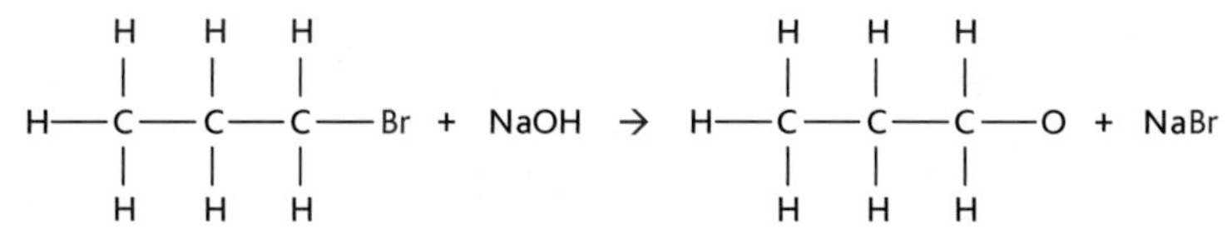

$$\mathrm{H{-}\overset{H}{\underset{H}{C}}{-}\overset{H}{\underset{H}{C}}{-}\overset{H}{\underset{H}{C}}{-}Br + NaOH \rightarrow H{-}\overset{H}{\underset{H}{C}}{-}\overset{H}{\underset{H}{C}}{-}\overset{H}{\underset{H}{C}}{-}O + NaBr}$$

fig. 2.11.13 **The production of propan-1-ol from 1-bromopropane.**

First you work out the molar masses of propan-1-ol and sodium bromide. $A_r(\mathrm{H}) = 1.0$, $A_r(\mathrm{C}) = 12.0$, $A_r(\mathrm{O}) = 16.0$, $A_r(\mathrm{Na}) = 23.0$ and $A_r(\mathrm{Br}) = 79.9$.

$$\text{Atom economy} = \frac{M_r\ \text{propan-1-ol}}{M_r\ \text{1-bromopropane} + M_r\ \text{sodium hydroxide}} \times 100\%$$

$$= \frac{(3 \times 12.0) + (8 \times 1.0) + (1 \times 16.0)}{(3 \times 12.0) + (7 \times 1.0) + (1 \times 79.9) + (23.0 + 16.0 + 1.0)} \times 100\%$$

$$= \frac{60 \times 100}{162.9} = \mathbf{36.8\%}$$

This tells you that most of the starting materials are turned into waste. Even if there was a 100% yield in the reaction, nearly two-thirds of the mass of the atoms used would be wasted unless the sodium bromide was also a useful product of the reaction.

HSW Choosing a catalyst

Ethanoic acid could be produced in industry as we produce it in the laboratory, by oxidation of ethanol. This however has a low atom economy as there are other products of the reaction. One method of producing ethanoic acid used industrially involved the direct oxidation of butane and naphtha (both from fossil fuels) and this also had a very poor atom economy of about 35%.

A better method is to react methanol with carbon monoxide:

$$CH_3OH + CO \rightarrow CH_3COOH$$

This process has an atom economy of 100% – there are no waste products.

It was first developed by BASF in 1960 using the reaction of methanol and carbon monoxide with two catalysts, cobalt and an iodide-based cocatalyst, at conditions of 300 °C and 700 atmospheres pressure. These conditions require a lot of energy input to achieve and the process can be hazardous at this temperature.

This was further developed by Monsanto in 1966. The same reaction was used, but using a new catalyst system of rhodium/iodide ion, which operated at milder conditions of 150–200 °C and 30– 60 atmospheres. **Figure 2.11.14** shows the Monsanto proceess as a flow diagram.

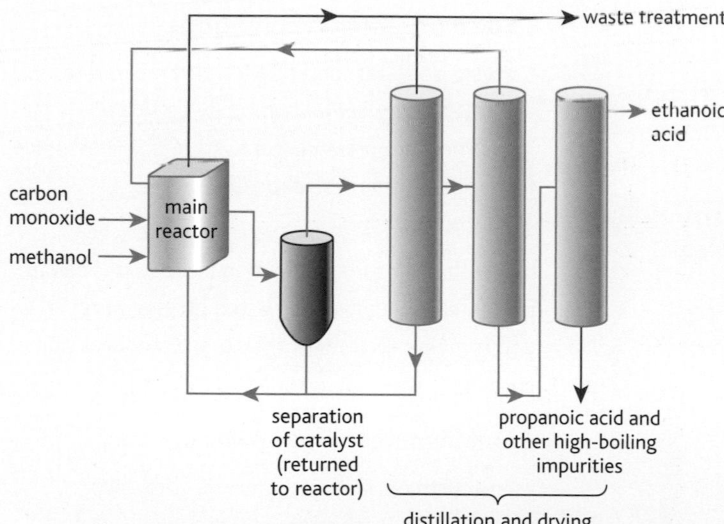

fig. 2.11.14 **The Monsanto process for the production of ethanoic acid.**

The disadvantages with this process are:

- Rhodium metal is very expensive – more expensive than gold.
- Rhodium and iodide ions form insoluble salts such as RhI_3, and the water content in the reaction vessel has to be kept relatively high to prevent this.
- A final distillation step is required to remove water, adding to the costs and energy demand. Any precipitation occurring removes the catalyst from the reaction mixture, which must be recovered and returned to the main reactor.
- Rhodium also catalyses side reactions such as:

$$CO(g) + H_2O(g) \rightarrow CO_2(g) + H_2(g)$$

Iridium was suggested as an alternative to rhodium. Initial studies by Monsanto had shown the iridium complex to be a less active catalyst than the rhodium complex. However, subsequent research showed that the iridium catalyst could be promoted using ruthenium, and this combination leads to a more active and more specific catalyst than the rhodium compound. This is now called the Cativa process. It releases less carbon dioxide into the atmosphere.

The switch from rhodium to the cheaper iridium allows the use of less water in the reaction mixture. This reduces the number of drying columns necessary, and decreases the formation of by-products. Compared with the original Monsanto process, less propanoic acid by-product is produced.

The Monsanto process continues to evolve. Currently the methanol comes from synthesis gas which is made from oil. Methanol can also be produced from wood, household waste or sewage. At some time this process may be developed so that it does not use fossil fuels.

Questions

1 Consider the addition reaction of bromine to propene. Work out the atom economy of this reaction.

2 Look at the flow diagram in fig. 2.11.14. What steps are taken to:
   a produce a pure product b reduce costs c protect the environment?

HSW Energy considerations in the chemical industry

The chemical industry uses a lot of energy, although in the past 50 years the average amount of energy required to produce a tonne of chemical has reduced by 50%. Energy is used in a chemical factory for:

- raising the temperature of reactants so that a reaction begins or continues
- producing electrical energy for electrolysis
- distillation to separate and purify or concentrate desired products
- heating to dry product material
- waste treatment.

The energy used in the separation, drying and waste management may be more than that used in the reaction stages.

Where does the energy come from?

Much of the energy used in the chemical industry comes from the combustion of fossil fuels, eg natural gas and coke. Sometimes superheated water is used for heating – this is water heated under pressure so the boiling temperature is higher than 100 °C.

Where electricity is required, eg for electrolysis, this may come from hydroelectric power or electricity purchased from a supplier at a preferential rate. A small amount of this electricity may be generated by wind power, solar power or other renewable resources.

fig. 2.11.15 Volkswagen AG car factory, Hannover, Germany

Reducing energy demand

Chemical manufacturers use various methods to reduce their use of energy, for both financial and environmental reasons.

Good housekeeping

Just as we waste energy around the home, in chemical factories energy can be wasted by leaking taps and pipes, insufficient insulation and poor maintenance.

Heat recovery

In many combustion and other exothermic processes, much of the heat energy released goes up the chimney and is wasted. In large industrial processes, heat energy is extracted from the waste gases by a **heat exchanger**. This works like a refrigerator, transferring energy from the flue gases to where it can be useful. A heat exchanger can provide hot water and heating in the plant. This energy can also be used to preheat reactants.

Choice of reaction conditions

Carrying out a reaction at a lower temperature will save energy but will normally be slower. Adjustments in pressure, catalysts, etc. will be made to compensate for this if possible.

It is often more economic to use pure oxygen rather than air for an oxidation process. If it has to be heated, energy is not wasted heating nitrogen, which makes up 80% of the air but produces no product.

Combined heat and power

Many chemical factories generate their own electricity rather than buying it in. This is often cheaper and avoids transmission costs.

Microwaves in the pharmaceutical industry

Microwaves can be used in the pharmaceutical industry for heating reactions. This started in the mid-1980s using domestic microwaves. Although the results were promising they were difficult to reproduce and the procedure was often dangerous.

fig. 2.11.16 **Using a microwave oven heats reactants more economically, reducing reaction times.**

Figure 2.11.16 shows a modern industrial microwave oven. Using a microwave oven allows a process that would require 48 hours in a conventional oven to produce an 84% yield in a very short time. Reactions that would need a heavy metal catalyst can proceed quickly even without the catalyst. There are environmental benefits to not using toxic heavy metals. Medicinal chemists are able to carry out in minutes reactions that would have needed overnight in a conventional oven.

Modern industrial microwave ovens are much more powerful than domestic microwave ovens. A domestic oven uses pulses of radiation, while a commercial microwave oven is continuous and much more powerful. Microwaves create an electric field which causes polar molecules such as water to line up with the field. Before they can do that the field switches to the opposite direction and the molecules swivel round and line up. The microwave energy is effectively converted into thermal energy. The polar molecules then heat up the molecules around them.

If a mixture is heated in a closed vessel using microwaves it is possible to carry out the reaction at approaching twice the boiling temperature of the solvent. Given that a 10 °C rise in temperature can double the rate of a reaction, the increase in a rate over say a 60 °C rise would be 106 times. Modern microwave heaters can deal with up to 1 dm^3 of reaction.

Examzone

You are now ready to try the second Examzone test for Unit 2 (Examzone Unit 2 Test 2) on page 254. This will test you on what you have learnt in Unit 2, chapters 6 to 11.

Questions

1 Suggest why a heat exchanger may be installed in a chemical plant that carries out an exothermic reaction.

2 Suggest reasons why a microwave oven is suitable for the pharmaceutical industry.

Examzone: Unit 1 Test 1 (chapters 1.1 to 1.4)

1 Define the following terms:

Relative atomic mass **(2)**

The Avogadro constant **(1)**

ppm **(1)**

(Total 4 marks)

2 Define the term **standard enthalpy of combustion**. **(3)**

(Total 3 marks)

3 Calculate the enthalpy change for the reduction of propanoic acid to propanal:

$CH_3CH_2COOH + H_2 \rightarrow CH_3CH_2CHO + H_2O$

given the following enthalpies of combustion / kJ mol^{-1}:

propanoic acid −1527; hydrogen −286; propanal −1821. **(3)**

(Total 3 marks)

4 The reaction of an acid with a base to give a salt is an exothermic reaction. In an experiment to determine the enthalpy of neutralisation of hydrochloric acid with sodium hydroxide, 50.0 cm^3 of 1.00 mol dm^{-3} HCl was mixed with 50.0 cm^3 of 1.10 mol dm^{-3} NaOH. The temperature rise obtained was 6.90 °C.

a Define the term **enthalpy of neutralisation**. **(1)**

b Assuming that the density of the final solution is 1.00 g cm^{-3} and that its heat capacity is 4.18 J K^{-1} g^{-1}, calculate the heat evolved during the reaction. **(3)**

c 0.0500 mol of acid was neutralised in this reaction; calculate $\Delta H_{neutralisation}$ in kJ mol^{-1}. **(2)**

d Suggest why sodium hydroxide is used in slight excess in the experiment. **(1)**

e Give a major source of error in this experiment and a suggested improvement. **(2)**

(Total 9 marks)

5 a Enthalpy changes can be calculated using average bond enthalpy data.

(i) The enthalpy change to convert methane into gaseous atoms is shown below.

$CH_4(g) \rightarrow C(g) + 4H(g)$
$\Delta H = +1664\,kJ\,mol^{-1}$

Calculate the average bond enthalpy of a C—H bond in methane. **(1)**

(ii) Use the data in the table below and your answer to **(a)(i)** to calculate the enthalpy change for

$2C(g) + 2H_2(g) + Br_2(g) \rightarrow CH_2BrCH_2Br(g)$

Bond	Average bond enthalpy/kJ mol^{-1}	Bond	Average bond enthalpy/kJ mol^{-1}
C—C	+348	H—H	+436
Br—Br	+193	C—Br	+276

(3)

b The standard enthalpy of formation of 1,2-dibromoethane, CH_2BrCH_2Br, is −37.8 kJ mol^{-1}.

Suggest the main reason for the difference between this value and your calculated value in **a(ii)**. **(1)**

(Total 5 marks)

6 a The first ionisation energy of potassium is +419 kJ mol^{-1} and that of sodium is +496 kJ mol^{-1}.

(i) Define the term **first ionisation energy**. **(3)**

(ii) Explain why the first ionisation energy of potassium is only a little less than the first ionisation energy of sodium. **(3)**

b Potassium forms a superoxide, KO_2. This reacts with carbon dioxide according to the equation:

$4KO_2(s) + 2CO_2(g) \rightarrow 2K_2CO_3(s) + 3O_2(g)$

Carbon dioxide gas was reacted with 4.56 g of potassium superoxide.

(i) Calculate the amount, in moles, of KO_2 in 4.56 g of potassium superoxide. **(2)**

(ii) Calculate the amount, in moles, of carbon dioxide that would react with 4.56 g of potassium superoxide. **(1)**

(iii) Calculate the volume of carbon dioxide, in dm^3, that would react with 4.56 g of potassium superoxide. Assume that 1.00 mol of a gas occupies 24 dm^3 under the conditions of the experiment. **(1)**

(iv) What volume of oxygen gas, in dm^3, measured under the same conditions of pressure and temperature, would be released? **(1)**

(Total 11 marks)

7 Complete the following table. **(2)**

(Total 2 marks)

Atom	Number of protons	Number of neutrons	Number of electrons
$^{191}_{77}Ir$			
$^{96}_{42}Mo$			

8 The element gallium consists of two types of atom of relative mass 69.0 and 71.0 respectively. The percentage abundance of the atoms of relative mass 69.0 is 60.2%.

a Calculate the relative atomic mass of gallium. **(2)**

b What name is given to these different atoms of gallium? **(2)**

(Total 4 marks)

9 The graph shows a plot of IE (ionisation energy) against the number of the electron removed for sodium. Explain the shape of this graph in terms of the electron structure of sodium. **(3)**

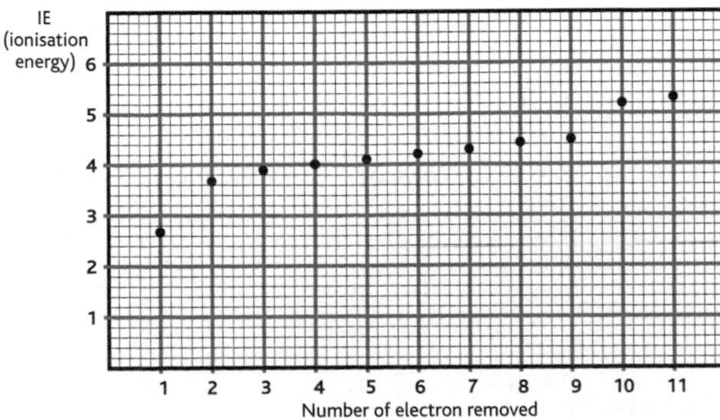

(Total 3 marks)

10 a (i) Write the equation for the reaction of magnesium metal with chlorine, showing state symbols. **(2)**

(ii) The product of this reaction is ionic. Use this information to explain why it has a relatively high melting temperature (714 °C). **(2)**

b Why is magnesium iodide more covalent than magnesium chloride? **(2)**

c Describe the bonding in magnesium metal. **(3)**

(Total 9 marks)

11 a Copy and complete the electron configuration for carbon.

$1s^2$ .... **(1)**

b Explain how successive ionisation energy data could be used to confirm that carbon is in group 4 of the periodic table. **(1)**

c Draw a dot and cross diagram for a molecule of carbon tetrachloride, CCl_4, showing **outer electrons only**. **(2)**

d Explain how the following are achieved in a mass spectrometer.

(i) Ionisation **(2)**

(ii) Deflection **(1)**

e (i) Define the term **relative isotopic mass**. **(3)**

(ii) Carbon consists of the isotopes ^{12}C, ^{13}C and ^{14}C. Chlorine consists of the isotopes ^{35}Cl and ^{37}Cl.

Use this data to calculate the maximum relative molecular mass of a molecule of carbon tetrachloride, CCl_4. **(1)**

(iii) Explain, in terms of sub-atomic particles, the meaning of the term isotopes. **(2)**

(iv) Give a large-scale use of mass spectrometry. **(1)**

(v) Why do isotopes of the same element have the same chemical properties? **(1)**

(Total 15 marks)

Examzone: Unit 1 Test 2 (chapters 1.5 to 1.7)

1 a An organic compound **X** contains 82.75% carbon and 17.25% hydrogen by mass.

(i) Calculate the empirical formula of **X**. **(2)**

(ii) Deduce from the mass spectrum below the relative molecular mass of **X**, giving a reason for your choice. Hence show that the molecular formula of **X** is C_4H_{10}. **(2)**

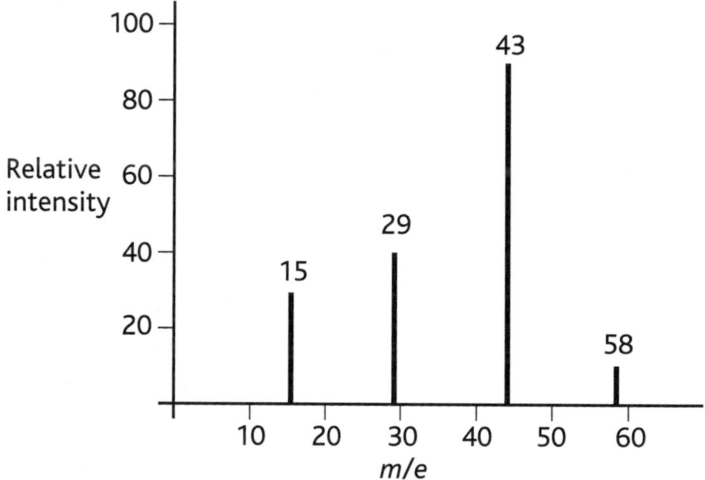

b There are two possible structures, **A** and **B**, for this molecule:

```
    H   H   H   H              H   H   H
    |   |   |   |              |   |   |
H — C — C — C — C — H      H — C — C — C — H
    |   |   |   |              |   |   |
    H   H   H   H              H  CH3  H

          A                        B
```

(i) Identify the species responsible for the peaks in the mass spectrum at 43, 29 and 15. **(3)**

(ii) Hence deduce which of the structures **A** or **B** is present giving a reason for your answer. **(2)**

c Complete combustion of **X** in oxygen gives carbon dioxide and water only.

(i) Write an equation for this combustion reaction. **(1)**

(ii) Calculate the total volume of the gaseous mixture produced when 2 cm^3 of gaseous C_4H_{10} is mixed with 15 cm^3 of oxygen and completely burned. All volumes are measured at room temperature and pressure. **(1)**

(iii) With reference to the substances in **c(i)**, explain why alternative fuels need to be developed. **(2)**

(Total 13 marks)

2 **a** **(i)** Explain the term **homologous series**. **(2)**

(ii) To which homologous series does ethene, C_2H_4, belong? **(1)**

b Draw the full structural formulae, showing all the bonds, for each of the following:

(i) The organic product of the reaction of ethene, C_2H_4, with aqueous potassium manganate(VII) and sulfuric acid. **(2)**

(ii) 3,4-dimethylhex-2-ene. **(2)**

(iii) A repeating unit of poly(propene). **(2)**

c Ethene reacts with hydrogen chloride gas to form C_2H_5Cl.

(i) What type of reaction is this? **(2)**

(ii) Give the systematic name for C_2H_5Cl. **(1)**

(iii) Explain why the reaction between ethane and hydrogen chloride gas has a high atom economy. **(1)**

(Total 13 marks)

3 **a** But-2-ene, $CH_3CH{=}CHCH_3$, exists as geometric isomers.

(i) Draw the geometric isomers of but-2-ene. **(2)**

(ii) Explain how geometric isomerism arises. **(1)**

b **(i)** Draw the structural formula of a compound which is an isomer of but-2-ene but which does not show geometric isomerism. **(1)**

(ii) Explain why the isomer drawn in **(i)** does not show geometric isomerism. **(1)**

(Total 5 marks)

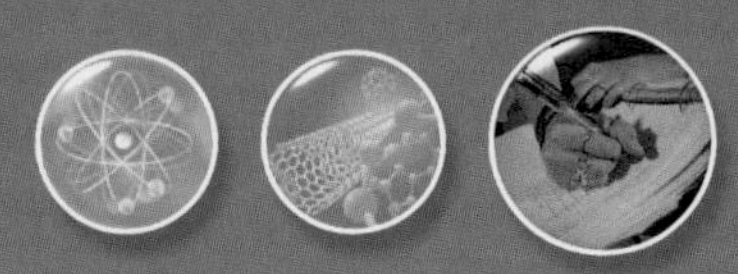

Examzone: Unit 2 Test 1 (chapters 2.1 to 2.5)

1 a Draw the ammonia molecule, NH_3, making its three-dimensional shape clear. Mark in the bond angle on your diagram. Explain why ammonia has this shape and this bond angle. **(4)**

b Explain whether ammonia is a polar molecule or not. **(2)**

c Explain, in terms of the intermolecular forces in both compounds, why ammonia has a higher boiling temperature than phosphine, PH_3. **(3)**

d **(i)** Explain, in terms of electrons, how ammonia can react with hydrogen ions to form ammonium ions, NH_4^+. **(2)**

(ii) State the number of protons and the number of electrons present in an ammonium ion. **(2)**

(Total 13 marks)

2 a Hydrogen iodide, HI has a boiling temperature of –35 °C, and hydrogen chloride, HCl a boiling temperature of –85 °C.

(i) Draw a dot and cross diagram of a hydrogen iodide molecule (showing outer shell electrons only). **(2)**

(ii) Explain why hydrogen iodide has a higher boiling temperature than hydrogen chloride. **(3)**

b Hydrogen iodide and hydrogen chloride react with water to form acidic solutions.

(i) Write the equation for the reaction of hydrogen iodide with water. **(1)**

(ii) State why the solution formed is acidic. **(1)**

c **(i)** Excess dilute hydrochloric acid reacts with a lump of calcium oxide, CaO to form an aqueous solution of calcium chloride.

Write the equation for this reaction, including state symbols. **(2)**

(ii) In a similar reaction with dilute sulfuric acid and a lump of calcium oxide, the reaction stops after a short time even though some calcium oxide remains. State why the reaction stops so quickly. **(1)**

(Total 10 marks)

3 a Methane and poly(ethene) are both hydrocarbons.

(i) State the type of bond between carbon and hydrogen atoms in the molecules of both compounds. **(1)**

(ii) State the type of **intermolecular** force present in **both** compounds. **(1)**

(iii) Explain why poly(ethene) melts at a higher temperature than methane. **(3)**

(Total 5 marks)

4 Domestic bleaches contain sodium chlorate(I), NaOCl.

a Write an **ionic** equation to show the disproportionation of the chlorate(I) ion. Use oxidation numbers to explain the meaning of the term disproportionation in this reaction. **(3)**

b Domestic bleaches are dilute solutions of sodium chlorate(I). The amount of ClO^- ions in a sample can be found by reacting it with excess acidified potassium iodide solution.

$ClO^- + 2I^- + 2H^+ \rightarrow I_2 + Cl^- + H_2O$

The iodine produced is then titrated with standard sodium thiosulfate solution.

- 10.0 cm^3 of a domestic bleach was pipetted into a 250 cm^3 volumetric flask and made up to the mark with distilled water.
- A 25.0 cm^3 portion of the solution was added to excess acidified potassium iodide solution in a conical flask.
- This mixture was titrated with 0.100 $mol\,dm^{-3}$ sodium thiosulfate solution, using starch indicator added near the end point.
- The mean titre was 12.50 cm^3.

(i) Give the colour change you would see at the end point. **(1)**

(ii) State why the starch indicator is added near the end point. **(1)**

(iii) The equation for the reaction between iodine and thiosulfate ions is

$2S_2O_3^{2-} + I_2 \rightarrow S_4O_6^{2-} + 2I^-$

Calculate the amount (in moles) of chlorate(I) ions in 1.00 dm^3 of the **original** bleach. **(5)**

(iv) Use the equation below to calculate the mass of chlorine available from 1.00 dm^3 of the **original** bleach. Give your answer to 3 significant figures. **(1)**

$ClO^- + Cl^- + 2H^+ \rightarrow Cl_2 + H_2O$

c Sodium thiosulfate can be used to remove the excess chlorine from bleached fabrics.

$S_2O_3^{2-} + 4Cl_2 + 5H_2O \rightarrow 2SO_4^{2-} + 10H^+ + 8Cl^-$

By considering the change in oxidation number of sulfur, explain whether chlorine or iodine is the stronger oxidising agent when reacted with thiosulfate ions. **(2)**

(Total 13 marks)

5 **a** The compounds lithium chloride, sodium bromide and potassium iodide can be distinguished from one another by the use of flame tests.

(i) Copy and complete the following table. **(3)**

Compound	Flame colour
Lithium chloride	
Sodium bromide	
Potassium iodide	

(ii) Explain the origin of the colours in flame tests. **(2)**

b These compounds can also be distinguished from one another by the use of concentrated sulfuric acid.

(i) State what would be **seen** when concentrated sulfuric acid is added to separate solid samples of each of these compounds:

lithium chloride

sodium bromide

potassium iodide. **(4)**

(ii) Write an equation, including the state symbols, for the reaction between solid lithium chloride and concentrated sulfuric acid. **(2)**

(Total 11 marks)

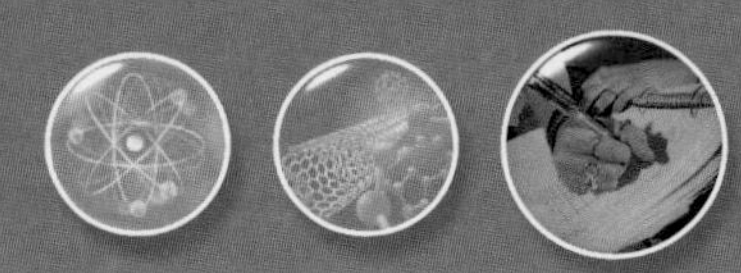

Examzone: Unit 2 Test 2 (chapters 2.6 to 2.11)

1 a (i) State two factors, other than a change in temperature or the use of a catalyst, that influence the rate of a chemical reaction. **(2)**

(ii) For one of the factors you have chosen explain the effect on the rate. **(2)**

b The Maxwell–Boltzmann distribution of molecular energies at a given temperature, T_1 is shown below.

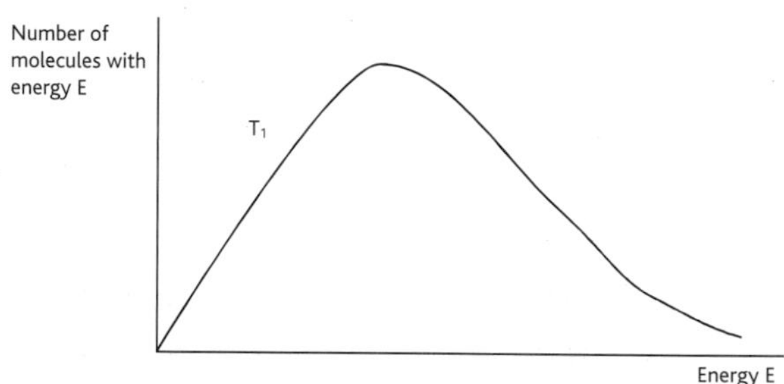

(i) Copy the graph and on the same axes draw a similar curve for a reaction mixture at a higher temperature, T_2. **(2)**

(ii) Place a vertical line marked E_a at a plausible value on the energy axis to represent the activation energy for a reaction. **(1)**

(iii) Use your answers to parts **b (i)** and **(ii)** to explain why an increase in temperature causes an increase in the reaction rate. **(3)**

(Total 10 marks)

2 a Define the term **standard enthalpy of formation**. **(3)**

b The dissociation of phosphorus pentachloride is a reversible reaction.

$PCl_5(g) \rightleftharpoons PCl_3(g) + Cl_2(g)$

(i) Use the values of enthalpy of formation given to calculate ΔH for the forward reaction. **(1)**

	ΔH_f / kJ mol^{-1}
$PCl_5(g)$	−399
$PCl_3(g)$	−306

(ii) Explain, with reasons, the effect that raising the temperature would have on the composition of the equilibrium mixture. **(2)**

(iii) Other than by changing the temperature, suggest how the amount of PCl_5 present at equilibrium could be increased. Give a reason for your answer. **(2)**

(Total 8 marks)

3 Four of the structural isomers of $C_4H_{10}O$ are alcohols. One of these isomers is butan-2-ol.

a Draw the structural formulae of two other alcohols with molecular formula $C_4H_{10}O$ and name each of these isomers. **(4)**

b A sample of butan-2-ol, C_4H_9OH was heated with a mixture of sulfuric acid and potassium dichromate(VI).

(i) State the colour change that would be observed during the reaction. **(1)**

(ii) Give the name of the organic product formed, and name the type of reaction occurring. **(2)**

(Total 7 marks)

4 In the preparation of the alcohol butan-2-ol, 13.7 g of 2-bromobutane was hydrolysed with 9.0 g of potassium hydroxide in aqueous solution. The following reaction occurred.

$$CH_3CHBrCH_2CH_3 + KOH \rightarrow CH_3CHOHCH_2CH_3 + KBr$$

a Calculate the amount (in moles) of each **reactant** in the above experiment, and use your answers to state which reactant was present in excess. **(4)**

b Calculate the maximum possible mass of butan-2-ol which could be obtained in the above experiment. **(4)**

c Give a reason why, in practice, this maximum yield is not obtained. **(1)**

d The reaction taking place can be classified as nucleophilic substitution. Explain the term **nucleophile** and identify the nucleophile in the reaction. **(2)**

e The above experiment was repeated under identical conditions, except that 2-iodobutane was used in place of 2-bromobutane. State and explain the effect that this change would have on the rate of reaction. **(2)**

(Total 13 marks)

5 This question concerns reaction sequences for the conversion of propan-1-ol into 2-bromopropane and butanoic acid.

a Propan-1-ol, $CH_3CH_2CH_2OH$, can be converted into 2-bromopropane, $CH_3CHBrCH_3$, in two stages.

$$CH_3CH_2CH_2OH \rightarrow CH_3CH{=}CH_2 \rightarrow CH_3CHBrCH_3$$

(i) Give the reagents and conditions necessary for the conversion of $CH_3CH_2CH_2OH$ into $CH_3CH{=}CH_2$ **(2)**

(ii) Give the reagent and write the mechanism for the conversion of $CH_3CH{=}CH_2$ into $CH_3CHBrCH_3$ **(4)**

(iii) Briefly explain why 2-bromopropane is the main product of this second reaction rather than 1-bromopropane. **(1)**

(Total 7 marks)

6 There is considerable concern about the depletion of the 'ozone layer'. Ozone, O_3 is constantly being formed in the upper atmosphere by oxygen molecules reacting with oxygen atoms. These atoms are formed when other oxygen molecules absorb high energy ultraviolet radiation and undergo homolytic bond fission:

$O_2 \rightarrow O^\bullet + O^\bullet$ [Reaction 1]

$O^\bullet + O_2 \rightarrow O_3$ [Reaction 2]

The ozone formed absorbs ultraviolet radiation and breaks down to oxygen molecules and atoms:

$O_3 \rightarrow O_2 + O^\bullet$ [Reaction 3]

Another reaction, which is not light dependent, also removes ozone:

$O_3 + O^\bullet \rightarrow 2O_2$ [Reaction 4]

Over time the ozone and oxygen concentrations reach a steady state.

a **(i)** What is meant by the term **homolytic bond fission**? **(2)**

(ii) Which of these reactions is likely to be an initiation step? **(1)**

(iii) Use the reactions above to explain what is meant by the term **steady state**. **(3)**

b Reaction 2 has an enthalpy change of $-100\,kJ\,mol^{-1}$ and Reaction 4 an enthalpy change of $-390\,kJ\,mol^{-1}$. Assuming that Reactions 1 and 2 are followed by Reaction 4, show how this results in the upper atmosphere warming up by effectively converting light energy into heat energy. Calculate the amount of heat produced per mole of oxygen molecules. **(4)**

c In the upper atmosphere chlorofluorocarbons, CFCs such as CCl_2F_2 break down to give chlorine atoms and these chlorine atoms react with ozone molecules:

$Cl^\bullet + O_3 \rightarrow ClO^\bullet + O_2$ [Reaction 5]

(i) Write an equation to show how the ClO radicals can react with oxygen atoms to form oxygen molecules. **(1)**

(ii) Use the equations given in earlier parts and that given for Reaction 5 to explain how the formation of only a few chlorine atoms can cause the removal of a large number of ozone molecules. **(2)**

(iv) Suggest a reason why the rates of all the reactions in the upper atmosphere are likely to be slow. **(1)**

d 2-hydroxy-4-methoxybenzophenone is used in lotions to protect against sunburn caused by ultraviolet radiation.

Suggest why the release of CFCs into the environment may lead to an increase in sales of lotions containing 2-hydroxy-4-methoxybenzophenone. **(2)**

(Total 16 marks)

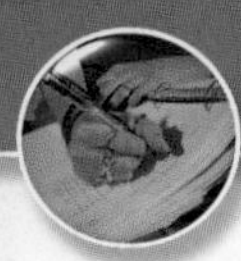

7 This question concerns the three isomers **A**, **B** and **C**, each of which has a relative molecular mass of 134.

a The mass spectrum of substance **A** is shown below. Identify the species responsible for the peaks labelled 1, 2 and 3. **(3)**

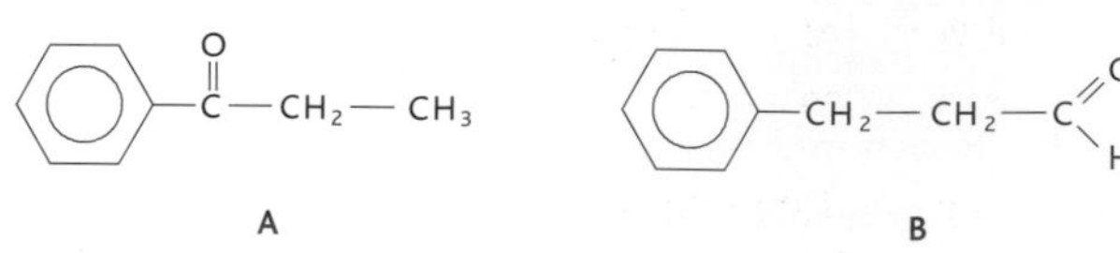

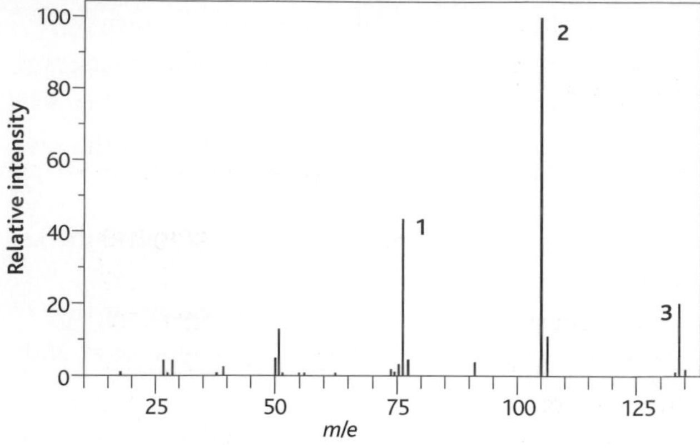

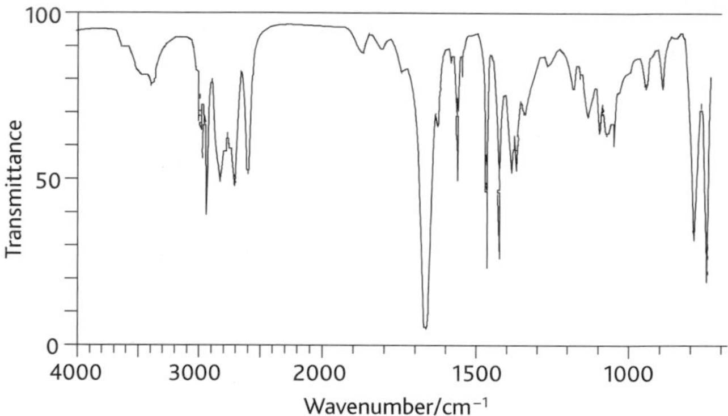

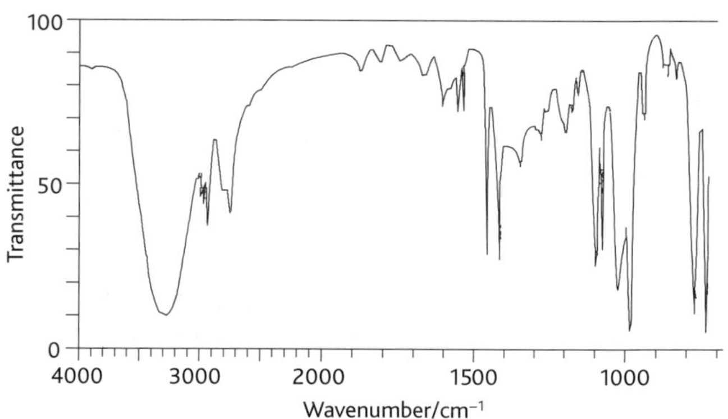

b The infrared spectra of two of these substances were also measured.

(i) Use the table and the spectra on the right to identify which spectrum is that of substance **C**. **(1)**

Bond	Wavenumber/ cm^{-1}
C—H (arenes)	3000–3100
C—H (alkanes)	2850–3000
C=O	1680–1750
O—H (hydrogen bonded)	3200–3570
O—H (not hydrogen bonded)	3580–3650
C=C (arenes)	1450–1600

(ii) Give **one** reason for your choice. **(1)**

(iii) Give one **other** reason why the other spectrum could **not** be that of substance **C**. **(1)**

c **(i)** State which of the substances **A**, **B** and **C** will react with bromine dissolved in hexane and state what would be observed. **(2)**

(ii) Give the structural formula of the organic product(s) obtained in **c**(i). **(1)**

d Greenhouse gases such as CO_2 absorb infrared radiation and it is this that is causing the Earth's atmosphere to warm up. Name another greenhouse gas, other than CO_2, which absorbs IR radiation. **(1)**

(Total 10 marks)

Index

The Periodic Table of Elements

Key

relative atomic mass
atomic symbol
name
atomic (proton) number

1	2											3	4	5	6	7	0 (8)
							1.0 **H** hydrogen 1										*(18)* 4.0 **He** helium 2
(1)	*(2)*											*(13)*	*(14)*	*(15)*	*(16)*	*(17)*	
6.9 **Li** lithium 3	9.0 **Be** beryllium 4											10.8 **B** boron 5	12.0 **C** carbon 6	14.0 **N** nitrogen 7	16.0 **O** oxygen 8	19.0 **F** fluorine 9	20.2 **Ne** neon 10
23.0 **Na** sodium 11	24.3 **Mg** magnesium 12	*(3)*	*(4)*	*(5)*	*(6)*	*(7)*	*(8)*	*(9)*	*(10)*	*(11)*	*(12)*	27.0 **Al** aluminium 13	28.1 **Si** silicon 14	31.0 **P** phosphorus 15	32.1 **S** sulfur 16	35.5 **Cl** chlorine 17	39.9 **Ar** argon 18
39.1 **K** potassium 19	40.1 **Ca** calcium 20	45.0 **Sc** scandium 21	47.9 **Ti** titanium 22	50.9 **V** vanadium 23	52.0 **Cr** chromium 24	54.9 **Mn** manganese 25	55.8 **Fe** iron 26	58.9 **Co** cobalt 27	58.7 **Ni** nickel 28	63.5 **Cu** copper 29	65.4 **Zn** zinc 30	69.7 **Ga** gallium 31	72.6 **Ge** germanium 32	74.9 **As** arsenic 33	79.0 **Se** selenium 34	79.9 **Br** bromine 35	83.8 **Kr** krypton 36
85.5 **Rb** rubidium 37	87.6 **Sr** strontium 38	88.9 **Y** yttrium 39	91.2 **Zr** zirconium 40	92.9 **Nb** niobium 41	95.9 **Mo** molybdenum 38	[98] **Tc** technetium 43	101.1 **Ru** ruthenium 44	102.9 **Rh** rhodium 45	106.4 **Pd** palladium 46	107.9 **Ag** silver 47	112.4 **Cd** cadmium 48	114.8 **In** indium 49	118.7 **Sn** tin 50	121.8 **Sb** antimony 51	127.6 **Te** tellurium 52	126.9 **I** iodine 53	131.3 **Xe** xenon 54
132.9 **Cs** caesium 55	137.3 **Ba** barium 56	138.9 **La*** lanthanum 57	178.5 **Hf** hafnium 72	180.9 **Ta** tantalum 73	183.8 **W** tungsten 74	186.2 **Re** rhenium 75	190.2 **Os** osmium 76	192.2 **Ir** iridium 77	195.1 **Pt** platinum 78	197.0 **Au** gold 79	200.6 **Hg** mercury 80	204.4 **Tl** thallium 81	207.2 **Pb** lead 82	209.0 **Bi** bismuth 83	[209] **Po** polonium 84	[210] **At** astatine 85	[222] **Rn** radon 86
[223] **Fr** francium 87	[226] **Ra** radium 88	[227] **Ac*** actinium 89	[261] **Rf** rutherfordium 104	[262] **Db** dubnium 105	[266] **Sg** seaborgium 106	[264] **Bh** bohrium 107	[277] **Hs** hassium 108	[268] **Mt** meitnerium 109	[271] **Ds** darmstadtium 110	[272] **Rg** roentgenium 111							

Elements with atomic numbers 112-116 have been reported but not fully authenticated

	140 **Ce** cerium 58	141 **Pr** praseodymium 59	144 **Nd** neodymium 60	[147] **Pm** promethium 61	150 **Sm** samarium 62	152 **Eu** europium 63	157 **Gd** gadolinium 64	159 **Tb** terbium 65	163 **Dy** dysprosium 66	165 **Ho** holmium 67	167 **Er** erbium 68	169 **Tm** thulium 69	173 **Yb** ytterbium 70	175 **Lu** lutetium 71
* Lanthanide series														
* Actinide series	232 **Th** thorium 90	[231] **Pa** protactinium 91	238 **U** uranium 92	[237] **Np** neptunium 93	[242] **Pu** plutonium 94	[243] **Am** americium 95	[247] **Cm** curium 96	[245] **Bk** berkelium 97	[251] **Cf** californium 98	[254] **Es** einsteinium 99	[253] **Fm** fermium 100	[256] **Md** mendelevium 101	[254] **No** nobelium 102	[257] **Lr** lawrencium 103